127

129 Here

p.266. Rates of behaviour.

p134 - Reliability

GARLAND SERIES IN ETHOLOGY

Series Editor: Gordon M. Burghardt

Primate Ecology and Human Origins:
Ecological Influences on Social Organization
Edited by: Irwin S. Bernstein and Euclid O. Smith

The Development of Behavior:
Comparative and Evolutionary Aspects
Edited by: Gordon M. Burghardt and Marc Bekoff

The Behavioral Significance of Color
Edited by: Edward H. Burtt, Jr.

The Behavior and Ecology of Wolves
Edited by: Erich Klinghammer

Handbook of Ethological Methods
By: Philip N. Lehner

Cat Behavior: The Predatory and Social Behavior
of Domestic and Wild Cats
By: Paul Leyhausen

Dominance Relations: An Ethological View of Human
Conflict and Social Interaction
Edited by: Donald R. Omark, Fred F. Strayer and Daniel G. Freedman

Species Identity and Attachment: A Phylogenetic Evaluation
Edited by: Aaron Roy

OTHER GARLAND BOOKS IN ETHOLOGY

Primate Bio-Social Development: Biological, Social,
and Ecological Determinants
Edited by: Suzanne Chevalier-Skolnikoff and Frank E. Poirier

Faces from New Guinea
By: Paul Ekman

The Dog: Its Domestication and Behavior
By: Michael W. Fox

Ethological Dictionary: In English, French, and German
By: Armin Heymer

Handbook of ETHOLOGICAL METHODS

Philip N. Lehner

Colorado State University

Garland STPM Press
New York & London

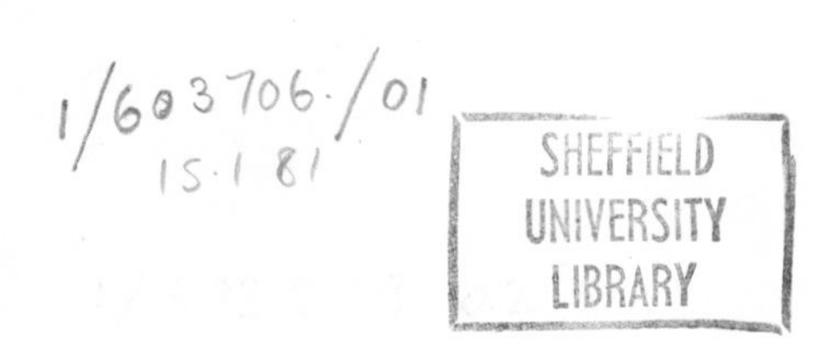

15 14 13 12 11 10 9 8 7 6 5 4 3 2 1

Library of Congress Cataloging in Publication Data
Lehner, Philip N 1940–
Handbook of ethological methods.
(Garland series in ethology)
Bibliography: p. 327
1. Animals, Habits and behavior of—Methodology—
Handbooks, manuals, etc. I. Title. II. Series.
QL751.L398 591.5 77-90468
ISBN 0-8240-7024-0

Published by Garland STPM Press
545 Madison Avenue, New York, New York 10022
Printed in the United States of America

This book is dedicated to
MY PARENTS
who have provided opportunity,
guidance, faith, and encouragement
for my several and varied endeavors.

Contents

Foreword

At the outset the study of animal behavior was an easy course to follow. We needed no special equipment. The outdoors was teeming with animals whose behavior or ecology we knew little about. We needed merely to plunge in with field notebook and binoculars, a fair measure of fortitude to put up with the elements, and a large measure of imagination to interpret what we were studying. The writings of Konrad Lorenz, Niko Tinbergen and others are full of examples of these early studies. They opened whole new worlds of understanding.

Ethology has now reached maturity. Those two ogres, the granting agency and the graduate school committee, are demanding a much higher level of hypotheses and experimental design than in the halcyon days when one could do field ethology without these constraints. Henry Eyring has said "Doing research is easy; it's finding the forefront of knowledge that is difficult." How true, if by doing research we mean the actual recording of observations. But reaching this stage is difficult. Until now the investigator has had to fend largely for himself, poring through behavior journals to learn how others have tackled problems. This is a tedious and often scatter-gun approach. Now Philip Lehner has succeeded in guiding the investigator in ethology from the initial look at a research problem to its final interpretation. This book will save the beginning investigator from many mistakes, oversights, and countless hours in designing his study. More senior investigators will find this book invaluable as well. Who among us has not bemoaned after completing a piece of research, "Why didn't I do this?" or "Why didn't I think of that?" Lehner has drawn together a prodigious amount of information and presented it in such a fashion that both veteran and novice can readily locate his needs.

Dr. Lehner is an avid field ethologist, currently studying the vocal repertoire of coyotes in Jackson Hole. He sits on a hillside armed with little more than a spotting scope, tape recorder, parabolic mirror, and trusty field notebook. With the magnificent panorama of the Tetons as backdrop, he can observe the daily drama of coyotes moving about beneath him and so link his recordings to the actual behavior taking place below him. At the other extreme Lehner has done the most rigorously controlled research on such things as the effects of pesticides on color discrimination in mallards, using the Skinner box, event recorders, and rigid control of light intensity. He remains a champion of starting out any research with exploratory descriptive studies, using the traditional field notebook and pencil and generous amounts of time spent asking questions about the animals under observation.

This is no coldly factual book of methodology. Lehner has sprinkled the book liberally with quotations from ethologists that show how human the scientist is. Older readers will savor these quotations as they recall past associations and difficulties. The newcomer to the field will find them reassuring and taking on added meaning as he advances in his research experience. Like any writing of real merit this book will stand rereading on numerous occasions, each reading bringing new insights and relevance.

Allen W. Stokes

Preface

The purpose of this book is to present a logical, meaningful, and practical approach to the study of animal behavior. This is not a simple task because there is much artistry in the conduct of animal behavior studies. Research can become very personal and observations difficult to describe. If you are skilled and blessed with a generous dose of serendipity, you gain an intimate harmony with the animals you are studying and begin to feel that a window of understanding has opened. You feel comfortably less alien and more attuned to, yet isolated from, what you are observing. You become one with the animals, just as an experienced rider demonstrates perfect synchrony with the rising and falling back of a galloping horse. You begin to observe with apparent insight into an animal's behavior. However,

> The pursuit of science is an intensely personal affair. Experimenters cannot always tell us how or why they do what they do, and the fact that their conclusions are sound so much of the time remains a puzzle. [SIDMAN 1960:vi]

Nevertheless, with Sidman's statement at hand I optimistically plunged into this project. The impetus for writing this book came from listening to undergraduates, graduate students, and lay persons who were interested in animal behavior. I heard a common question (one I had previously voiced myself): "How do you study the behavior of animals?" That is a difficult question; but I hope this book goes a long way toward answering it. In essence, I have written the book I wish I had when I was just beginning my studies of animal behavior. No attempt has been made to write an all-inclusive treatise. Rather, it is introductory with sufficient depth and direction to allow the reader to pursue additional topics independently.

I hope this book will be useful to a diversity of people in several ways. First, it can be effective as a textbook for laboratory and techniques classes in animal behavior. Second, it should serve as a useful "companion volume"—a "companion" to a regular textbook and a "companion" to the researcher in the field. Third, it is a composite of practical information which I hope will guide the weekend ethologist, as well as serve as a useful reference for advanced ethologists. To hope that the book accomplishes all that is to be ambitiously optimistic, and to actually attain those goals is next to impossible. However, if this book is read by many, used by some, and stimulates a few, I will be satisfied.

My feelings in undertaking this task were similar to those of Niko Tinbergen when in the Preface to his classic *The Study of Instinct* he wrote:

> I am fully aware that this first attempt at a synthetic treatment is incomplete and unsatisfactory in many respects. But somebody had to make the attempt. I cannot imagine a better result of my venture than that it should lead to better presentations in the future.

I am indebted to the large number of teachers, students and colleagues who, over the years, have shared their knowledge and ideas with me. Although they are too numerous to mention here individually, they know who they are, and I thank them. John Alcock, Marc Bekoff, Gordon Burghart, and Tim Clark provided valuable criticism on the entire manuscript. Wayne Aspey, Fritz Knopf, and Hugh Drummond critiqued the statistical and description sections. I wish to thank these people for the time and energy they spent and the constructive criticism they provided in greatly improving the manuscript; however, I remain solely responsible for any errors of omission or commission. Finally, thanks go to my wife Barbara and my children Jason and Jenny, who put up with my harried schedule and unpredictable disposition. They shared the burden and trusted me when I promised: "Not much longer."

Philip N. Lehner

Handbook of ETHOLOGICAL METHODS

1 Introduction

One does not meet oneself until one catches the reflection from an eye other than human.
[EISELEY 1964:24]

In choosing to study the behavior of animals you are setting off on a voyage across waters that are often rough and, in some areas, poorly charted. This book is basically a compilation of practical information that is intended to help smooth the waters and assist you in charting your course. I have taken the liberty of infusing it with personal and borrowed philosophy. These philosophical interludes are meant to cement together concrete blocks of cold facts and invite you to stop and ponder what you have learned and what it means.

The most disheartening (and sometimes disarming) question that an uninitiated ethologist can face is: "So what?" You can be confronted with that question at technical meetings or cocktail parties, and answers vary from informative discourses to obscene threats. Some of the discussions found in this book may help you to wrestle with that question in your own mind.

The study of behavior encompasses all of the movements and sensations by which animals and men mediate their relationships with their external environments—physical, biotic and social. No scientific field is more complex, and none is more central to human problems and aspirations. [ALEXANDER 1975:77]

One of the greatest challenges you face is to keep your own studies of animal behavior in perspective. Why did you choose that particular species or concept to study? (Chap. 1, Sec. D; Chap. 2, Secs. A, B) What is the focus of your study? Three dimensions of this question are discussed below in Chapter 1, Section A. What is al-

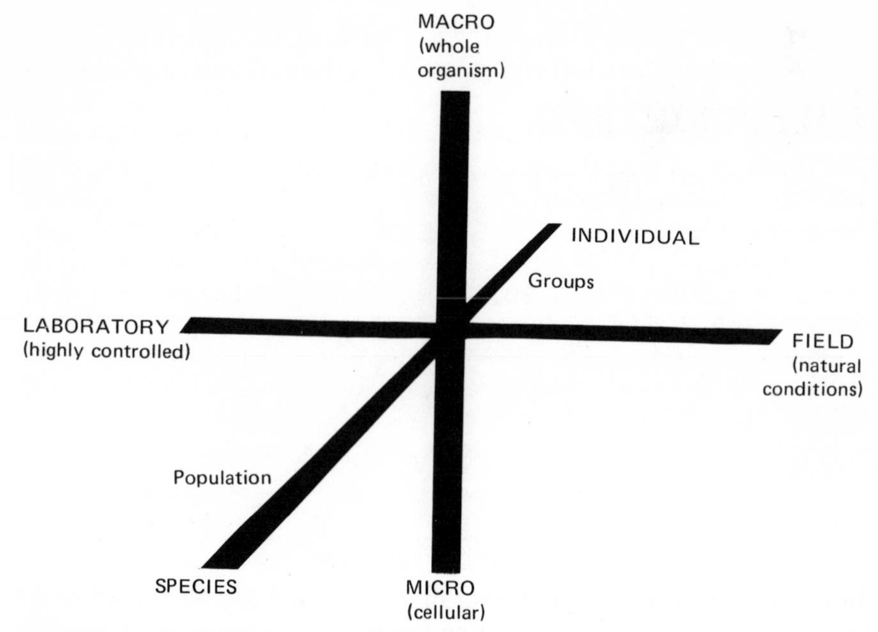

Figure 1-1 Three dimensions of ethology.

ready known about the subject (Chap. 3, Sec. C)? Are you replicating someone else's research, testing someone else's hypothesis, or are you answering your own questions? Will your results be limited to the individuals you are observing or will you consider them a representative sample of a larger population?

Earlier I said that the study of animal behavior is a fascinating voyage; remember to continually assess where you are, where you have been, and where you are going.

A. WHAT IS ETHOLOGY?

Everyone seems to have their own definition of ethology. A verbal exchange among Konrad Lorenz, Niko Tinbergen, Theodore Schneirla, and others on the definition of ethology makes interesting reading and history (Schaffner 1955:77–78). Eisner and Wilson (1975:1) define ethology as the:

> . . . study of whole patterns of animal behavior under natural conditions, in ways that emphasize(d) the functions and the evolutionary history of the patterns.

However, for our purposes, let us focus on Lorenz's (1960*a*) definition: "Ethology can be briefly defined as the application of orthodox biological methods to the problems of behavior." As a basis for this definition, Lorenz *(op. cit.)* described the genealogy of ethology—". . . biology . . . is its mother . . . whereas . . . for a father, a very plain zoologist Charles Darwin." Ethology's heritage should be kept firmly in mind.

Ethology is a nearly limitless discipline which operates basically along three dimensions (Fig. 1-1). These are continuous dimensions along which studies can be focused. This book will concentrate on methods applicable to the study of the whole organism at the individual or group level under field conditions.

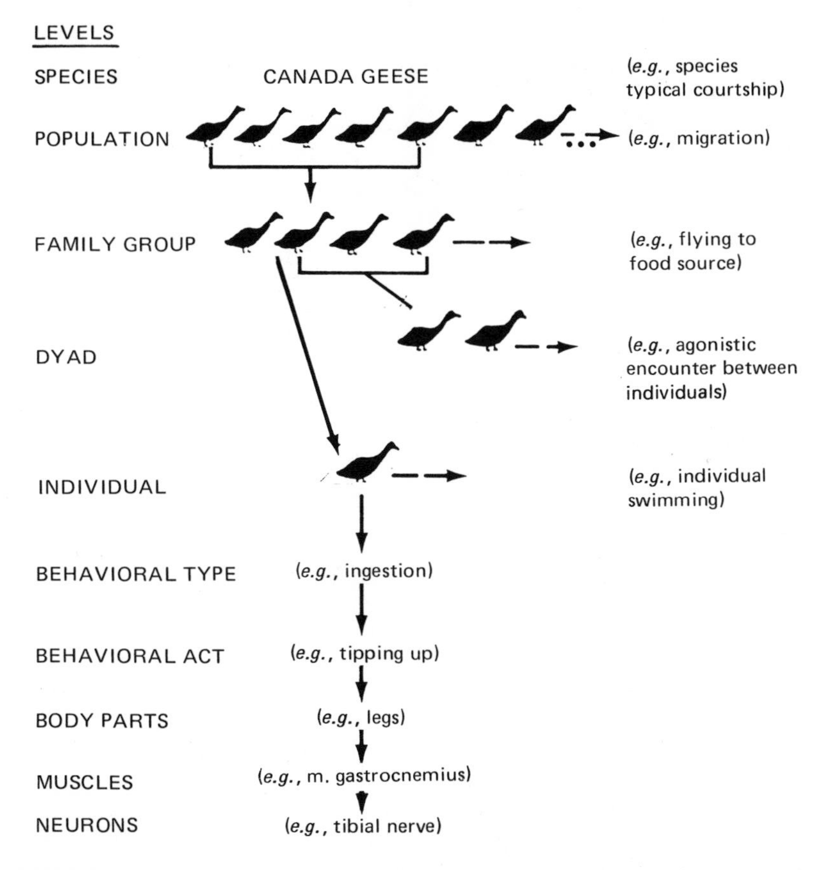

Figure 1-2 Levels of behavior in Canada geese. At each level there can be intraspecific and interspecific interactions (indicated by — →) with other family groups, individuals, and so on.

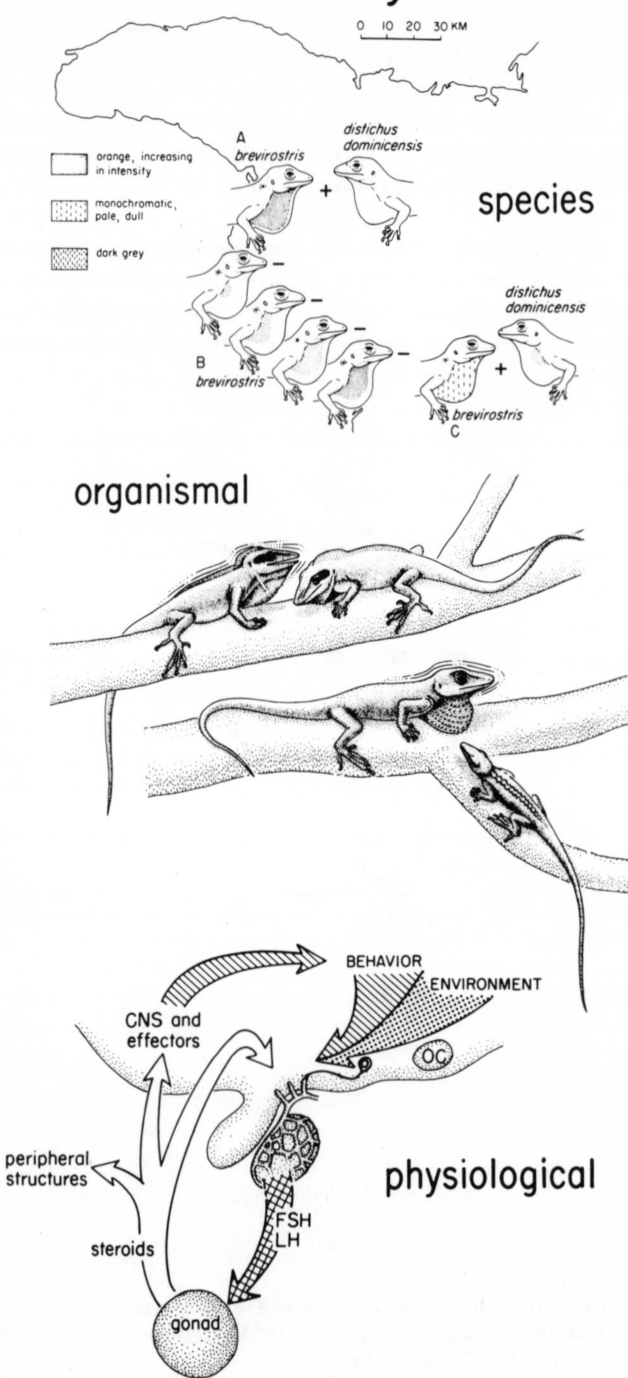

Levels of Analysis
0 10 20 30 KM
orange, increasing in intensity
monochromatic, pale, dull
dark grey
A
brevirostris
distichus
dominicensis
+
species
−
−
−
−
B
brevirostris
distichus
dominicensis
+
brevirostris
C
organismal
BEHAVIOR
ENVIRONMENT
CNS and effectors
OC
peripheral structures
physiological
FSH
LH
steroids
gonad

1. Levels of Behavior

Behavior occurs at various organizational levels along the individual–species and macro–micro dimensions (Fig. 1-1), a perspective which the ethologist must obtain and maintain (Menzel 1969). Figure 1-2 gives an example of how these levels can be viewed for Canada geese *(Branta canadensis).*

The concept of levels of organization is important to the ethologist, and he should be able to "zoom in" and "zoom out" (Menzel 1969) to-and-from the aspect of behavior he is studying (Fig. 5-2). We can approach a real understanding of behavior only if we put it in perspective. Crews (1977) illustrated this point in his study of the reproductive behavior of the American chameleon *(Anolis carolinensis)* (Fig. 1-3).

The ethologist must decide at what level he will be conducting research and how his study will integrate into what is already known at the other levels.

B. WHAT TO STUDY?

For some people this question is seemingly unimportant since they have focused most of their attention on a chosen species. Choice of species, however, is only a partial answer to the question: "What to study?" Others want the adventures of Jane Goodall, George Schaller, Konrad Lorenz, and Niko Tinbergen. They do not really care what

Figure 1-3 Traditionally, species interactions, the "species" level of analysis, have been studied in the field, while physiological mechanisms underlying species-typical behavior patterns have been investigated in the laboratory. Recent research with *Anolis carolinensis* demonstrates that these ordinarily separate lines of inquiry can be integrated into a single program of research at the "organismal" level of analysis and thus lead to a better understanding of causes and functions of reproductive behavior. These levels of research are illustrated in the figure. At the species level, pure populations of *A. brevirostris* (population *B*) exhibit clinical variation in dewlap color, whereas populations to the north and south, which are sympatric with *A. distichus*, have uniform dewlap colors. At the organismal level, aggressive posturing between male *A. carolinensis* (on the top branch) involves erected nuchal and dorsal crest, lateral compression of body, black spot behind eye, engorged throat, and lateral orientation, whereas during the courtship display (bottom branch), the male has a relaxed body posture and extended dewlap, and the crest and eye spot are absent. The physiological level shows some of the principal intrinsic and extrinsic factors mediating reproductive events in vertebrates and their feedback relationships. (Adapted from Crews, D., 1977, The annotated anole: Studies on the control of lizard reproduction, *Amer. Sci.* 65:432, Fig. 4.)

they study as long as it is adventurous. However, the serious student of ethology must be prepared for uncomfortable hours under adverse conditions when even the most avid ethologist would admit to physical misery. Ethology has often been glamorized, but it is not always glamorous (See Lott's *Protestations of a Field Person,* reprinted on p. 25).

Ethologists have historically focused on the animal's behavior *per se*, whereas psychologists have traditionally concerned themselves with the relationships among motivation, stimuli, and behavior. Experimental psychologists generally view learning and behavior according to two paradigms: 1) stimuli *eliciting* behavior (classical conditioning) and 2) *emitted* behavior being followed by reward or punishment (operant conditioning). Both types of learning occur in the real world. Hence we can view the behavior of an animal according to the model in Figure 1-4.

The model is separated into three components:

Stimuli—an organism may react to exogenous (external) and/or endogenous (internal) stimuli (Glickman 1973).

Behavior—a BEHAVIOR will be elicited and another ongoing BEHAVIOR may be inhibited.

Behavioral Effect—the BEHAVIOR may result in *Positive Reinforcement* in which case *Positive Feedback* will increase or maintain the probability of the behavior occurring again; it may result in *Punishment* which causes *Negative Feedback* and decreases the probability of that behavior occurring again; finally the BEHAVIOR may have no apparent effect on the probability of the behavior occurring again.

For example, we put a bowl of dog food near a sleeping dog. It is hungry (endogenous stimulus; *e.g.*, lowered blood-sugar level), therefore it is receptive to the odor of the dog food (exogenous stimulus), awakens (opening its eyes and raising its head), gets up, approaches the food, and eats (BEHAVIOR sequence). Sleeping (BEHAVIOR) was inhibited (or at least no longer stimulated). The food tastes good (*i.e.*, provides positive gustatory stimuli) and begins to decrease the dog's hunger (Positive Reinforcement; *e.g.*, elevated blood-sugar level) thereby maintaining (Positive Feedback) the feeding (BEHAVIOR). When satiation occurs the feeding (BEHAVIOR) will gradually have No Effect and the BEHAVIOR will gradually stop. If we had put an aversive stimulus, such as quinine (Punishment) in the food, Negative Feedback would have decreased the probability that the dog would have taken another bite (BEHAVIOR).

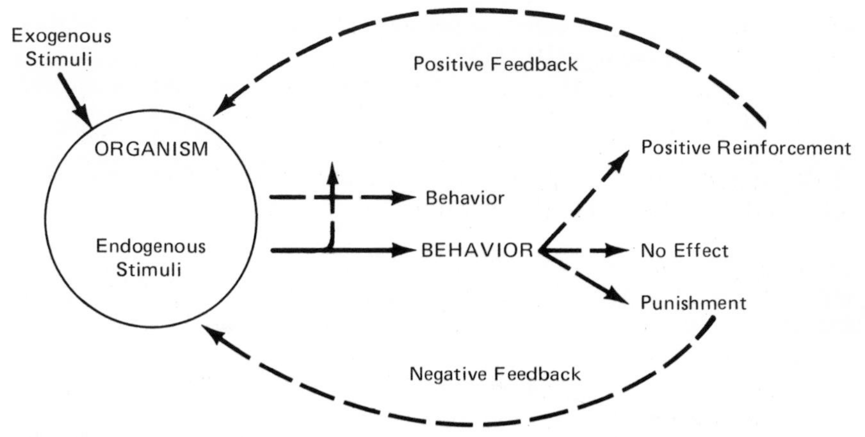

Figure 1-4 A model of animal behavior. The dashed lines indicate possible outcomes of behavior. (See text for explanation and an example.)

The model is useful, not only in envisioning the immediate modification of behavior through the behavioral effect, but also the animal's behavior within an evolutionary context.

An animal may perform behaviors that are primarily instinctual (*e.g.*, courtship) because of an inherent reward which the animal receives as positive feedback. Whether this is pleasure as we know it, or what Lorenz calls "feeling good" (Nisbett 1977:138), does not really matter in the model. The assumption is that there is immediate positive feedback that sustains the behavior in the individual, and long-term feedback (natural selection) that maintains it in the gene pool and hence in the population.

Applying the model to the evolution of behavior, we can envision positive reinforcement increasing individual fitness and punishment decreasing it. That is, the evolutionary process of natural selection will positively reinforce adaptive behaviors by allowing the genes from those individuals to reproduce themselves in future generations. Conversely, disadvantageous behavior will be punished by decreasing the output of those individual's genes. Obviously the ultimate punishment is death, which ends gene transfer in that individual.

This is, of course, an oversimplification of behavior, but it is profitable to begin by viewing behavior within a simple framework before leaping into complex, in-depth analyses, and models (p. 319). Nevertheless, keep in mind Hinde's warning:

Behaviour is diverse: an attempt to squeeze it into a system involving only a few explanatory concepts is liable to lead to one of two results—either facts which do not fit will be ignored, or the concepts will be stretched until they become valueless. [HINDE 1957:116]

Behavior is *what* an animal does. However, ethologists do not restrict themselves to examining only *what* an animal does. Nielsen (1958) stated that ethology also includes the study of *when, how* and *why*, and I would add to this list the study of *where*.

What—A description of the behavior of the animal.

When—The temporal component of the behavior. This can include the occurrence of the behavior with respect to the animal's lifetime, the season, time of day, or position in a sequence. The duration of a behavior and its contribution to an animal's time budget are also considered under the study of *When*.

How—This includes the motor patterns used to accomplish a goal-oriented behavior (*e.g.*, flying from one tree to another) as well as the underlying physiological mechanisms (an aspect not covered in this book). Studies of the relevant stimuli associated with behavior are included in this category. The evolutionary and phylogenetic determinants of behavior are also studied to answer *How* questions. For example: How did flight evolve in birds?

Why—Two basic concepts underlie the study of the *Why* of behavior. These are *motivation* and *ecological adaptation*. These separate, but related, concepts are generally treated in different disciplines: motivation in psychology and ecological adaptation in behavioral ecology. However, the latter is sometimes incorporated into ethological studies, and the former has figured prominently in conceptual behavior models (*e.g.*, displacement behavior).

Where—This is the spatial aspect of a behavior. Studies include where a behavior occurs geographically or relative to other animals or environmental parameters. Keep in mind that spatial characteristics are three dimensional.

Many ethologists subscribe to Tinbergen's (1963) categorization of four areas of study in ethology: function, causation, ontogeny, and evolution.

Function—This can include the study of *proximate* and/or *ultimate* function, or what Hinde (1975) has called the "weak meaning" and "strong sense" of function, respectively. The proximate function refers to the immediate effect of the behavior on that animal, other animals, or the environment. Through many observations correlations can be established which lead to conclusions of cause and effect (*i.e.*, proximate function). The ultimate function refers to adaptive significance of the behavior in terms of improving the individual's fitness, and how natural selection operates to maintain the behavior.

Causation—What are the mechanisms that underly the behavior? What are the contexts in which it occurs, and what are the exogenous and endogenous stimuli that elicit the behavior? (See Figure 1-4.)

Ontogeny—How does the behavior develop in the individual? What maturational and learning processes are important in the development of the behavior?

Evolution—How did the behavior develop in the species? This includes a phylogenetic comparative approach in which the behavior is studied in closely related species to reveal differences which may reflect evolutionary change (Lorenz 1951; Kessel 1955). This often leads to the development of an "ethocline" which presents the differences in a behavior among several species along an evolutionary continuum (Evans 1957).

Tinbergen's four areas of study can be grouped into two sets which address questions of "proximal" or "ultimate" causation (Wilson 1975), or as Alcock (1975) stated, questions of "how" or "why," respectively. The matrix below illustrates levels of study at which research on the two sets of questions is often directed, as well as how the two sets can be divided into the study of causes and origins.

Research directed at	1. Level of individual 2. "Proximal causation" (Wilson 1975) 3. "How questions" (Alcock 1975)	1. Level of individual or population 2. "Ultimate causation" (Wilson 1975) 3. "Why questions" (Alcock 1975)
Origins	Ontogeny	Phylogeny (Evolution)
Causes	Control (Causation)	Function

In addition, Marler and Hamilton (1966) suggested the following five broad areas of investigation of animal behavior that overlap those of Tinbergen: *motivation, ecology, social communication, phylogeny*, and *ontogeny*.

The above should give you a general idea of the various aspects of animal behavior that you can study. The following sections discuss the steps an ethologist takes in studying animal behavior.

C. SCIENTIFIC METHOD

Ethology is a science, and as such, serious studies should adhere to established guidelines. These guidelines are clearly expressed as follows:

Scientific research is systematic, controlled, empirical, and critical investigation of hypothetical propositions about the presumed relations among natural phenomena. [KERLINGER 1967]

The "orthodox biological methods" mentioned in Lorenz's definition of ethology include the scientific method. The scientific method (Fig. 1-5) is a logical approach to research in all sciences including ethology (see Hailman 1975, 1977 for an opposing viewpoint). This book is organized around the scientific method modified for use in ethology (Chap. 1, Secs. C, D).

Now that I have focused in on the cold-blooded approach that will serve as the foundation for this book, let me state that this does not exclude the artistry inherent in ethological studies. I also want to champion the researcher who gets to know his animals, who becomes one with nature, and who feels empathy with the animals he is studying.

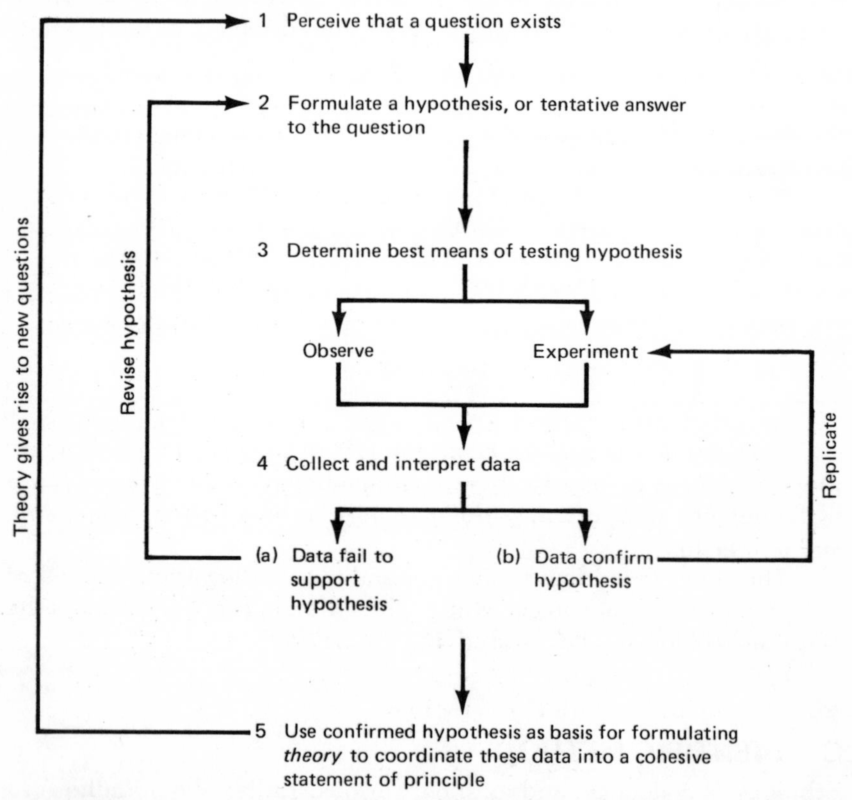

Figure 1-5 Flow diagram of scientific inquiry (from Jessop 1970). Serendipity (fortuitous discovery) can operate at all stages of the process.

Several researchers have described their state of mind and thought processes while studying animals in their natural habitat. I would especially recommend reading Tinbergen (1958) and Schaller (1973). Dethier (1962) also gives an illustrative and entertaining description of the development of his laboratory research on blowflies. The ethological researcher should be more than a collector of data; he should seek to "understand" his animal subjects at a level higher than quantitative analysis can provide.

When we watch animals at different levels on the evolutionary scale, as when Seitz watches fishes, when Dr. Tinbergen watches gulls, when Dr. Lorenz watches ducks, when Dr. Schneirla watches ants, or when I watch doves, the observer can get a feeling of what is going to happen next, which is compounded in different degrees of the intellectual experience of relationships that are involved on the one hand, and, on the other hand, of building yourself into the situation. [LEHRMAN 1955]

Darling (1937) stated that in order to gain insight into behavior, the observer must become "intimate" with the animals under observation. Lorenz (1960*b*) suggested that an even higher level of empathy with the animals under study is necessary for a true understanding of their behavior.

It takes a very long period of watching to become really familiar with an animal and to attain a deeper understanding of its behaviour; and without the love for the animal itself, no observer, however patient, could ever look at it long enough to make valuable observations on its behaviour. [LORENZ 1960*a*:xii]

The ethologists quoted above were not encouraging rampant anthropomorphism at the expense of objectivity (Carthy 1966). Rather, they were stating that genuine interest in the animal *per se* will foster greater insight into its behavior than will mechanistic (machine or machinelike) data collection.

D. ETHOLOGICAL APPROACH

As mentioned previously, the ethological approach is the result of fitting the scientific method to ethology.

Ethology . . . is characterized by an observable phenomenon (behavior, or movement), and by a type of approach, a method of study (the biological method). . . . The biological method is characterized by the general scientific method. [TINBERGEN 1963:411]

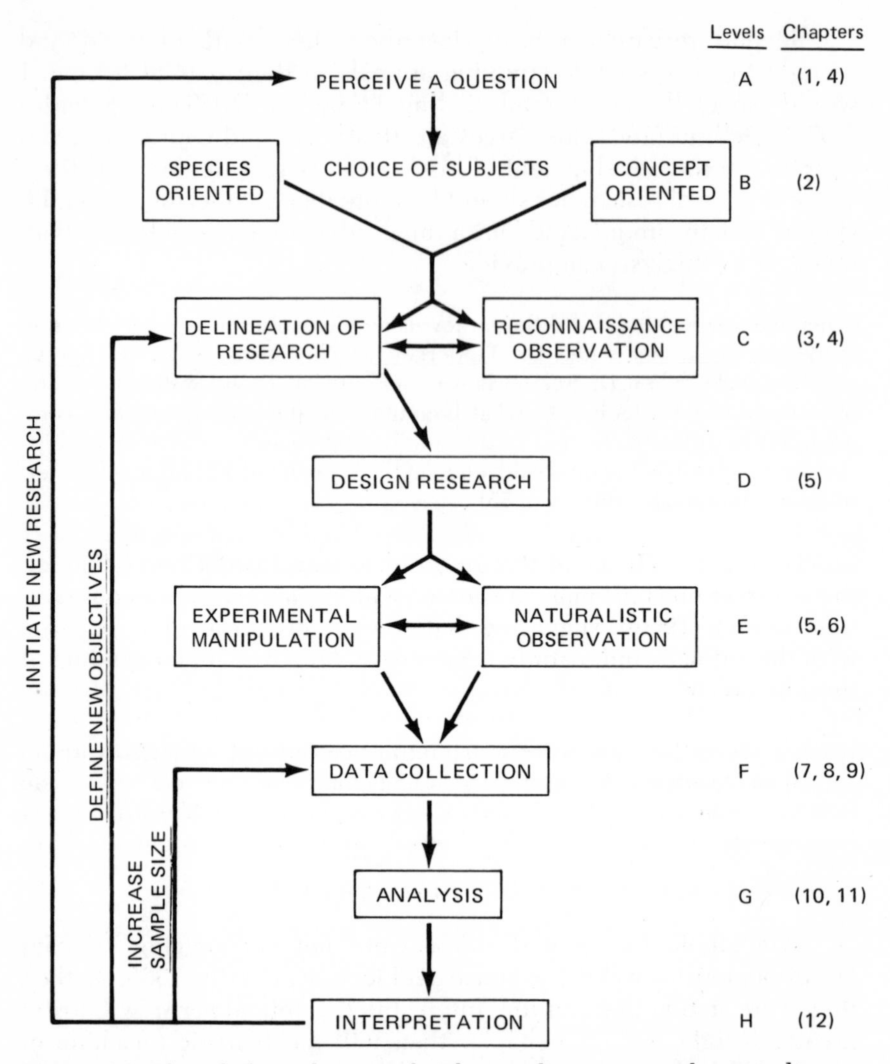

Figure 1-6 The ethological approach. The numbers in parentheses indicate the chapters in this book which discuss the respective steps in the ethological approach.

The ethological approach consists of a careful stepwise procedure by which descriptive data are gathered and hypotheses are tested (hypothetical-deductive method). It expands on the four basic operations described by Delgado and Delgado (1962): 1) the isolation of objects or variables to be measured, 2) the establishment and definition of units of measurement, 3) the comparison of actual events and units, and 4) the interpretation of the meaning of the observed events. Even at its earliest stages (*i.e., ad libitum* observation) the approach includes ill-defined hypothesis testing. Most people's approach to life is through constant hypothesis testing. For example, the very first time we go to work we travel a specific route which we hypothesize will get us there. After a series of successful trips our hypothesis that our office is located in Room 300 in the building at the corner of Oak and Center streets gains considerable support. However, should our boss move our office while we are on vacation, or the building be leveled by fire, we will have to reject our hypothesis and either revise it or formulate a new one.

More particularly, ethologists often perceive questions which arise from the rejection of our long-standing theories about the way animals should behave.

> . . . biologists are drawn to study events that seem to contradict what we have been taught to expect on the basis of our knowledge of non-living things. It is this discrepancy between what an animal "ought to do" and what it is actually seen to do that makes us wonder. Like a stone released in mid-air, a bird ought to fall; yet it flies away. [TINBERGEN 1972:20]

That is, we constantly seek the order which we believe exists in the universe, and when events occur which seem to upset our preconceived ideas of the ways things should be, we either ignore the situation or as scientists seek explanations through study.

E. OBSERVATION VS. EXPERIMENTATION

These two approaches will be discussed in some detail in later sections; however, at this point I want to stress the importance of both approaches. They are designated as naturalistic observation and experimental manipulation in Figure 1-6; but they are not synonymous with a field vs. laboratory dichotomy. They are not mutually exclusive but rather are two areas along a continuum (Fig. 1-1) and should be used to compliment each other. The following excerpts from Tinbergen illustrate how his study of the begging behavior of herring-gull chicks evolved from naturalistic observation to experimental manipulation:

> When the chicks are a few hours old they begin to crawl about under the parent, causing it to shift and adjust its position every so often. Sometimes the parent stands up and looks down into the nest, and then we may see the first begging behaviour of the young. They do not lose time in contemplating or

studying the parent, whose head they see for the first time, but begin to peck at its bill-tip right away, with repeated, quick, and relatively well-aimed darts of their tiny bills. They usually spread the wings and utter a faint squeaking sound. . . . Now and then the chicks peck at the food on the ground, but more often they aim at the parent's bill, and although this aiming is not always correct, it rarely takes them more than three or four attempts until they score a hit. . . . Their remarkable "know-how," not dependent on experience of any kind never fails to impress one as an instance of the adaptedness of an inborn response. It seems so trivial and common at first sight, but the longer one watches it the more remarkable it appears to be. . . . The reaction is innate, and it is obviously released by very special stimuli which the parent bird alone can provide, and which enable the chick to distinguish the parent's bill-tip from anything else it may encounter in its world.

We naturally were interested in the nature of these stimuli. In the literature we had found some observations which seemed to show that here again was a reaction dependent on only very few "sign stimuli." . . . Heinroth . . . wrote (1928) that his Herring Gull chicks had the habit of pecking at all red objects, especially when they were kept low, so that they could peck downwards. . . . The special sensitivity to red was further demonstrated by the fact that Goethe could elicit responses by red objects of various kinds, and of an appearance that was rather different from a Herring Gull's bill: such as cherries, and the red soles of bathing shoes!

It seemed to us worthwhile to go into this problem a little deeper. That the chicks were responding to the red patch was obvious; however, as the bill without red did also elicit some response, there must be more in a parent bird's bill that stimulated the chick. Also, the downward tendency had to be explained. As regards opportunity for experimental work, the reactions to cherries and bathing shoes showed that it should be easy to design dummies capable of eliciting responses. Further, the very fact that reactions to crude dummies were not rare, showed that the chick's sensory world must be very different from ours, for we would never expect a bathing shoe to regurgitate food.

Therefore, when in the summer of 1946 no war conditions prevented us any longer from working in the field, I took my zoology students out for a fortnight's work in one of the Herring Gull colonies on the Dutch Frisian Isles. We carried with us an odd collection of Herring Gull dummies and thus started a study which was to occupy and fascinate us during four consecutive seasons. [TINBERGEN 1960:178, 184–186]

The astute researcher oscillates between the two approaches utilizing the strengths of each to better test his hypotheses.

In 1972, Tinbergen (p. 20) cautioned:

I believe that Ethology is at the moment swinging away from observation, and that it is time for another move, a return towards a better balance between the two tendencies.

The "balance between the two" (observation and experimentation) is the key to good ethological research. Both approaches should include quantification, but *quantification and statistical analysis must not become the overlord of good quality observation.* Sometimes initial observations cannot be quantified, but their value is not necessarily greatly diminished.

The current belief that only quantitative procedures are scientific and that the description of structure is superfluous is a deplorable fallacy, dictated by the "technomorphic" thought habits acquired by our culture when dealing preponderantly with inorganic matter. [Lorenz 1973:152]

Historically, observation was thought by many to be the realm of ethologists and experimentation was believed to be found only in psychologists' laboratories. In his summary of the Seventh International Ethological Conference, Roeder (1961) diagrammed this dichotomy as follows:

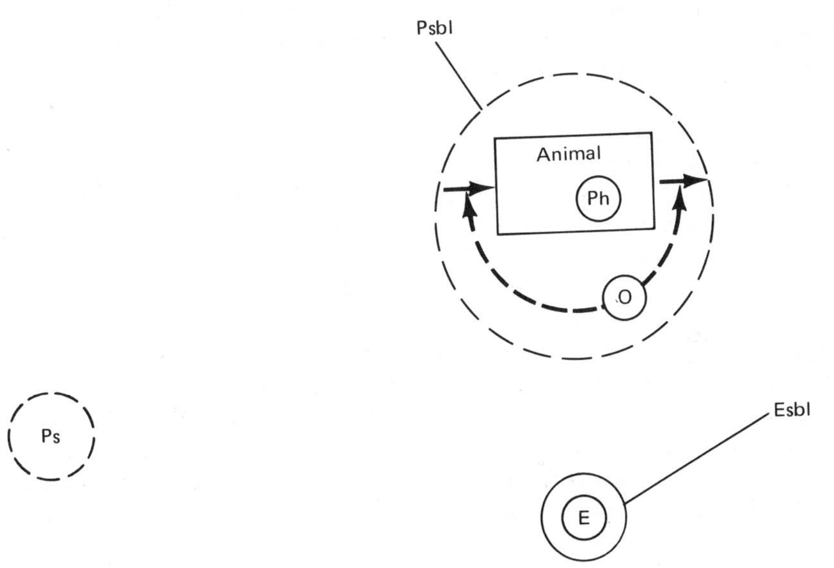

Ph—physiologist looks inside the animal
O—"optomotor boy" observes the input and the animal's output
E—ethologist sits in a blind *(Esbl)* and observes the animal
Ps—psychologist puts a blind *(Psbl)* around the animal

Lorenz expressed his suspicion that experimental psychologists were being guided solely by quantification without observing the behavior of animals placed in operant-conditioning chambers.

I can never help a shrewd suspicion that the worshiper of quantification and despiser of perception may occasionally be misled into thinking that two goats plus four oxen are equal to six horses. Counting pecks of pigeons in Skinner Boxes without observing what the birds inside really do, might occasionally add up to just this. [LORENZ 1960*a*:72]

Lorenz's comment somewhat overstated the case, but is still valid in many instances today. However, my purpose here is not to campaign for more direct observation in the experimental psychologists' methods (Hutt and Hutt, 1974, have effectively done this), but rather to insist that ethologists recognize the necessity for observation, experimentation, and quantification.

In 1963, while discussing the history of ethology, Tinbergen expressed concern that the balance was shifting too far towards experimentation.

We must hope that the descriptive phase is not going to come to a premature ending. Already there are signs that we are moving into an analytical phase in which the ratio between experimental analysis and description is rapidly increasing. [TINBERGEN 1963:412]

The study of animal behavior has come of age. Experimentation and quantification are no longer the tools only of psychologists and observation the only tool of ethologists. The disciplines have long since overlapped, and there is no longer a territorial battle claiming that methodologies belong solely to one discipline or the other (Hutt and Hutt 1970; Willems and Raush 1969). Ethologists and psychologists no longer work in their private vacuums, but rather they are united in the quest for knowledge about behavior. See Burghart (1973) for an extensive account of the development of ethological and psychological concepts into a more holistic approach. Both ethologists and psychologists should be committed to a balanced approach, one that includes naturalistic observation and experimental manipulation, both of which should include quantification.

2
Choice of Subjects

Ethologists come to study particular species for different reasons, but generally travel via two routes: 1) their special interest in a particular *species*, or 2) the species that is suitable for investigating a particular *concept* (see also Gans 1978). These two routes are intertwined so that a researcher may become interested in pursuing a concept after studying various aspects of a species' behavior. Likewise, a researcher may initially study a species in pursuit of answers to a conceptual problem and become fascinated with other aspects of the species' behavior.

A. SPECIES-ORIENTED RESEARCH

Many ethologists who settle on a particular species to study are naturalists who discover a fascinating species while spending time in the field. The following excerpt from *Curious Naturalists* describes how Tinbergen began his many years of research on the digger wasp:

> On a sunny day in the summer of 1929 I was walking rather aimlessly over the sands, brooding and a little worried. I had just done my finals, had got a half-time job, and was hoping to start on research for a doctor's thesis. I wanted very much to work on some problem of animal behaviour and had for that reason rejected some suggestions of my well-meaning supervisor. . . .
>
> While walking about, my eye was caught by a bright orange-yellow wasp the size of the ordinary jam-loving *Vespa*. It was busying itself in a strange way on the bare sand. With brisk, jerky movements it was walking slowly backwards, kicking the sand behind it as it proceeded. The sand flew away with every jerk. I was sure that this was a digger wasp . . .
>
> I watched these wasps at work all through that afternoon, and soon be-

came absorbed in finding out exactly what was happening in this busy insect town. . . .

As I was watching the wasps, I began to realize that here was a wonderful opportunity for doing exactly the kind of field work I would like to do. Here were many hundreds of digger wasps—exactly which species I did not know yet, but that would not be difficult to find out. . . .

My worries were over; I knew what I wanted to do. This day, as it turned out was a milestone in my life. For several years to come I was to spend my summers with these wasps, first alone, then with an evergrowing group of co-workers, most of them zoology students of the University of Leiden. [TINBERGEN 1958:5–8]

Dethier, who spent many years pursuing the behavioral biology of the blowfly, describes how he settled upon the blowfly as a research animal in the following excerpt from his delightful book *To Know a Fly*:

When choosing an experimental animal, therefore, why settle for anything so prosaic as the laboratory rat, so giddy as the guinea pig, so phlegmatic as the frog, so reptilian as the chicken, so cousinly as the chimpanzee? Why not choose an excitingly different creature like the aardvark or the dugong? Why not choose the fly?

With so many kinds of flies in nature's burgeoning storehouse of life, how does one choose a proper species for study? The answer is simple. Let the species choose you. This was how our laboratory came to work with the black blowfly fifteen years ago. [DETHIER 1962:7–8]

The rest of Dethier's account of his early years of research on the blowfly make heuristic and enjoyable reading.

Von Frisch, a nobel laureate along with Lorenz and Tinbergen, began his many productive years on the study of bees through his attempt to "set the record straight" regarding the color vision of bees. Von Frisch describes the launching of his career of bee study in the preface to his fascinating book, *The Dancing Bees*:

. . . some forty-five years ago . . . a distinguished scientist, studying the colour sense of animals in his laboratory, arrived at the definite and apparently well-established conclusion that bees were colour-blind. It was this occasion which first caused me to embark on a close study of their way of life; for once one got to know, through work in the field, something about the reaction of bees to the brilliant colour of flowers, it was easier to believe that a scientist had come to a false conclusion than that nature had made an absurd mistake. Since then I have been constantly drawn back to the world of the bees and ever captivated anew. I have to thank them for hours of the purest joy of discovery, parsimoniously granted, I admit, between days and weeks of despair and fruitless effort. [VON FRISCH 1953:iii]

Some researchers have begun their studies in animal behavior on a single species and then become interested in a limited type of behavior which they have generalized to other species. For example, Allee recounts how he began studying freshwater isopods and then became interested in social behavior in general.

Almost forty years ago as a graduate student in zoology I was engaged in studying the behavior of some common small fresh-water animals called isopods . . . day after day I put lots of five or ten isopods into shallow water in a round pan. . . . When a current was stirred in the water the isopods from the streams usually headed against it; but those from ponds were more likely either to head down current or to be indifferent in their reaction. . . .

Rather cockily I reported after a time to my instructor that I had gained control of the reaction of these animals to a water current. By the judicious use of oxygen in the water, I could send the indifferent pond isopods hauling themselves upstream, or I could induce the stream isopods to going with the current. I had not reckoned with another factor that presently caught up with me.

After a winter in the laboratory it seemed wise as well as pleasant to take my pan out to a comfortable streamside one sunny April day and there check the behavior of freshly collected isopods in water dipped from the brook in which they had been living. To my surprise, the stream isopods, whose fellows all winter had gone against the current, now went steadily downstream or cut across it at any angle to reach another near-by isopod. When I used five or ten individuals at a time, as I had done in the laboratory, they piled together in small close clusters that rolled over and over in the gentle current. Only by testing them singly could I get away from this group behavior and obtain a response to the current; and even this reaction was disconcertingly erratic.

It took another year of hard work to get this contradictory behavior even approximately untangled; to find under what conditions the attraction of the group is automatically more impelling than keeping footing in the stream; and that was only the beginning of the road that I have followed from that April day to this time, continuing to be increasingly absorbed in the problems of group behavior and other mass reactions, not only of isopods, but of all kinds of animals, man included.

As the years have gone on, aided by student and other collaborators and by the work of independent investigators, I have tried to explore experimentally the implications of group actions of animals. [ALLEE 1938:5–7]

Allee pursued his interest in the social behavior of animals and became an early and influential ethologist.

Ethologists often become deeply interested in particular species and pursue different lines of questions as they arise while studying other aspects of the species' behavior. The cliche that "research generally provides more questions than it answers" keeps some ethologists studying one, or a few, species for their entire career. This is often efficient in that they can build on past experience with the species and are able to make effective use of their time.

> It is obvious that the best contributions have come from people who have given years of their life to careful, patient observation of one species. [TINBERGEN 1953:129]

B. CONCEPT-ORIENTED RESEARCH

Although Lorenz came to study birds, particularly waterfowl, as an outgrowth of his apprenticeship with Heinroth, he selected species with an eye to unraveling analogous and homologous relationships in patterns of behavior. Nisbett describes Lorenz's choice of subjects this way:

> However his interest in an animal may have arisen in the first place—and this may in part have been by the interplay of chance and curiosity—his chosen subjects did in fact form a coherent and rational array. The different species fell into several groups. First, there were those which were in their own right the central objects of his study; initially the jackdaws, then the herons, and now the geese. Second, were the closely related species: ravens for comparison with jackdaws, or mallard ducks to watch out of the corner of an eye while looking at geese. These showed not only what the ducks had in common with their geese cousins but also what they had developed differently: he could ask himself 'why?' Heron society was markedly different from that of jackdaws or geese; again 'why?' Then there were the species unrelated to jackdaws or geese, but which had similar elements of behaviour. This allowed him to look for patterns of behaviour to which evolution came independently in different species. [NISBETT 1977:44]

Concept-oriented research is also illustrated by Alcock's (1973) use of red-winged blackbirds to test L. Tinbergen's (1960) "search image hypothesis." Earlier work with the species had suggested that they might selectively search in patches where they had found food previously, but more importantly for Alcock they were omnivorous and available. Therefore, he used this species in an experimental food maze to answer the question ". . . do birds learn *where* they are likely to find food and come to use locational cues to direct their searching and/or do birds learn *what* food they are likely to find and come to search preferentially for that item on the basis of visual cues associated with the food?" The bird in this case was the red-winged blackbird. Other researchers have tested L. Tinbergen's hypothesis on carrion crows *(Corvus corone)* (Croze 1970), domestic chicks *(Gallus gallus)* (Dawkins 1971), and great tits *(Parus major)* (Royama 1970).

1. August Krogh Principle

Researchers who pursue answers to conceptual questions or relatedly concentrate their efforts on a particular type of behavior attempt to study species that best represent the concept or type of behavior under study. In 1929 August Krogh stated that "For a large number of problems there will be some animal of choice, or a few such animals, on which it can be most conveniently studied." Krebs (1975) has labeled this statement "The August Krogh Principle."

The indigo bunting was selected by Carey and Nolan (1975) to test the "Verner-Willson-Orians hypothesis" that polygyny (one male mating with two or more females) would evolve in avian species where critical resources are distributed sufficiently unequally to offset the disadvantages of reduced parental attention and possibly increased attraction of predators and depletion of food resources. Preliminary study of an indigo bunting population in Indiana had led Carey and Nolan to predict that the population would be polygynous and therefore provide a good opportunity to test the hypothesis.

Experimental psychologists traditionally use rats, pigeons, or selected primates as their subjects in studying learning theories that they believe will then be applicable to (or at least worthy of testing on) man. Two reasons are generally given for pursuing concepts of learning on such few species. First, many psychologists suggest that learning processes are basically the same in all species, differing in amount rather than kind. These species are being used as biological models in essentially the same way that the drug industry uses rats and mice to test drugs designed for human consumption. Second, there is a wealth of background data already available on these species. It is questionable whether the August Krogh Principle was invoked when the rat was selected for the early psychology studies. It was probably convenience that provided and continued the momentum that established the rat as the psychologist's choice subject (Beach 1950).

Several years ago while cooling off in a local pub (after a hot afternoon observing yellow-headed blackbirds in a mosquito-infested marsh), Dr. Gordon Orians explained to J.R. Watson and myself that he had become very interested in the slugs that foraged in his backyard garden in Seattle. His interest was more in testing ecological theory than in saving his vegetables. Later Cates and Orians reported on their test of the hypothesis that ". . . early successional plant species make a lesser commitment of resources to defend against herbivores, and

Table 2-1. Credit-debit sheet of some characteristics to be considered when selecting a subject species.

Characteristic	Questions	Credit	Debit
Suitability	Is the species suitable for the concept being studied (August Krogh Principle)? Does it engage in interesting behavior which you can observe repeatedly?		
Availability	Is the species found locally or will you have to travel to study it in the field? If it is found in a foreign country, what are the political ramifications?		
	If you want to observe the species in the wild, is it accessible in its habitat? Can observations be easily made? Is it nocturnal or diurnal?		
	If you want to bring the animal to you can it be easily obtained? Is it on the rare and endangered species list? Can it be easily captured?		
Adaptability	How will the animal adapt to life in captivity? Can you provide for its special needs? Is its habits compatible with yours?		
Background Information	What is already known about the species? Is there a reasonable backlog of data on which to build? How does it help you answer the questions above and anticipate other problems?		
Summary	Total the credits and debits, assess the financial commitment, and accept or reject the species as your subject.		

should then provide better food resources for generalized herbivores than later successional and climax plants" (Cates and Orians 1975:410). What would be a good research animal on which to test these predictions? Cates and Orians (1975:411) explain:

> Generalized herbivores are required for testing these predictions: we have used in our experiments two species of slugs, both known to graze on a wide variety of plants, and both abundant in and around Seattle, where they are active most of the year.

2. Suitability vs. Availability

It would be ideal to study the species we want, either because of our interest in the species *per se* or because it is the most appropriate species for studying a concept (August Krogh Principle). However, this is often not possible or desirable due to a multitude of factors; suitability and availability are only two of these factors. For example, Leuthold (1977:13) states that "Ease of observation is probably the main factor that has influenced the choice of species [among African ungulates] for field studies."

In choosing a subject species, you would find it valuable to construct a credit-debit sheet of the desirable and undesirable characteristics of the particular species for your study. Table 2-1 contains a partial list of characteristics to be considered.

However you come to choose your subject animals, do it well. Your investment in studying that species may be years or your entire career. Your colleagues will associate you with the species whether you like it or not. Select wisely and spend your time well.

3 Reconnaissance Observation

A. HOW TO OBSERVE

An early step in the study of animal behavior involves intensive reconnaissance observation. This may occur before you have decided what aspect of behavior to study and probably before you have formulated any hypothesis (Lorenz 1935). This early stage in which you become familiar with the animal's behavior ". . . is the most arduous and demanding aspect of behavioral study" (Marler 1975:2). It is extremely important, for no successful research can be launched without this background knowledge. One source of questions and hypotheses are these initial observations.

Having myself always spent long periods of exploratory watching of natural events, of pondering about what exactly it was in the observed behaviour that I wanted to understand before developing an experimental attack, I find this tendency of prematurely plunging into quantification and experimentation, which I observe in many younger workers, really disturbing, unless, as happens to some, they do, from time to time, return, more purposefully than before, to plain, though more sophisticated watching. [TINBERGEN 1951:vi]

1. Watching vs. Observing

In response to Nisbett's questioning him about why he studied animals rather than man, Lorenz replied:

The naive justification is that you like doing it, you love animals and you gloat over them in a stupid way. If they do not give you that simple pleasure, not even an Asiatic yogi would have the patience to look at animals for as long as is necessary to observe them. [NISBETT 1977:9]

Many people watch animals but few really observe them. Most visitors to zoos stop and watch some animals. When they leave they can tell you about most of the animals they saw and some that they watched; but few can relate any in-depth *observations* of particular behaviors. Bird watchers typically see a bird, identify it, and rush on hoping to add more species to their lists. Observers take the time to study the behavior of the birds, describing in their notebooks the intricate details of individual and social behaviors and perhaps recording the birds' vocalizations for later reference and enjoyment. Intense observation is generally more rewarding than superficial watching. It is also necessary in the study of animal behavior. Observation may be as much a state of mind, of awareness, as it is a clearly defined technique. This is reflected in the quotes on pages

An observer must be more than a visual recorder; he must also be aware of input to his other senses and must think. One must be disciplined enough to know when to be a machine-like recorder of data and when to contemplate what is happening or has happened. The astute observer often develops and tests hypotheses mentally while keeping animals under observation. Obviously, attention to rapidly occurring behavior is not consistent with theorizing, hence priorities must be established.

Priorities for activity while observing various animals are generally the same, as follows:

1. Record data accurately and completely.
2. Check equipment to make sure it is working properly and repair it if necessary.
3. Think about what is happening, put it in perspective, formulate hypotheses, think about them, discard the easily disproven ones, and write down those that call for additional thought and testing. This constant consideration of hypotheses is a procedure to be used only when making initial reconnaissance observations for the formulation of questions, objectives and hypotheses. It should be avoided as much as possible during actual data collection (p. 108).
4. Satisfy basic bodily needs. Plan your sampling periods so that your basic bodily needs do not suddenly appear to be priority number one.

Periods spent in the field are likely to be some of the most treasured times of your existence. It should be both productive and enjoyable, but it won't be constant fun. Lott (1975) reflected the feelings of many ethologists when he wrote *Protestations of a Field Person*:

"Welcome back! Have a good vacation?"

"I wasn't on vacation. I was in the field."

Well that's not what I mean. Being in the field isn't a vacation; it's hard work, a hard life, and besides . . .

But hold your tongue. People who spend months at a time noting the behavior of animals in odd corners of the world are usually greeted that way. We're happy with our work, of course, but for several reasons it doesn't qualify as a vacation.

Sometimes just living there is a problem. If you can't find or afford a convenient house, camping out becomes living in a tent by the second week. And the kind of stick-to-your-ribs food that stores well in a burlap bag or metal box soon starts to stick in your throat. Some colleagues and I went to visit Patti Moehlman's burro study in Death Valley a couple of years ago, and brought along some steak. Patti had been without refrigeration for weeks. Her response was succinct and eloquent: "GOLL-EE REDMEAT." She ate enraptured, purging for a time the taste and texture of globs of margarine and peanut butter on week-old bread, and Vienna sausage taken neat, still cold, and coated with gelatin. So welcome was the steak that it hardly mattered that the water from a desert cloudburst streamed into our plates as we ate crouching under a picnic table.

But food and housing are far from the worst of it. You can get used to eating almost anything and sleeping almost anywhere. The worst of it is that you get to be a little bit batty.

To be more specific, you get to be sort of manic-depressive. You experience mood swings that increase as a direct function of the number of seven-day weeks you've spent on the project. The research has its ups and downs, of course, but they are nothing compared to the emotional roller coaster the researcher is riding. The most salient symptom is that your evaluation of the study gets to be wildly unrealistic.

High noon may find your pulse racing as your mind forms the kind of modest, contained, but penetrating remarks that persuade the National Academy of Sciences plenary session that the Nobel Committee did indeed know what it was doing when it cited your analysis of the distribution of deer droppings, the grumpiness of goose gatherings, or the ballistics of bison bellowing as the intellectual link that completed the conceptual chain from molecule to mastodon. Your lips move a little as you take on a set of bored, cynical journalists who came to the press conference to play it for laughs and a chance to get in a dig at the granting agency that spent nearly $1,750 in support of your research. A basic stock of fine ironic wit, a dash of captivating candor, an irresistibly lucid illumination of The Link in layman's terms and they are first sobered, then entranced. When you release them from your spell, they will sprint to their typewriters and set their two forefingers to banging out near poetry in praise of basic research and (blush) you.

That evening you may be so sunk in shame that you want to change not only your study but your name. How could you have committed yourself to a study so barren and one for which you are so ill prepared? What will you say to that granting agency when they ask what became of more than $1,750 intended to support *significant* basic research? If you take the entire blame you'll never get another chance. Besides it wasn't *all* your fault; but how do you make them understand that fate has thwarted you at every turn, that your field glasses fogged up during nearly three goose gatherings, that the microphone salesman was lying, *lying* when he told you how far away it would pick up bison bellows? Yes, who will bear witness that your failure was not really your fault now that God has turned his face from you?

And so you go on, ever more sublimely happy, ever nearer suicide. During your more lucid moments you realize, of course, that you're getting to be a little bit batty, and you come to crave some stabilizing influence to dampen your oscillations. Contact with an old friend becomes so welcome that you hold your tongue even if he says something stupid like, "Welcome back! Have a good vacation?" [See also Hailman's (1973) discussion of fieldism]

The following quotation carries a hidden message: ". . . as the *work* [italics mine] by Cain and Sheppard has shown . . ." (Tinbergen 1958). Whether intended or not, the word *work* is indicative of what ethological studies can, at times, become. Even though the overall experience is enjoyable and the results rewarding, it can become tedious, and you are often confronted with having to convince yourself that the end justifies the means. Schaller's description of part of his study on the Serengeti lion is illustrative.

My existence revolved around lions, I was wholly saturated with them, talked and wrote about them, and thought about them. . . . A few times, though, I saw too much of lions. Once Bill . . . and I decided to track a lion continuously by radio for several weeks. . . . The first few days were rather pleasant. . . . As the days passed this delight vanished, and we went about our task with grim determination. . . . We stayed with the male for twenty-one consecutive days and suffice it to say that for once I had a surfeit of lions. [SCHALLER 1973: 90–91]

Estes (1967:45) admitted to near boredom when he wrote, "A week later, at 9:20 P.M., while I was out alone, the rallying call of hyenas again distracted me from *the rather tedious job of recording gnu activity patterns* [italics mine]."

The descriptions of behavior of wild animals that you read in the literature are often the result of days and weeks and years of careful stalking and painstaking observations. Often hours are spent in a blind under less than ideal conditions, with inclement weather making you physically uncomfortable and your view of the animals poor, and the inactivity of the animals is frustrating. Your binoculars get beaten about and rained and snowed upon, and the pages of your field notes become limp and stuck together. Field research can be trying at times, but you can make the best of it by being physically and mentally prepared. Expect Murphy's Law: "If anything can go wrong it will" to take effect from time to time. Allow for some slack in your schedule to absorb days when you cannot collect data because of poor weather or equipment breakdowns. Often, you release much frustration merely by recording the disasters in your field notes, realizing that in years to come you will remember even those days fondly as you reflect on the data-rich days with pride.

In conclusion, successful collection of data through observations necessitates your: 1) having the proper equipment (*e.g.*, binoculars and telescopes), 2) understanding the various ways to describe behavior, 3) having a well-designed system for recording your field notes, and 4) knowing when your data are sufficient. The observations we are discussing here are either *ad libitum* or initial reconnaissance observations on which a future well-designed study will be based. Nevertheless, the skills used are the same; only the relative emphasis on the data collected is likely to be different.

2. Field Notes

Field notes are often the best, and sometimes the only, record you have of your activities and observations in the field. Good field notes are the end result of developing a skill into an art, and the basis for learning those skills is a knowledge of fundamentals.

> Note taking . . . is an *art* requiring sensitivity and skill. Perfection is unattainable, a goal to be sought and pursued through selective training and persistent practice just as beauty or reality are sought by the sculptor, the musician or the poet. [EMLEN 1958:178]

The system of taking notes that you decide upon will determine their value to you and other researchers in the future. Most ethologists with whom I have spoken use some variation of the system developed by Dr. Joseph Grinnell of the Museum of Vertebrate Zoology, University of California, Berkeley. The usual format is to divide the notebook into three sections: 1) journal, 2) species accounts, and 3) catalog. The species-accounts section is used to keep records of your observations by species in contrast to the journal, where your observations are recorded by date and time. The catalog section is used to record specimens collected in the field. The discussion to follow is based upon Grinnell's system, but will be concerned only with the journal section.

Good-quality white, lined paper should be used. It should be a bond, having a rag content of 50–100%. The size should be approximately 6¼ inches by 8½ inches, (16 centimeters by 21½ centimeters), preferably looseleaf and kept in a three-ring binder. Many fieldworkers use two notebooks. One is taken into the field for the day's records, and the other contains past records for the year and is kept safely in camp or at home. Past journals should be kept by year. Field notes are extremely valuable, and loss or destruction in the field must be avoided.

Black waterproof ink is usually recommended, a hard pencil being second best. You should always carry a pencil, however, for both a backup and for writing in heavy mist or rain; ink is not waterproof until it is dry.

The following three inks are recommended: 1) Pelikan Drawing Ink, 2) Higgins Eternal Black Ink, and 3) Koh-I-Noor Rapidograph Ink.

A sample field note (journal) page is shown below:

Page no.
Name
Locality
Date

Other observers
Weather
Habitat type
Time into field

Time: Observations and remarks

Time: Observations and remarks

Time out of field

Page number: Number consecutively for the year.

Locality: Specific locality, direction and estimated distance from known point (*e.g.*, E shore of Cobb Lake, 6 km NE of Ft. Collins, Colorado); section, township and range are also useful information.

Date: Date of observations, travel, meetings, etc.

Other observers: Provides for later verification and elaboration (Remsen 1977).

Weather: Precipitation, percent cloud cover, wind speed and direction, temperature, etc. INSERT CHANGES IN YOUR NOTES AS THEY OCCUR. Temperature can be measured by carrying a small, metal-shielded thermometer with you into the field. For most studies it will be sufficient, and worthwhile (*e.g.*, Mrosovsky and Shettleworth 1975), to estimate measures of the other weather variables. Beaufort's 13 categories are useful for determining approximate wind speed:

0. *Calm:* Movement of the air is less than 1 mile (1.6 Km) an hour. Smoke rises vertically; the sea is mirror-smooth.
1. *Light air:* 1–3 miles (1.6–4.8 Km) an hour. The drift of smoke indicates the direction of the breeze.
2. *Light breeze:* 4–7 miles (6.4–11.3 Km) an hour. Leaves begin to rustle.
3. *Gentle breeze:* 8–12 miles (12.9–19.3 Km) an hour. Leaves and twigs in motion; crests on waves begin to break.
4. *Moderate breeze:* 13–18 miles (20.9–29.0 Km) an hour. Small branches move; dust rises; many whitecaps on waves at sea.
5. *Fresh breeze:* 19–24 miles (30.6–38.6 Km) an hour. Small trees in leaf begin to sway.
6. *Strong breeze:* 25–31 miles (40.2–49.9 Km) an hour. Large branches begin moving.
7. *Moderate gale:* 32–38 miles (51.5–61.1 Km) an hour. Whole trees in motion.
8. *Fresh gale:* 39–46 miles (62.8–74.0 Km) an hour. Twigs break off.
9. *Strong gale:* 47–54 miles (75.6–86.9 Km) an hour. Foam blows in dense streaks across the water at sea.
10. *Whole gale:* 55–63 miles (88.5–101.4 Km) an hour. Trees uprooted; huge waves build up with overhanging crests.
11. *Storm:* 64–75 miles (103.0–120.7 Km) an hour.
12. *Hurricane:* Wind velocities above 75 miles (120.7 Km) an hour.

Habitat Type: General topography and vegetative cover; note prominent physiographic features.

Time into field: Time at which you begin your field-related activities (*e.g.*, leave home or camp for study area or to pick up colleague, etc.)

Time: Time of observation, both beginning and end; use military time (*e.g.*, 1345–1510). Midnight = 0000; Noon = 1200; 6:00 P.M. = 1800.

Observations and remarks:

A. Record the species, number, age, and sexes if possible.
B. Describe behavior as accurately, clearly, concisely, and completely as possible (see discussion below).
C. Include remarks such as unusual occurrences (Short 1970), thoughts, and ideas about the behavior.

RECORD OBSERVATIONS AT ONCE—DO NOT TRUST TO MEMORY

Time out of field: It is desirable to record time at which you leave the study site, as well as when you reach home or camp.

General Comment: At the end of each day's field observations it is often desirable to head a new page General Comments and list:

Route of travel
Hours of observation
Weather
Species observed
General impressions about the day's observations

Some fieldworkers recopy their field notes after a day's observations (*e.g.*, Schaller 1973; Remsen 1977; P. Johnsgard, personal communication). Others believe that your first impressions are the most accurate and that in recopying you are prone to edit them, thereby making them less accurate. Regardless, remember that this is a record of your fieldwork—*be complete*. It is a source of data and hypotheses and is the basis for future research—*be accurate*. It may be used by other researchers for similar purposes—*be clear and concise*. Last, but not least, it will be a diary and a source of memories; insert thoughts you will enjoy having again, twenty or forty years later.

3. Equipment

a. Blinds

The purpose of a blind (or hide) is to allow observation of animals with as little disturbance as possible. That is, you hope you are observing behavior which is unaffected by your presence, whether you are perceived or not by the animals. In situations where animals are accustomed to human presence or where you can habituate them to yourself, no blind is necessary. At the other extreme are species with whom great caution must be employed.

Blinds are of two general types—*natural* and *artificial*—and may be sunken or built at or above ground level. Knowledge of the animal's reactions to novel objects at various places in their environment is necessary for selecting the proper blind. Must you select a natural blind with as little disturbance as possible or can you construct an artificial blind and allow the animals time to habituate themselves to it? Since coyotes in the National Elk Refuge were accustomed to seeing vehicles along the dirt roads, Ryden (1975) rented a bright yellow van to use as a blind for her observations. The same ploy was used by Kucera (1978) in his study of mule deer in Big Bend National Park, and by Walther (1978) in his observations of oryx in Serengeti National Park.

Artificial blinds are generally designed to blend in with the habitat as closely as possible. An unobtrusive blind is less likely to disturb the animals and attract curious humans.

Description and discussions of blinds and their use can be found in Hanenkrat (1977) and outdoor photography literature, such as Baufle and Varin (1972), Ettlinger (1974), and Marchington and Clay (1974). Pettingill (1970) describes the construction of two types of typical artificial blinds (Fig. 3-1).

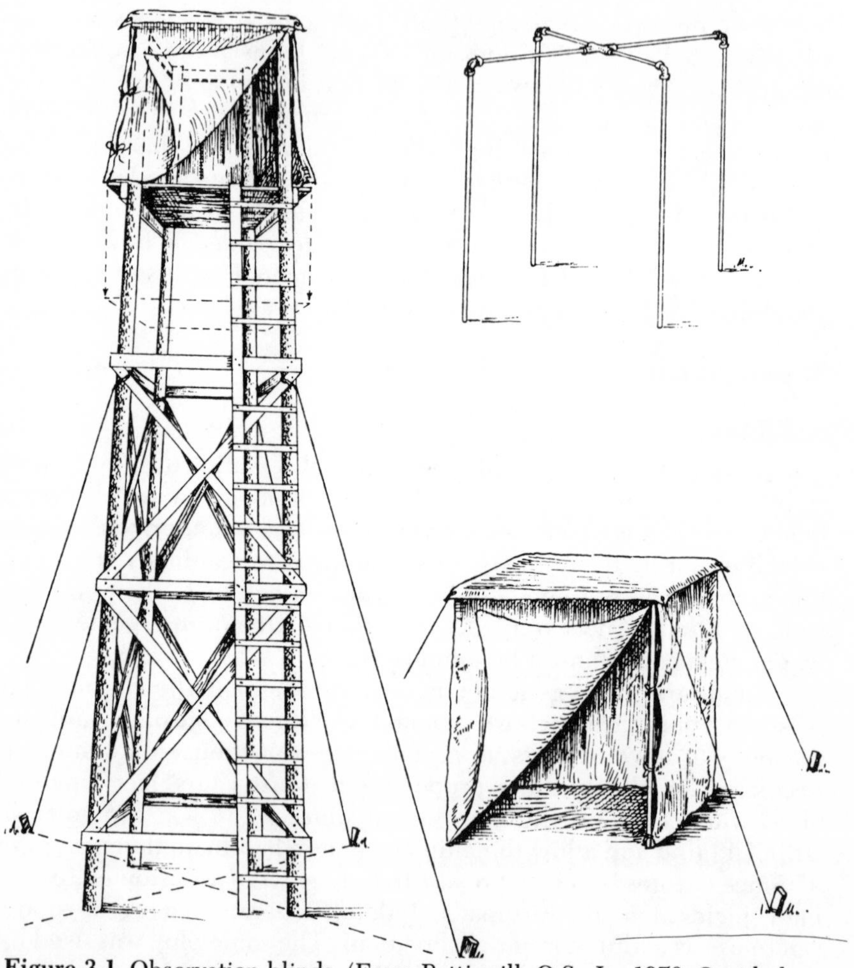

Figure 3-1 Observation blinds. (From Pettingill, O.S., Jr., 1970, Ornithology in laboratory and field, 4th ed., Burgess. Publ. Co., Minneapolis, p. 427, Pl. 29.)

The following characteristics should be considered when selecting a natural blind or constructing an artificial blind:

1. Behavior of animal
   a. Reaction to strange objects; approach—avoid.
   b. Spatial distribution of behavior patterns; are you at the right spot to observe the behavior?
2. Observational capability
   a. Number and size of openings for observation and filming.
   b. Capacity for number of observers anticipated.
3. Permanence
   a. Can it be permanent for several weeks or years? Will the animals be there during your observation periods?
   b. Should the blind be temporary and movable? How often and how rapid will the moves be? A camouflage suit can be considered the most movable artificial blind.
4. Climatic conditions
   a. Severity; must it be built to withstand severe wind and precipitation?
   b. Prevailing winds may be important for locating the blind downwind to reduce your olfactory and auditory stimuli from reaching the animals.
   c. Comfort and movability are often conflicting objectives. Remember that you can function effectively as an observer-recorder only if you are reasonably comfortable.

b. Binoculars and Spotting Scopes

One of the most important pieces of equipment to the ethologist is a good pair of *binoculars*. In fact, as Tinbergen (1953:132) has said, they "... are almost indispensable."

Humans are a vision-dependent species; therefore the ethologist tends to rely very heavily on what he sees. A very large percentage (perhaps over 95%) of what we record as observations are what we see. Equipment that will provide us with better vision or in other ways make the animals more visible, such as a strategically located blind, will pay off immensely.

Binoculars are constructed of lenses and prisms. The important components for our discussion are labeled in Figure 3-2.

Important characteristics are designated by the numbers commonly used to describe binoculars, such as:

7	×	35
↑		↑
magnification		diameter of objective lens (mm)

Magnification tells you the number of times greater than normal that an object being viewed will appear. Although increased magnifi-

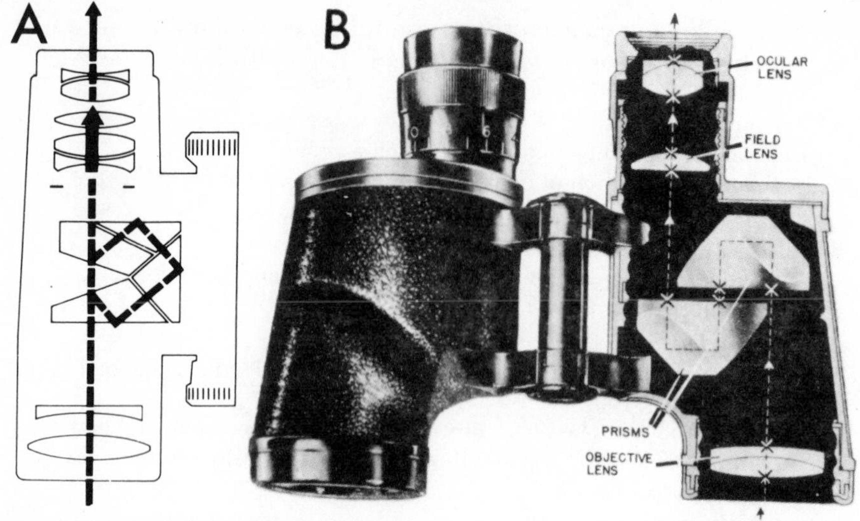

Figure 3-2 *A*, light passing through the prisms of one side of a roof-prism binocular, an advancement discussed in the text. *B*, porro-prism binocular, standard field. Each side has a two-lens ocular system, an objective lens and two prisms. X's mark the 10 optical surfaces where undesirable reflections occur. It is these surfaces that are coated to reduce reflection. A wide-field binocular has an additional lens in the ocular system, making a total of 12 surfaces to be coated (from Reichert and Reichert 1961).

cation would appear to be desirable, it often carries with it some problems. Generally, the following occurs with increased magnification:

1. The field of view becomes smaller (see below).
2. Clarity is lost, since the more powerful the lens the more the imperfections in the lens are also magnified.
3. Light transmission is lowered.
4. Increased blurring results from movement of the binocular.

The first three problems can, of course, be overcome by the manufacturer, but this will result in higher-priced binoculars.

Objective lens diameter affects the amount of light that enters the binocular. What is important, however, is the amount of light which finally passes through the *ocular lens* (light transmission or *relative brightness*).

Everything else being equal, the larger the objective lens the greater the relative brightness. To compare light transmission for uncoated binoculars use the following formula:

$$\left(\frac{\text{diameter of objective lens (mm)}}{\text{magnification}}\right)^2 = \text{relative brightness}$$

For example, 7 × 50 binoculars have about twice the relative brightness of 7 × 35 binoculars.

$$\left(\frac{50}{7}\right)^2 = 51 \qquad \left(\frac{35}{7}\right)^2 = 25$$

One important disadvantage of large objective-lens binoculars is that they are heavy (this weight is overcome in roof-prism binoculars, discussed later).

You can check for relative brightness by holding the binoculars eight to ten inches in front of your eyes and pointed at the sky. The circle of light which you see in the ocular lens is the *exit pupil*. Relative brightness increases as the square of the exit pupil:

$$\text{relative brightness} = (\text{exit pupil})^2 = \left(\frac{\text{diameter of objective lens}}{\text{magnification}}\right)^2$$

A large exit pupil can be wasted on a human eye pupil which is constricted on a bright day. The human pupil may be constricted to less than 3 mm on bright days whereas the binocular's exit pupil may often exceed this (Table 3-1). Therefore, the binocular usually becomes the primary limiting factor only under conditions of low light intensity.

Another measure of light transmission which takes into account relative brightness, as well as the percentage of light transmitted, is *relative light efficiency*. This is calculated according to the formula:

$$\text{relative light efficiency} = 2 \times \text{relative brightness} \times \text{percentage of light transmitted}$$

One factor which has a great affect on the percentage of light transmitted is *coating*. Light passing through untreated binoculars loses about 5% of its brightness at each surface it passes through. This can be considerable when ten surfaces are normally involved (Fig. 3-2*B*). Coating of the surfaces can reduce this figure to 0.5% at each surface. It generally increases the relative light transmission by 50%. Therefore, it is wise to insist that all the optics are coated in any binocular you purchase or intend to use.

Field of view is another characteristic of binoculars which is very important to the ethologist. This is limiting to different degrees depending on the type of research to be conducted. However, everything else being equal, you should select binoculars with the largest

Table 3-1. Some specifications for representative binoculars (from Reichert and Reichert 1961).

Model	Diameter Exit Pupil (in mm) *(EP)*	Relative Brightness *(RB)*	Relative Light Efficiency *(RLE)* Percentage of Coating			Field Feet at 1000 Yds. (in ft)	Field Angular Degrees	Relative Field (in ft) *(RF)*
			20%	80%	100%			
Binoculars—Standard Field								
6 × 15	2.5	6.2	6.8	8.7	—	370	7	2220
6 × 30	5	25	27	35	37.5	450	8.5	2700
7 × 18	2.6	6.8	7.5	9.5	—	325	6.2	2275
7 × 35	5	25	26.5	35	37.5	380	7.3	2660
7 × 50	7.1	50	55	70	75	380	7.3	2660
8 × 30	3.75	14	15.5	20	—	330	6.3	2640
Binoculars—Semiwide Field and Wide Field								
6 × 24	4	16	—	22	—	636	12	3816
6 × 25	4.2	17.5	—	—	26.2	577	11	3462
7 × 35	5	25	27.5	35	37.5	525	10	3675
8 × 30	3.75	14	15.5	20	21	390	7.4	3120
8 × 30	3.75	14	15.5	20	21	450	8.5	3600
8 × 40	5	25	27.5	35	37.5	375	7.2	3000
9 × 35	3.9	15	—	21	23	390	7.4	3510
10 × 50	5	25	27.5	35	37.5	370	7	3700

field of view. Usually the field of view is expressed in feet at 1,000 yards (900 meters); that is, the width of the scene you can see at 1,000 yards (900 meters). The field of view is controlled primarily by the *field lens* (Fig. 3-2*B*). The normal field of view can be increased (semiwide or wide field) by the manufacturer by using a different ocular system, especially a larger field lens. This, of course, also increases the price of binoculars.

Generally, the greater the magnification the smaller the field of view. Reichert and Reichert (1961) devised a formula to calculate the *relative field* for different types of binoculars.

$$\text{relative field} = \text{magnification} \times \text{field at 1,000 yards}$$

The larger the relative field the closer the binocular approaches the ideal—that is, high magnification and large field of view.

Binocular focusing systems are of two types. With *individual eyepiece focus* the observer focuses each eyepiece separately while keeping the other eye closed. If you focus them while viewing a distant object, your binoculars will be in focus at all distances beyond about 30 feet (9.1 meters). You must refocus for closer objects. Binoculars constructed with individual eyepiece focus are generally better sealed against moisture and dirt than are center focusing binoculars.

When adjusting *center focus* binoculars the observer uses the center focus wheel to focus the left eyepiece while the right eye is closed, and then the right eyepiece is focused while the left eye is closed. Now the observer can focus the binoculars for varying distances by using the center focus alone. This is an advantage over individual eyepiece focus binoculars; however, center focus binoculars are not as easily sealed against moisture and dirt.

It should be clear from this overview of binoculars that the choice of the proper binocular is always going to be a trade-off of advantageous and disadvantageous characteristics. These are summarized in Table 3-2.

An advance in prism binoculars has been the replacement of porro-prisms (Fig. 3-2*B*) with roof-prisms (Fig. 3-2*A*). Leitz was the first to introduce roof-prism binoculars, and they still produce high-quality optical instruments. The optical characteristics of roof-prism binoculars, although essentially the same as those for porro-prism binoculars, are in some instances better. Representative figures are given in Table 3-3.

The greatest advantages of roof-prism binoculars are their ease of handling and low weight. These result from their slim shape and small size compared to porro-prism binoculars (Fig. 3-2). They are more

Table 3-2. Summary of advantages and disadvantages that accrue from features in binoculars.

Features	Advantages	Disadvantages
1. Greater magnification	Viewing animals more closely	a. Smaller field of view b. Lower light transmission c. Poorer image clarity d. Increased movement of binoculars e. Greater weight
2. Larger objective lens	More light transmission	Heavier
3. Larger field of view	Increased size of scene	Increased price
4. Coating	Increased light transmission	Increased price
5. Individual eyepiece focus	Better sealing against moisture and dirt	Individual focus of eyepieces at close distances
6. Center focus	Use of center focus alone for varying distances	Less well sealed against moisture and dirt

expensive than porro-prism binoculars, but only the individual consumer can properly weigh convenience against cost.

The Department of Defense has developed *starlight binoculars* that are effective under conditions of low light intensity (Fig. 3-3*A*). They work on the principle of photomultiplication, the same as the starlight scope (Fig. 3-3*B*), but they are not nearly as effective. These binoculars can sometimes be obtained on short-term loan through the Civil Affairs Section of the U.S. Army.

The only *spotting scopes* that are useful to the ethologist are prism scopes which are constructed on the same principle as prism binoculars (Fig. 3-2). Therefore their operation is basically the same as one half of a prism binocular. The same characteristics of brightness and field of view are applicable to scopes and are calculated in the same way.

Table 3-3. Some representative optical specifications for roof-prism binoculars.

Model	Diam. Exit Pupil (in mm) *(EP)*	Relative Brightness *(RB)*	Field Feet at 1000 yds. (in ft)	Field Angular Degrees	Relative Field (in ft) *(RF)*
6 × 18	3	9	420	8	2520
7 × 21	3	9	372	7.6	2604
8 × 24	3	9	366	7	2928

Figure 3-3 *A*, Starlight binoculars. *B*, Starlight scope.

Since scopes are used for greater magnification than binoculars, an important characteristic is *resolving power* (clarity). It is difficult to manufacture optics with high magnification which also produce a clear image. Reichert and Reichert (1961) described a method for measuring resolving power of a scope which is then useful for comparing scopes:

$$\text{resolving power} = \frac{\text{maximum distance at which you can distinctly see specific details through the scope}}{\text{maximum distance at which you can distinctly see them with the unaided eye (20} \times \text{20).}}$$

For example, if you can distinctly see a small bird at a maximum distance of 20 feet (6.1 meters), but with a 20-power scope you can see it distinctly at a maximum of 360 feet (108 meters), then:

$$\text{resolving power} = \frac{360}{20} = 18$$

In a very high-quality scope the resolving power should equal the magnification. Although the optical characteristics vary between manufacturers, the following figures are common:

Magnification	Resolving Power
20 ×	18
30 ×	22
40 ×	20

A feature which is often very useful is a "zoom" lens. This can be purchased as a permanent feature of the scope or through a variable-power eyepiece.

Scopes should be tested and compared before a choice is made for use or purchase. As with binoculars, there is the inevitable compromise between magnification, brightness, and clarity. Table 3-4 provides some characteristic specifications.

Scopes are greatly affected by two factors: Heat waves from the ground and movement. The first can be overcome best by observing from the highest point available. If possible, use the top of a hill or rock outcropping. If you are not concerned about concealment, just getting up on the roof of a vehicle will often reduce the effect of heat waves considerably.

Movement is reduced best by the use of a sturdy tripod (Figs. 5-2, 8-15*B*). These come in many designs and price ranges. There are four questions to keep in mind when you examine a tripod for use or purchase:

1. How rugged is it?
2. Does the height adjust to a comfortable eye level?
3. Do the legs extend smoothly and rapidly for quick extension and occasional adjustment?
4. How portable is it?

Table 3-4. Some specifications for representative spotting scopes (Reichert and Reichert 1961).

Model	Diameter Exit Pupil (in mm) *(EP)*	Relative Brightness *(RB)*	Relative Light Efficiency *(RLE)* Percentage of Coating			Field Feet at 1000 Yds. (in ft)	Field Angular Degrees	Relative Field (in ft) *(RF)*
			20%	80%	100%			
20 × 40	2	4	4.5	5.6	6	107	2.1	2140
16 × 50	3.1	9.6	—	—	14.5	200	3.8	3200
20 × 50	2.5	6.2	7	8.8	9.4	118	2.3	2360
15 × 60	4	16	17.5	22.5	24	150	3	2250
20 × 60	3	9	10	12.5	13.5	112	2.1	2240
20 × 60	3	9	—	—	13.5	170	3.3	3400
30 × 60	2	4	4.5	5.6	6	80	1.5	2400
40 × 60	1.5	2.2	2.4	3.1	3.3	61	1.2	2440

Gunstock mounts can be a useful compromise between a good tripod and no tripod at all. They can be purchased, adapted from an old gunstock, or carved and custom fitted (Fig. 3-4). Also, window mounts are extremely useful if most of your observations will be made from a vehicle.

Photography through binoculars and spotting scopes is mentioned here only because it is possible, not because it is recommended. Although manufacturers do provide adapters for mounting a camera on binoculars or scopes, I recommend that it be avoided if possible. It is best to treat your observational and photographic equipment as two different and important types of equipment. The problems of assembly and disassembly, coupled with the generally poorer quality of photographs, does not warrant the extensive use of binoculars and scopes for photography. If you choose to investigate this type of photography further, I suggest you consult Reichert and Reichert (1961), who extol its virtues.

A combination binocular-camera (BINO/CAM) is marketed by Tasco Sales, Inc., 1075 N.W. 71st Street, Miami, Florida 33138. It combines a 7 × 20 wide-angle (field of view-508 feet) binocular with a 110 camera. The camera uses film cartridges, has six F stops (5.6–32) on the 112-mm telephoto lens, a 1/200 shutter speed and a built-in tripod adapter. The unit, which sells for approximately $300, may find limited use by ethologists but is too new to have received much use as yet.

Figure 3-4 Andy Sandoval observing bighorn sheep through a spotting scope mounted on a gunstock.

The U.S. Army *starlight scope* is a useful, though not very portable, device developed by the Department of Defense for night observations (Fig. 3-3*B*). It uses a battery-operated photomultiplier to increase the brightness of the image several thousand times. The image is then projected onto a small screen inside the eyepiece. Its limitations are its size, focal distance and availability. Starlite scopes can sometimes be obtained on short-term loan through the Civil Affairs Section of the Army. If you are planning to borrow one, be prepared to assume liability for an $18,000 instrument.

Javelin Electronics (6357 Arizona Circle, Los Angeles, California 90045) manufacturers starlight scopes which are smaller and more practical for ethological work (*e.g.* Waser 1975a). They are available in different models which vary in physical dimensions, light amplification, and magnification and have options for use with still and motion-picture cameras, as well as video recorders.

Several types of special *miscellaneous observational devices* have been developed by individual researchers to meet their particular needs. For example, Parker (1972) and Smith and Spencer (1976) developed mirror and pole devices which allowed them to look into high birds' nests. Descriptions of various devices are scattered throughout the technical and popular literature. However, time can often be better spent applying your own ingenuity to developing a device to meet your needs than it is searching for something suitable in the literature. You can also probably save yourself a great deal of money.

B. HOW TO DESCRIBE BEHAVIOR

At the heart of the modern approach to the analysis of behavior in animals is the problem of description. [MARLER 1975:2]

1. Empirical vs. Functional Descriptions

As you first observe the behavior of an animal you will likely be confused by the complexity of what the animal does; but in time some order will appear in the types of behavior engaged in, the contexts in which they appear, and the movements and postures that are involved (Marler 1975). Familiarity with an animal's behavior and insight into its function are continuing processes that generally lead to revision of both hypotheses and terminology.

Perhaps it will surprise no one, but differentiating the behavioral vocabulary of a given species in a given setting is, like learning a human language, a never-ending task (and delight). After one acquires first the rudiments and then broad comprehension, one continues to gain recognition and facility with the more subtle nuances of the language and the shaded meanings of more complex constructions. After 18 years of observing macaques in fairly similar observational settings, I continually encounter new subtleties of behavior or detect a special import in the linkage of two elements previously thought of as behaviorally independent. [ROSENBLUM 1978:20]

Nevertheless, in time, it will be necessary to describe what you have observed in terms which are clear yet unassuming.

The problem of description (and naming behaviors) is resolved through experience in observing the animal's behavior and your ability to select terminology that will assist, not hinder, future analysis.

Your descriptions should inform others of your observations in an objective way without bias to your own experiences or personal beliefs. *Anthropomorphism,* the attribution of human characteristics to nonhuman animals, is often considered one of the gravest sins that an ethologist can commit (Carthy 1966). However, avoiding anthropomorphism is difficult, if not impossible, since it can be argued that we cannot have knowledge of anything which we have not ourselves experienced either directly or indirectly. As Rioch (1967) has remarked, we are both limited and directed by our vocabulary (symbolic behavior) in describing observed behavior.

I will readily admit that observation has one great drawback; it is hard to convey to others. Experimental conditions can be reproduced, pure observation unfortunately cannot. Therefore it does not have the same objective character. The observer who studies and records behavior patterns of higher animals is up against a great difficulty. He is himself a subject, so like the object he is observing that he cannot be truly objective. The most "objective" observer cannot excape drawing analogies with his own psychological processes. Language itself forces us to use terms borrowed from our own experience. [LORENZ 1935:92]

Lorenz (1974) has gone on to suggest that in some instances the use of terms like "falling in love," "friendship" or "jealousy" is not anthropomorphic, but rather refers to functionally determined concepts. Where and how does the beginner draw the line? The best approach is to avoid the borderline entirely and deal only with empirical descriptions, discussed below.

There are two basic types of behavioral description:

Empirical description—description of the behavior in terms of body parts, movements and postures (*e.g.*, baring the teeth).

Functional description—incorporation of reference to the behavior's function, proximally or ultimately (*e.g.*, bared-teeth threat).

These types are nearly synonymous (see example below) with the two types used by Hinde (1970): 1) description by spatiotemporal patterns of muscular contraction, including patterns of limb and body movement, and 2) "description by consequence," respectively. Wallace (1973) calls Hinde's first type "description by operation."

The type of description selected will depend in part on your knowledge of the animal's behavior and type of study you wish to pursue. Descriptions can be thought to lie along a continuum of information conveyed. At some stage, conveying additional information generally entails drawing conclusions from data about function. After careful study the researcher may be able to use a term which more clearly describes the context of a behavior. For example, W.J. Smith (1968) studied the use of the "kit-ter" call by the eastern kingbird *(Tyrannus tyrannus)* and concluded from observational data that it provides information relative to the caller's indecision in flying vs. staying put, flying towards vs. flying away, or flying vs. landing. Hence, he labeled the call the "Locomotory Hesitance Vocalization."

As another example, let us say we are walking through a stubble field, and 50 m ahead of us a mourning dove flies up out of the stubble and lands in a tree 50 m to our right. We can describe the behavior of the dove in flight as:

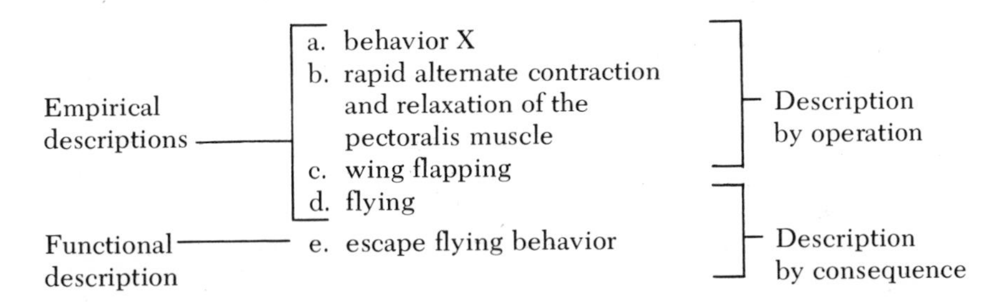

Describing (*i.e.*, naming) the behavior as "behavior X" provides us with no information unless we have access to a definition of the code being used. Rapid alternate contraction and relaxation of the pectoralis muscle tells us something about the mechanics of the behavior but does not provide the ethologist with much useful information. Wing flapping creates an image in the mind of the ethologist; but we still do not know if the dove was standing and flapping its wings (perhaps an intention movement) or actually flying. By describing the behavior as flying we get a clearer picture of the behavior and still are not assuming anything about underlying motivation. By describing it as "escape flying behavior" we are assuming that the dove was responding to a stimulus from which it wanted to escape. We do not really know if that was the true function of the flight or if, for example, it had finished feeding and was merely flying to the tree where it could rest with relatively greater safety.

This example illustrates that the same behavior may be used in several contexts. Mounting may occur in sexual or dominant-subordinate contexts in dogs, just as urination may be marking or merely elimination (Bekoff 1979). Functional descriptions should be avoided, except when the function is intuitively obvious (see below) or supported by data, since they can be confusing and misleading (Marler 1975) and lead to changes in terminology as a study progresses (Tinbergen 1959).

The distinction between empirical and functional descriptions is not always clearcut, so that the problem is generally resolved in terms of the observer's intent. For example, does "sniffing" imply searching for olfactory stimuli or merely wiggling the nose and vibrissae. This type of confusion over the observer's intent is clarified through the definition of behavior units, discussed on page 68).

Some descriptive terms are clearly functional but are readily accepted, since the goal of the behavior appears intuitively obvious. For example, the terms "nest building" and "egg retrieval" are accepted in ethological parlance as describing a type of behavior.

The type of behavior, as well as the type of data being collected, often force the use of both empirical and functional descriptions. Hinde (1970) suggests that since threat and courtship behavior in birds involves both relatively stereotyped motor patterns and an orientation with respect to the environment, both description by operation and consequence are necessary.

Eisenberg (1967) provided a list of behaviors for rodents (Table 3-5) that included both empirical and functional descriptions for convenience of presentation. It is useful to examine the list and identify those terms that are borderline, as well as those that are clearly empirical or functional.

2. Ethogram

An *ethogram* is a set of comprehensive descriptions of the characteristic behavior patterns of a species (Brown 1975). It results from many hours of observation (in some cases audio recording) and description, and it should be the starting point for any ethological research, especially species-oriented research. When concept-oriented research is conducted, the researcher may compile an ethogram of only those behaviors within, or closely related to, the category in which he is interested. However, the more we restrict our view of the animal's

Table 3-5. List of rodent behaviors utilizing both empirical and functional terms (from Eisenberg 1967).

General Maintenance Behavior

- Sleeping and resting
  - Curled
  - Stretched
    - On ventrum
    - On back
  - Sitting
- Locomotion
  - On plane surface
    - Diagonal
    - Quadrupedal saltation
    - Bipedal walk
    - Bipedal saltation
    - Jumping
  - Climbing
    - Diagonal coordination
    - Fore and hind limb alteration
  - Swimming
- Care of the body surface and comfort movements
  - Washing
    - Mouthing the fur
      - Licking
      - Nibble
    - Wiping with the forepaws
    - Nibbling the toenails
  - Scratching
  - Sneezing
  - Cough
  - Sandbathing
    - Ventrum rub
    - Side rub
    - Rolling over the back
    - Writhing
  - Stretch
  - Yawn
  - Shake
  - Defecation
  - Urination
  - Marking
    - Perineal drag
    - Ventral rub
    - Side rub
- Ingestion
  - Manipulation with forepaws
  - Drinking (lapping)
  - Gnawing (with incisors)
  - Chewing (with molars)
  - Swallowing
  - Holding with the forepaws
- Gathering foodstuffs and caching
  - Sifting
  - Dragging, carrying
  - Picking up
    - Forepaws
    - Mouth
  - Hauling in
  - Chopping with incisors
  - Digging
  - Placing
  - Pushing with forepaws
  - Pushing with nose
  - Covering
    - Push
    - Pat
- Digging
  - Forepaw movements
  - Kick back
  - Turn and push (forepaws and breast)
  - Turn and push (nose)
- Nest Building
  - Gathering
  - Stripping
    - Biting
    - Jerking
    - Holding
  - Pushing and patting
  - Combing
  - Molding
  - Depositing
- Isolated animal exploring
  - Elongate, investigatory
  - Upright
  - Testing the air
  - Rigid upright
  - Freeze (on all fours)
  - Escape leap
  - Sniffing the substrate
  - Whiskering

Continued on next page

total behavioral patterns, the greater the probability of misinterpreting results.

Table 3-5. Rodent behaviors (continued)

Social Behavior

- Initial contact and contact promoting
  - Naso-nasal
  - Naso-anal
  - Grooming
  - Head over-head under
  - Crawling under and over
  - Circling (mutual naso-anal)
- Sexual
  - Follow and driving
  - Male patterns
    - Mount
    - Gripping with forelimbs
    - Attempted mount
  - Copulation
    - Thrust
    - Intromission
    - Ejaculate
  - Female patterns
    - Raising tail
    - Lordosis
  - Neck grip
  - Postcopulatory wash
- Approach
  - Slow approach
  - Turn toward
  - Elongate
- Agonistic
  - Threat (proper) (remains on all four legs)
  - Rush
  - Flight
  - Chase
  - Turn away
  - Move away
  - Bite
- Agonistic (continued)
  - Locked fighting (mutual)
  - Fight (single)
  - Defense (on back)
  - Side display
    - Shouldering
    - Sidling
    - Rumping
  - Uprights
    - Class I (upright threat)
    - Class II
    - Locked upright
  - Striking, warding
  - Sparring
  - Tail flagging
  - Kicking
  - Attack leap
  - Escape leap
  - Submission posture
  - Defeat posture
  - Tooth chatter
  - Drumming
  - Pattering (with forepaws)
  - Tail rattle
- Miscellaneous patterns seen in a social context
  - Sandbathing
  - Digging and kick back
  - Marking
    - Ventral rub
    - Side rub
    - Perineal drag
  - Pilo-erection
  - Trembling

The need for a broad, observational approach cannot be stressed too much. The natural tendency of many people, particularly of young beginners, is to concentrate on an isolated problem and to try to penetrate into it. This laudable inclination must be kept in check or else it leads to an accumulation of partial, disconnected results, to a collection of sociological oddities. A broad, descriptive reconnaissance of the whole system of phenomena is necessary in order to see each individual problem in its perspective; it is the only safeguard for a balanced approach in which analytical and synthetical thinking can cooperate. This, of course, is true not only of sociology, it is true of each science, but in ethology and sociology it is perhaps forgotten more often than in other sciences. [TINBERGEN 1953:130]

a. Repertoire and Catalog

A *catalog* of an animal's behaviors is a list of all that we have observed, listened to, or have knowledge of. This is only a portion of the animal's *repertoire*—all the behaviors that the animal is capable of performing. We call the catalog an *ethogram* when we believe that it closely approximates the complete repertoire. The size of the repertoire will, of course, vary from species to species as well as between individuals, depending on sex, age, and experience.

One decision that you must make during reconnaissance observations is when to stop. When do you have sufficient information to ask incisive questions, formulate precise hypotheses, and design a sound research project? At what point do you have a reasonably complete ethogram for the animal(s)?

If we were to observe an individual animal continuously for an extended period of time and record the behaviors that it showed, we could then plot the cumulative number of observed behaviors by the time (Fig. 3-5A).

An asymptote is reached after many hours of observation (arrow, Fig. 3-5A), beyond which few additional behaviors are seen for each unit of time spent observing. This asymptote may take tens, hundreds, or thousands of hours to reach, depending on the species studied. Also, if only one type of behavior is under study, then the time to the asymptote will be shortened. Fagen and Goldman (1977:268) concluded that ". . . familiarity with an animal's behavior will tend to require years of experience if the animal is a mammal or bird with a complex repertory. But if the animal's behavior is simple and relatively stereotyped such familiarity may be gained in a few months."

Hailman and Sustare (1973) described an interesting laboratory exercise in "the analytical power of biological observations." The object was to deduce the "behavioral organization" of a talking, stuffed toy elephant—Horton (manufactured by Mattel, Inc., Hawthorne, California). The first step consisted of listening to Horton's total vocal repertoire by one's pulling the string and listing the vocalizations emitted. These data were transferred to a cumulative graph (Fig. 3-6) and examined for an asymptote to determine if the entire repertoire had been recorded after 100 successive vocalizations.

Another way to look at the behavioral repertoire of an animal is through the time devoted to particular behaviors (i.e., time budget) or by frequency of occurence (Hutt and Hutt 1970). Since the frequency of occurrence varies for the behaviors in an animal's repertoire, a plot of cumulative percentages of total time spent in the various behaviors against their rank order by frequency of occurrence will show a curve

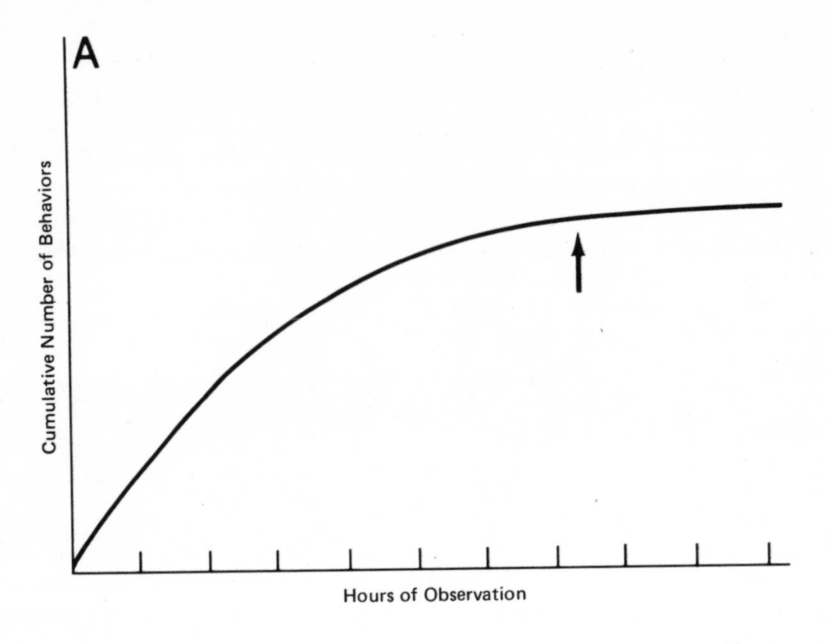

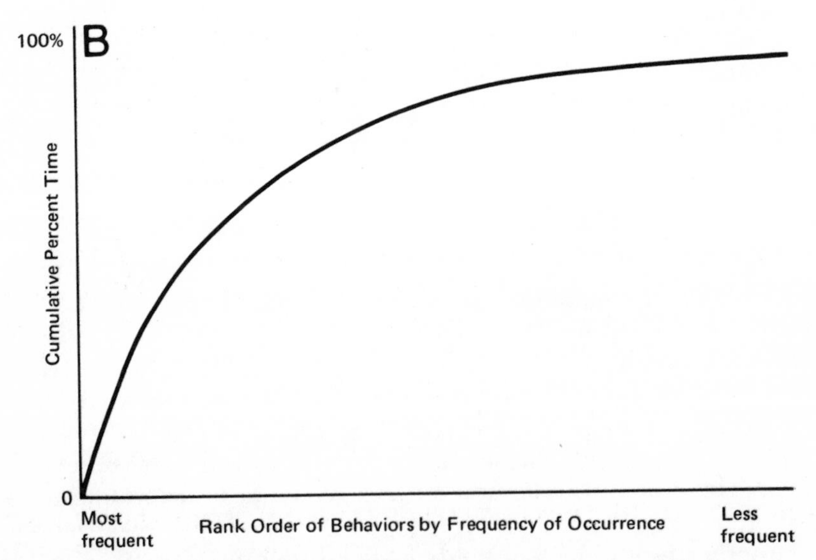

Figure 3-5 *A*, a hypothetical example of cumulative number of behaviors plotted against hours of observation. The arrow indicates the approximate asymptote. *B*, conceptual representation of an animal's repertoire plotted in cumulative percent of the time spent in the various behaviors. (*A* and *B* adapted from Hutt, S.J., and C. Hutt, 1974, Direct observation and measurement of behavior, courtesy of Charles C. Thomas, Springfield, Ill., p. 36, Fig. 3.)

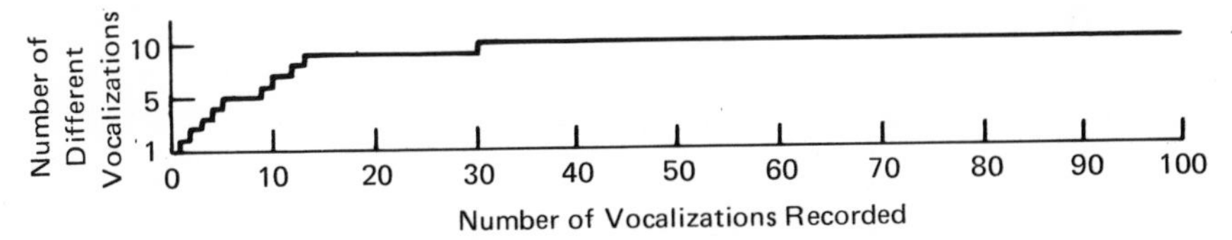

Figure 3-6 The cumulative number of different vocalizations as a function of the total number of vocalizations recorded (from Hailman and Sustare 1973).

which reaches an asymptote at the less frequently occurring behaviors (Fig. 3-5*B*).

Fagen and Goldman (1977) researched methods of analyzing behavioral catalogs and concluded that most distributions (types of behavioral act/number of acts observed) could be described by a logarithmic regression slope of approximately 0.3 (*e.g.*, Fig. 3-7).

Since the regression line has no finite asymptote, it does not allow the observer to predict repertoire size. However, this procedure did encourage Fagen and Goldman (1977:263) to recommend the following rule: "A ten fold increase in the total number of acts will, on the average, double the number of behaviour types in the catalog".

We can estimate our *sample coverage* (Fagen and Goldman 1977) by calculating the probability that the next behavioral act will be a new type. If $\hat{\theta}$ approaches 1, the probability of observing a new behavioral act is low.

$$\hat{\theta} = 1 - \frac{N_1}{I}$$

N_1 is the number of behavior types seen only once and I equals the total number of acts seen. When N_1 is small relative to I then θ will approach 1. The closer that θ approaches 1, the more complete the sample coverage. Altmann (1965) observed 5,507 acts in rhesus monkeys and saw 32 behavioral types only once.

$$\hat{\theta} = 1 - \frac{32}{5507} = 0.9942$$

This indicates that Altmann's sample coverage was essentially complete. Fagen and Goldman (1977) caution that this method emphasizes the significance of rare behavioral acts. They provide a more complex procedure for estimating the *repertory fraction* which properly weighs the frequency of occurrence of different behavioral types.

Of primary importance to any ethological study and particularly to a consideration of catalogs and repertoires is a determination of a behavioral unit. The size of the catalog will vary according to the way

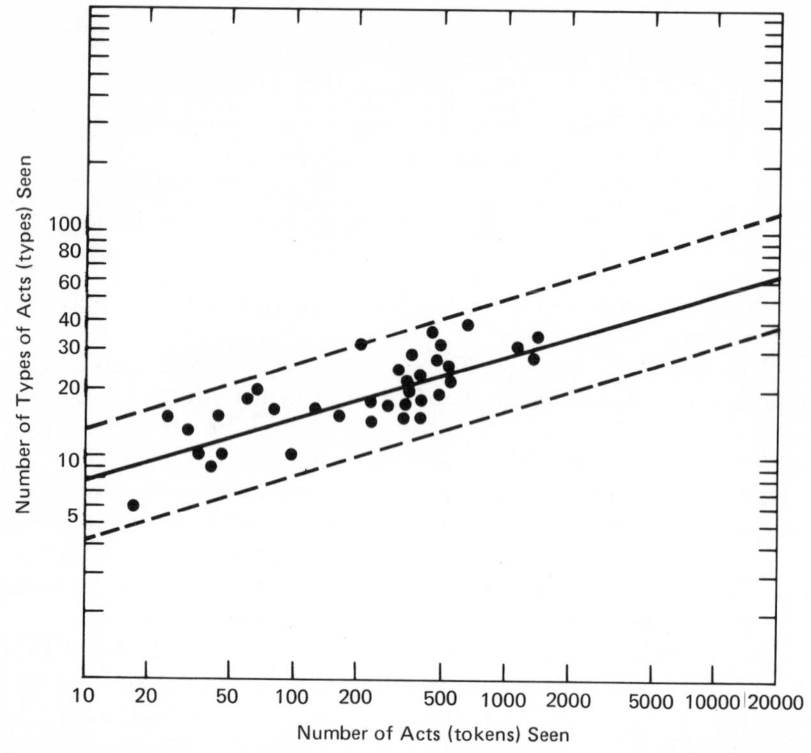

Figure 3-7 Plot of cat behavior data with fitted regression line $Y = 4 \cdot 01X^{0.29}$ (solid line) and 95 percent confidence bonds for regression line (dashed lines). (From Fagen, R.M., and R.N. Goldman, 1977, Behavioral catalogue analysis methods, *Anim. Behav.* 25(2):263, Fig. 1.)

in which behavioral units have been defined. The more inclusive (*i.e.*, lumping several different behavioral acts, such as threat and submission, into a single behavioral unit—agonistic behavior), yet mutually exclusive (*e.g.*, agonistic vs. ingestive), the smaller the catalog will be. The selection and definition of behavioral units will vary according to the objectives and logistics of the individual study (Chap. 4, Sec. A).

C. SECONDARY SOURCES

Before embarking on your study you should learn as much about your subject animal, especially its behavior, as you can. This means collecting information, in addition to your initial reconnaissance observations (p. 24), from all the secondary sources at your disposal, including available literature, data from other researchers' efforts, and films.

1. Literature

The following discussion will include selected sources of literature on animal behavior. It should be supplemented with material which can be provided by the reference librarian at the libraries where your personal library searches are conducted.

Access to journal articles can be sought in three ways. First, refer to a recent article on the topic and look at the articles the author lists in his bibliography. Each of those articles will also include references which can be used. This will allow you to pursue articles on a given topic back through many years of published articles.

Second, you can use the many abstracting services that provide references to journal articles on specified topics. Here are some of the abstract guides which are useful in searching for animal behavior literature:

Animal Behaviour Abstracts
Applied Ecology Abstracts
Biological Abstracts
Bioresearch Index
Current Advances in Ecological Sciences
Dissertation Abstracts International
    Section B. The Sciences and Engineering
Ecological Abstracts
International Abstracts of Biological Sciences
Key Word Index of Wildlife Research
Psychological Abstracts
Recent Literature of Mammalogy
Wildlife Abstracts
Wildlife Review
Zoological Record

Third, you can scan the yearly indexes in the professional journals where you suspect articles on your specific topic may have been published; also search the subject indexes in textbooks on animal behavior. A list of selected journals containing ethological articles is in the Appendix. The several *Current Contents* reproduce the tables of contents from professional journals divided into broad disciplinary areas and may be useful in keeping you abreast of new articles. Also, you can read the titles and abstracts of papers given at professional meetings and write the authors for more details or expanded versions.

Commercial literature searches are provided by several organizations and agencies. Many libraries have a *computer terminal* that gives access to over 50 independent literature search services throughout North America. In most cases, however, the data banks maintained by these commercial services include references back to only about 1970. Professional societies, may provide literature searches for a fee such as the American Society of Mammalogists' Citation Retrieval System. The Smithsonian Institution will provide abstracts of current government-funded projects on specified topics through their Scientific Information Exchange.

Library holdings of books on selected topics can best be determined through their subject card catalog. Efficient access through the card catalog is best obtained by first referring to the *Library of Congress Subject Heading List* under "Animals, Habits and Behavior of" for the subject heading that best describes your specific area of interest. In addition you should look at the *Superintendent of Documents Monthly Catalog* for federal publications in your area of interest. Federal publications often constitute a considerable portion of a library's holdings.

2. Other Researchers

Contact other researchers who are working (or have worked) on concepts or species in which you are interested. Discuss your proposed research relative to your own ideas. Do not parasitize and ask questions that imply "Tell me everything you know about—" Most researchers are glad to discuss projects but are unwilling to conduct minilectures for people who have not already attempted to learn as much as they can through other sources. Also, realize and respect the fact that most professionals are cautious about divulging the details of their research until it is in published form.

3. Films

Films are another source of preliminary information about an animal's behavior. Viewing films before embarking on a project is a strategy perfected by successful football coaches. The more familiar you are with the animal, the greater the probability of success in your project.

Other researchers are often willing to discuss not only their research and your proposed project with you, but will often lend motion-picture footage to responsible investigators.

Another source of films is through film libraries, where unedited footage or polished productions can be rented for a nominal fee. Four of these sources are listed on the next page:

Encyclopaedia Cinematographica
Archive
The Penn State University
211 Mitchell Building
University Park, PA 16802

Audio Visual Services
The Penn State University
17 Willard Building
University Park, PA 16802

Rockefeller University Film Service
Box 72
1230 York Ave.
New York, NY 10021

UCLA Media Center
Instructional Media Library
405 Hilgard Ave.
Royce Hall No. 8
Los Angeles, CA 90024

4
Delineation of Research

A. CONCEPTUALIZING THE PROBLEM

A separate chapter has been devoted to the concepts discussed below because of their relatively great importance. Proper delineation of research is the cornerstone of any successful study. A common fault among many beginning researchers is their inability to state clearly the questions they are attempting to answer and the objectives they are striving to meet. Before the actual research can begin you must decide what you are trying to accomplish. As mentioned earlier (p. 10), most of our activities in ethology will be concerned with hypothesis-testing in a broad or specific manner (discussed later). Broadly defined, hypotheses arise from a process of observation→question→hypothesis. Questions and hypotheses are a natural product of our thought processes and cannot be turned off and on. Therefore, while we are reading, observing animals, or listening to other researchers, we are constantly generating questions and formulating hypotheses. These thought processes should be allowed to run free during our initial reconnaissance observations, but they must be held in check during actual data collection when bias may creep in (p. 128).

1. What Are the Questions?

Ethologists are never really at a loss for questions. We are constantly generating questions as we pursue our own research. Reflecting on his many years of research on herring gulls, Tinbergen (1960:xiv) stated:

> Soon after starting such a study of a social bird's community life, one begins to realize how little one knows. Even an hour's careful observation of the

goings-on in a gullery faces one with a great number of problems—more problems, as a matter of fact, than one could hope to solve in a lifetime.

The real problem arises when we attempt to isolate and define a question, or group of related questions.

A clear statement of the *question* (problem) in scientific research is perhaps one of the most difficult steps (see Bronowski 1973). It is also a continuing process.

It may even take an investigator years of exploration, thought, and research before he can clearly say what questions he has been seeking answers to. Nevertheless, adequate statement of the research problem is one of the most important parts of research. [KERLINGER 1967:18]

A proper question, one suitable for research, asks: "What relationship exists between two or more variables?" The answer to the question is the goal of the research. That is, the goal (objective) of the research is to answer that question.

2. Stating Objectives

The objective of research, as stated above, is to answer one or more questions.

For example, an ethologist may be studying social behavior in domestic chickens. It is noticed that the frequency of agonistic encounters appears to vary with flock size. But does it vary systematically? A suitable research *question* is: What is the relationship between flock size and frequency of agonistic encounters? The *objective* of the research is to determine that relationship. The objective of experimental studies, such as that just mentioned, is to answer a question through tests of a hypothesis.

3. Stating Research Hypotheses

As discussed previously (p. 10), hypotheses may be of a broad undefined nature or specific and defined in such a way that they can be experimentally tested. Specific hypotheses are of two types: *research hypotheses* and *statistical hypotheses*. Statistical hypotheses are the outgrowth of research hypotheses and are the basis for statistical tests. They will be discussed in the chapter on analysis (p. 227).

Research hypotheses are conjectural statements about behavior and other related variables. They are what you perceive to be the "true situation." From the example in the previous section, your research hypothesis might be that agonistic encounters increase as flock size increases. This is the relationship that you believe exists between the behavior (agonistic encounters) and an independent variable (flock size). In 1966, Hamilton introduced the following research hypothesis:

> The current explanation of V-formations in bird flocks is that they establish favorable air currents, reducing flight energy requirements. An alternative hypothesis is proposed here, that this flock structure is a form of orientation communication, enabling the individuals of these flocks to take maximal advantage of the collective orientation experience of the group. [HAMILTON 1966:64]

In 1970, Lissaman and Shollenberger provided evidence in support of the former hypothesis, but Hamilton's alternative hypothesis remains untested.

5
Design of Research

A. NATURALISTIC OBSERVATION VS. EXPERIMENTAL MANIPULATION

We previously discussed reconnaissance observation (Chapter 3) as a means of gathering background knowledge about an animal before designing a research project. It helps not only to generate questions and define objectives, but also to determine what aspects of behavior can be measured, what manipulations are feasible, and the degree of variability that is to be expected. This initial observation occurs at level *C* in the ethological approach (Fig. 1-6), and it may occur before or simultaneously with the delineation of research (Chapter 4). Once the research has been delineated, it must be designed (level *D*), and an approach—naturalistic observation or experimental manipulation—selected (level *E*).

Ethological research can be broken down into *naturalistic observation*, in which the ethologist is collecting facts that may aid in understanding phenomena and/or lead to the formulation of hypotheses, and *experimental manipulation*, which:

> . . . usually consists of making an event occur under known conditions where as many extraneous influences as possible are eliminated and close observation is possible so that relationships between phenomena can be revealed. [BEVERIDGE 1950:20]

Naturalistic observation refers to the approach in which the observer studies the behavior of animals as it occurs naturally, with as little human intrusion as possible. Naturalistic observation does not have to be conducted in the wild. Often an environment can be created in the laboratory which closely approximates the natural habitat of the animal (*e.g.*, insects, fish). This type of approach has been called *descriptive* in that the objective is to document the natural history of a group or population (see Fig. 1-2) with an emphasis on behavior. This was the approach taken by Tinbergen in his initial scientific study of herring gulls, building on observations made throughout the earlier years of his life.

> Throughout the years of my boyhood watching the life in the large gullery was complete happiness . . . watching the snow-white birds soaring high up in the blue sky . . . It was this sentiment that sent me back to the gulls in later years, when I returned with a matured scientific interest, intent on exploring the secrets of their community life. [TINBERGEN 1960:xiii]

Tinbergen's initial objective could be stated as: To describe the community life of a herring-gull colony. His long-term studies, of course, later evolved into experimental and comparative studies conducted by himself and his students.

Naturalistic observation was the basic approach an used by Kruuk (1972) in his study of the spotted hyena *(Crocuta crocuta)* in the Serengeti National Park. The approach is reflected in the behavioral questions he asked:

1. How is the species organized socially, and how does this compare with other carnivores?
2. What are its foraging habits, and how do these compare with the feeding habits of its carnivorous relatives?
3. How are the questions in 1 and 2 related?

Naturalistic observation was the approach used by Nice (1937, 1943) in her study of song sparrows, Evans (1957) in his studies of wasps, Geist (1971) in his study of mountain sheep, Schaller (1972, 1963) in his studies of the mountain gorilla and African lion, and Darling (1937) in his study of red deer.

Experimental manipulation (discussed in the next chapter) and naturalistic observation complement each other, and most long-term research programs oscillate between the two approaches (Tinbergen 1951). Also, although we treat these two approaches as a dichotomy in ethology, it is often difficult to determine where one ends and the other begins.

Menzel, in his discussion of the two approaches in primate studies (additional discussion is found in Mason 1968), stated:

I am, in fact, convinced that naturalistic and experimental studies can be compatible with each other; that their respective methods can be applied to any situation (instead of being linked to a given situation such as laboratory or field); that both types of information are necessary to a meaningful general science of primate behavior; that any sharp division between naturalistic and experimental methodology is not only undesirable but impossible; and that, finally, disputes as to which method or situation is intrinsically best are nonsensical. [MENZEL 1969:78]

The further splitting of both naturalistic observation and experimentation into field and laboratory studies is discussed in the next chapter. Whichever approach is selected, a prescribed plan for collection of data is necessary (Chapter 7). If experimental manipulation is selected in order to test a hypothesis, then a careful consideration of variables and experimental designs is necessary.

B. VARIABLES

A *variable* is a property that takes on different values. In testing hypotheses there are two types of variables:

Independent variable—the property that changes (or is manipulated) and is believed to affect the dependent variable (also called the manipulated variable and active variable).

Dependent variable—the property that is believed to be affected by a change in the independent variable (also called the measured variable and assigned variable).

"An *independent variable* is the presumed cause of the *dependent variable*, the *presumed* effect" (Kerlinger 1967:39). Behavior will most often be treated as the dependent variable; however, it can be an independent variable as well (*e.g.*, the effect of one animal's behavior on another's).

The independent variable is designated as X and the dependent variable as Y. Some relationship is presumed to exist between the two such that changes in Y can be predicted from changes in X (Fig. 5-1). Line A describes a positive correlation and line B a negative correlation. Correlations can be much more complex than the linear correlations shown in Figure 5-1, but discussion of those correlations is beyond the scope of this book.

Variables can be divided into *quantitative variables* and *qualitative variables*. Quantitative variables vary in *amounts*. Qualitative variables vary in *kinds*. The type of variable studied will in part

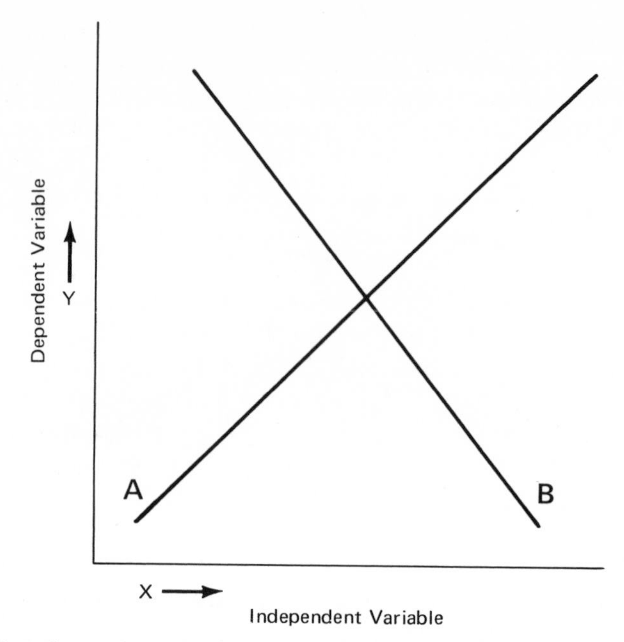

Figure 5-1 Relationship between dependent and independent variables. Slope *A* shows a positive correlation and slope *B* a negative correlation.

determine the scale of measurement (p. 109) you can use. That is, qualitative variables require a nominal scale, whereas ordinal, interval, or ratio scales can be used with quantitative variables.

The choice of the proper dependent variable (*e.g.*, behavior) to measure is extremely important. Most experimental designs employed by ethologists measure only one dependent variable at a time. The selection of the most appropriate dependent variable to measure should be based on four factors:

Sensitivity—likeliness of showing an effect due to a change in the independent variable.

Reliability—expected to provide the most consistent results.

Distribution—whether the observations in each treatment level would be normally distributed; parametric statistics can be applied only if this condition is met.

Practicality—if two or more dependent variable candidates emerge after consideration of the above three factors, choose one or more that are most easily measured; how much cost can you afford in terms of equipment, time and manpower?

All experiments are also affected by *nuisance variables*. These are undesired sources of variation which may affect the dependent variable and bias the results. Psychologists recognized their inability to control all the variables and coined the term *intervening variable* (Tolman 1958) to account for internal, directly unobservable psychological processes that in turn affect behavior; Kerlinger (1967:44) calls intervening variables "in-the-head" variables. All potential sources of undesired variation are nuisance variables (*e.g.*, previous experience; perceptual ability).

Nuisance variables can be controlled in four ways:

1. Hold the nuisance variable constant for all subjects.
2. Assign subjects randomly to the experimental conditions.
3. Include the nuisance variable as one of the treatments in the experimental design.
4. Treat the effects of the nuisance variable statistically through the use of covariance.

Examples of variables to be considered will be discussed in the next chapter.

C. BEHAVIOR UNITS

The choice of appropriate behavior units to be measured is at once one of the most important and difficult decisions to be made (Barlow 1977). As we shall see later (Chap. 5, Section C-5) this may be crucial to the validity of your research. The choice of an appropriate behavior unit is generally based on experience, tradition, logistics, and intuition.

> Even where there is agreement about what general kind of description is appropriate, there may be disagreement between "lumpers" and "splitters" about what should be counted as a *unit of behavior* (italics mine) for quantitative purposes. Where some tally fights, songs and journeys, others tally blows, notes and footsteps. [BEER 1977:158]

In the report of his field study of social communication in rhesus monkeys, S.A. Altmann (1965) states that ". . . categorizing the units of social behaviour involves two major problems: when to split and when to lump." He goes on to point out that there are natural units of behavior, "Thus, the splitting and lumping that one does is, ideally, a reflection of the splitting and lumping that the animals do."

1. Classification of Behavior Units

Behavioral units can be classified in several ways. The logical approach is to work from the general to the specific. First, determine the *level of organization* along the species-individual dimension (Fig. 1-2) and then classify the behavior units according to type.

One classification scheme according to type was developed by J.P. Scott (1950) (see Table 5-1). He suggests that, "The list provides a convenient guide for the description of behavior in a new species, but in any particular case certain types may be absent" (Scott 1963:23).

The type of behavior to be studied will be dictated by the questions being asked and the approach selected. In naturalistic-observation studies we usually gather data on many types of behavior, while in studies employing experimental manipulation we usually are concerned with only one type.

Within the general type, the behavior can be further classified according to *complexity* and *social interaction* by following a scheme prepared by Delgado and Delgado (1962) as a result of their studies of modifications in the social behavior of monkeys. Their classification of behavior units ". . . evolved from a system of definition" and was used within the framework of an explanatory model of behavior which they also developed. Briefly, their classification scheme is as follows:

A. Simple behavior units
   1. Individual
      (i) *Static or postural units* can be defined and identified by static relations (*e.g.*, sleeping alone)

Table 5-1. Scott's (1950) classification of behavior.

General Types of Adaptive Behavior	Definition
Ingestive	Eating and drinking
Investigative*	Exploring social, biological, and physical environment
Shelter-seeking*	Seeking out and coming to rest in the most favorable part of the environment
Eliminative	Behavior associated with urination and defecation
Sexual	Courtship and mating behavior
Epimeletic*	Giving care and attention
Et-epimeletic*	Soliciting care and attention
Allelomimetic*	Doing the same thing, with some degree of mutual stimulation
Agonistic	Any behavior associated with conflict, including fighting, escaping, and freezing

*Note that these are functional terms which must be supported by data before they are applied in any particular study (see p. 43).

(ii) *Dynamic or gestural units* can be defined and identified only by a sequence of spatial relations (*e.g.*, climbing a wall)
   (a) *localized*—involving only part of the system (*e.g.*, moving legs)
   (b) *generalized*—involving a change of position of the whole system in relation to its environment (*e.g.*, walking on ground)

2. Social
   (i) *Static* (*e.g.*, monkeys sleeping, embracing each other)
   (ii) *Dynamic* (*e.g.*, a monkey chasing another)

B. Complex behavioral units
   1. Simultaneous
   2. Sequential
   3. Syntactic—the significance of the behavioral unit may vary with context
   4. Roles (*e.g.*, groomer-groomed, threatener-threatened)
      (i) Active
      (ii) Passive

Delgado and Delgado's scheme of classification of behavioral units can be illustrated diagrammatically as follows:

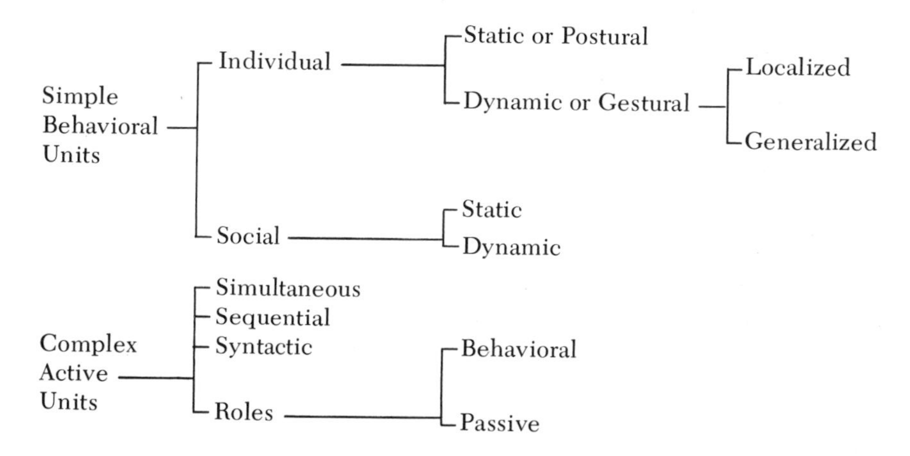

Next, the particular *behavior pattern* can be isolated. For example, a duck may move from one place to another (locomotion) by using one of four behavior patterns: walking, swimming, diving, or flying. These patterns could be broken down further according to social and environmental variables such as flocking and height and speed of flight.

We can then focus in on a *behavioral act* within a given behavior pattern. For example, flying can be broken down into the acts of taking off, flight, and landing.

Acts can be further classified into *component parts*. Taking off can be divided into movements of various parts of the body (*e.g.*, head, wings, legs), anatomical structures (*e.g.*, of muscles and bones), and neurological activity. Study of the internal physiological component parts of behavior are beyond the scope of this book; however, several excellent references are available (Eibl-Eibesfeldt 1975; Fentress 1977; Kandel 1977).

To summarize—in isolating behavior units for study we can work from general categories down to specifics:

1. Behavior types
2. Complexity and social interactions
3. Behavior patterns
4. Behavioral acts
5. Component parts

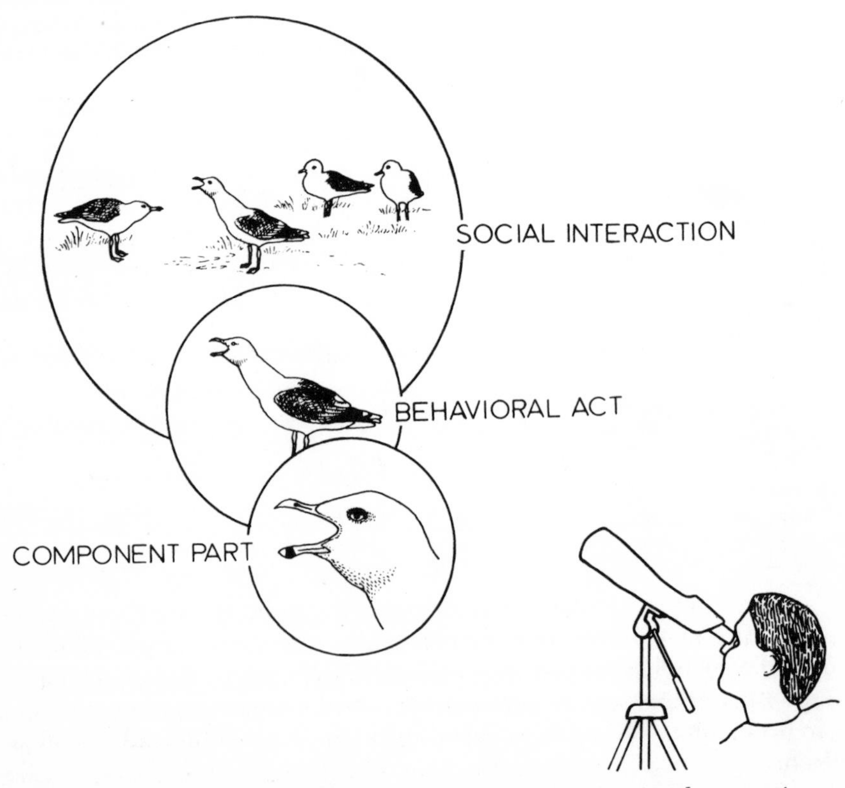

Figure 5-2 An observer focusing in and out between the social interactions, behavioral act, and component parts of a herring gull's oblique long call. (Drawing by Dan Thompson; based on Tinbergen 1959, 1960.)

The observer should not only concentrate on behavior units at the particular level selected for intensive study, but he can also gain insight while maintaining perspective through the process of "focusing-in" on and "focusing-out" from social interactions to component parts of the behavior. In Figure 5-2, the observer is focusing in on the alarmed herring gull described by Tinbergen:

> Some of these movements and postures are not difficult even for the human observer to appreciate, though the detection of most of them requires careful study. There are a multitude of very slight movements, most, if not all, of them characteristic of a special state of the bird. The student of behavior is to a high degree dependent on his ability to see and interpret such movements. In the beginning, he will notice them unconsciously. For instance, he will know very well on a particular occasion that a certain gull is alarmed, without realising exactly how he knows it. Upon more conscious analysis of his own perception (an important element in behaviour study), he will notice that the alarmed gull has a long neck. Still later, he will see another sign, the flattening of the whole plumage, which makes the bird look thinner. Upon still closer study, he will see that the eye of an alarmed bird has a very special expression, due to the fact that it opens its eyes extremely wide. [TINBERGEN 1960:7]

In addition, Drummond (in press) states that "If we examine the very broad range of phenomena regarded as behaviour patterns, it is apparent that the identity of each one resides in certain regularities, in those properties which are common to all instances, and that the regularities lie within a limited number of domains." He goes on to list the five following *domains of regularity*:

1. *Location* of the animal in relation to its environment (see below)
2. *Orientation* of the animal to the environment (see section on displays, p. 206)
3. *Physical topography* of the animal (see section on displays, p. 206)
4. *Intrinsic properties* of the animal (*e.g.*, changes in color, temperature, and electrical and chemical properties)
5. *Physical effects* induced in the environment by the animal (see p. 87)

2. Spatial and Temporal Aspects

Previously it was stated that an ethologist studies the what, when, how, where, and why of behavior. Here we are particularly concerned with the when and where questions. These two dimensions of behavior may be an important aspect of our research question or they may be ignored relative to other aspects. Examples of the study of spatial and temporal patterns will be given in a later chapter.

These are, of course, dimensions which are inherent in any behavior. They are also usually *relative dimensions*. This will become clearer as we consider the questions below which focus from the general to the specific dimensions.

Spatial questions:

1. In what geographical location is the animal found?
2. In what habitat is it located?
3. At what spot (vertically and horizontally)?
4. What is its location relative to other members of the same species?
5. How do its movements vary relative to where it (and other members of its species) are located?
6. How do the movements of one part of the body correlate with movements of other body parts (*e.g.*, synchronization of leg movements in locomotion)?

Temporal questions:

1. When is the individual observed (year, month, day)?
2. How does its occupation of a particular location correlate with season?
3. How does its behavior vary on a daily cycle?
4. How is its daily activity broken up into component behaviors (time budgets)?
5. What is the duration of occurrences of specific behaviors (or bouts of behavior)?
6. What is the relative timing of parts of the body in any particular behavior (*e.g.*, synchronization of leg movements in locomotion)?
7. What is the age of the animal(s)?

It can be seen that the spatial and temporal questions merged together in question six. These two dimensions are interrelated and are separated by the ethologist for convenience only.

3. Definitions of Behavior Units

It is obvious that we cannot measure what we cannot define. It is equally true that the way we define and record behavioral elements will be affected by the types of measurement we wish subsequently to apply to them. [HUTT and HUTT 1970:33]

Once a behavior unit is chosen it must be clearly *described* (p. 43) and *defined*. This is necessary in order to provide a clear picture of the behavior to other researchers. Sometimes descriptions are more elaborate than the actual criteria used by the researcher to determine when a behavior unit is occurring.

For the purpose of increasing reliability (both intra- and inter-observer), most behaviors are *operationally defined*. That is, we say when *X* and *Y* occur together then we will say that behavior Z (a composite of *X* and *Y*) occurs. *X* and *Y* may only be partial components of behavior Z. But we have chosen to define operationally the occurrence of Z as the simultaneous occurrence of *X* and *Y* because 1) we consider them the "most important" components. 2) *X* and *Y* are easily observed. 3) *X* and *Y* are the least arbitrarily defined components, and 4) *X* and *Y* have been used as criteria by other researchers.

As an example, the following are operational definitions for eight of the nineteen mutually exclusive behavior patterns Fernald (1977) observed in his study of adult male cichlid fish, *Haplochromis burtoni*:

Approach—swim quickly toward, then stop short near another fish.
Bite—bite opponent.
Chase—rapidly pursue fleeing opponent.
Frontal threat—spread opercula, lowered chin and spread pelvic fins while facing opponent.
Side threat—sideward presentation of spread opercula, fins, and distended chin to an opponent.
Mouth-to-mouth contact—two males grasp one another by the mouth and push, pull and bite.
Border fight—confrontation between two territorial males at the site of their common border.
Court—sideways quivering in front of a female with anal fin spread.

a. States and Events

Determining the duration of a behavior is often very difficult, not because we do not have the instruments, but rather we often do not have the necessary skill and observational experience. It is often difficult to determine when a behavior begins and when it ends. Operational definitions and descriptions of a behavior should be precise enough to allow other observers to make similar measurements with high reliability.

After observing animals for only a short period of time it becomes obvious that most behaviors can be divided into two categories (Altmann 1974) based on their duration:

State—the behavior an animal (or group) is engaged in; an ongoing behavior (*i.e.*, a robin flying).
Event—a change of states; it approaches an instantaneous occurrence (*e.g.*, a robin taking off).

Basically, a *state* is a behavior which you can time with a stopwatch; an *event* generally occurs so rapidly that you just count its occurrence. Sackett (1978) refers to these as *duration meaningful* and *momentary* behaviors, respectively.

Table 5-2. Measures for states and events.

Type of Measure	Definition	Usual Application
Total frequency	Number of occurrences per sample unit	Events, states
Partial frequency	Unknown percentage of total occurrences per sample unit	Events, states
Rate	Number of occurrences per unit time	Events
Duration	Amount of time per behavior unit	States

Events and states can be measured in various ways. The most frequent measurements are listed in Table 5-2.

Throughout an animal's life it is always cycling through states and events. In studying animal behavior we are merely sampling selected states and events, either as they occur "naturally" or are induced by the experimenter.

b. Bouts

The term *bout* is generally applied to: 1) a repetitive occurrence of the same behavior (*e.g.*, a bout of pecking) or 2) a relatively stereotyped sequence of behaviors that occur in a burst (*e.g.*, a courtship-display bout). Analysis of the latter type of bout will be considered in the next section.

Criteria for defining bouts of behavior are of two basic types:

1. *Change in behavior.* "If more than one behaviour is being observed, then a bout of one behaviour is said to end when a different type of behaviour begins" (Machlis 1977:9).
2. *Intervals between occurrences.* "A criterion interval X_t is chosen to separate one bout from another. All intervals equal to or greater than X_t are classified as between bout intervals and all those less than X_t as within bout intervals" (Machlis 1977:9).

The example below illustrates how the use of different criteria affects the analysis of bouts.

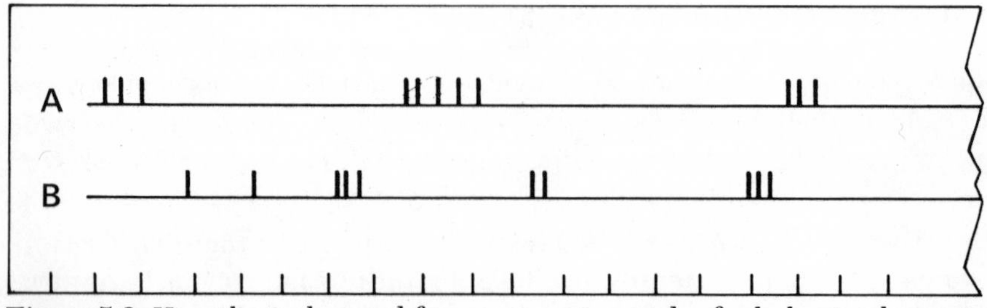

Figure 5-3 Hypothetical record from an event recorder for behavioral events *A* and *B*. Time marked every 10 seconds.

Behavior *A* appears to occur in discrete bouts that seem obvious from the record. Both bout criteria (above) could be applied equally well to defining these obvious bouts. Behavior *B*, however, will fall differently into bouts depending on the criterion chosen. If we choose criterion 1 then there are two bouts. If we use criterion 2 with a bout criterion interval (BCI) of 30 seconds, then there are three bouts; if we use a BCI of 10 seconds then there are five bouts.

Bout criterion intervals are often set arbitrarily according to qualitative or quantitative properties. For example:

> Bout: a consecutive series of songs which may vary in minor ways but nevertheless conforms to a particular song type. [MULLIGAN 1963:276]

> If mouth activities of the same kind followed each other in intervals of less than 16 seconds, they are considered to be a "bout." [HEILIGENBERG 1965:164]

Rosenblum (1978) discussed two dimensions of human behavior which he found useful in delineating bouts: 1) a change in the level (intensity) of motor output and 2) change in orientation. Bout criteria should only be formulated after the observer has become acquainted with the animal's behavior.

After becoming familiar with the social behaviors of laboratory rats, Grant and Mackintosh (1963) selected three seconds as the maximum amount of time that could elapse between behavioral acts of the same rat in order to consider them part of the same sequence. Dane and Van der Kloot (1964) studied interindividual courtship display sequences in groups of male goldeneye ducks *(Bucephala clangula)*. From their observations, they set a maximum time of five seconds which could elapse between displays of two ducks in order to consider them a stimulus-response sequence.

A more objective method of defining the BCI is to examine the frequency histogram of intervals between behaviors or the log survivorship curve (Fig. 5-4).

The log survivorship curve describes the probability of an event occurring relative to the time elapsed since the last event. When behaviors occur in bouts the slope of the curve is steep initially and then becomes gradual as the intervals lengthen. The curve is often considered to break into two portions: 1) a steep section of short within-bout intervals and 2) a gradual section of longer between-bout intervals. The break point between the steep and gradual sections of the curve can be merely approximated by visually inspecting the curve (Slater 1974) (Fig. 5-5).

But how do you know what constitutes an 'event' (i.e. bout) in the first place? M.B.

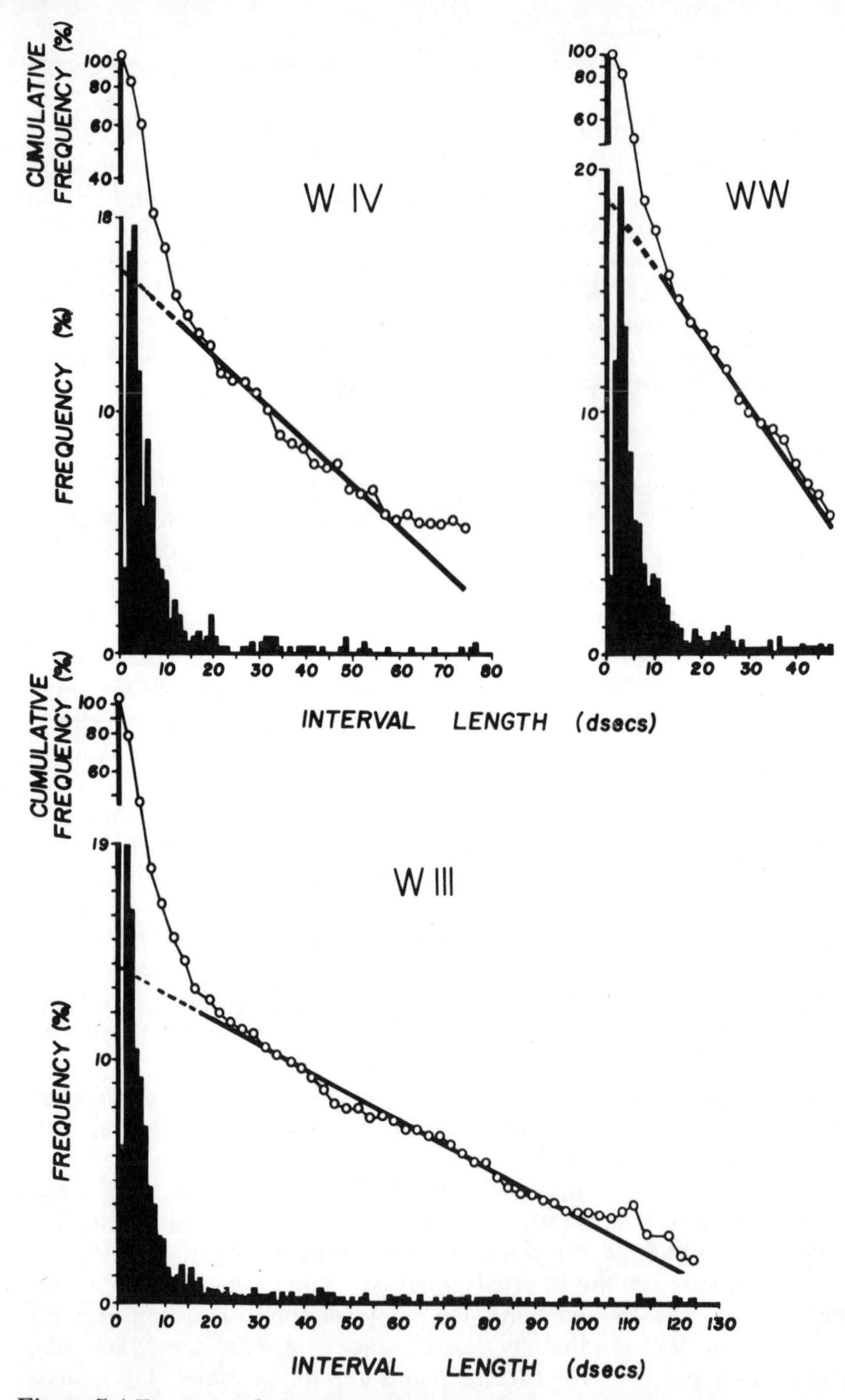

Figure 5-4 Frequency histograms of intervals between pecks and log survivorship curves. (From Machlis, L., 1977, An analysis of the temporal patterning of pecking in chicks, *Behaviour* 63(1–2):20, Fig. 4.)

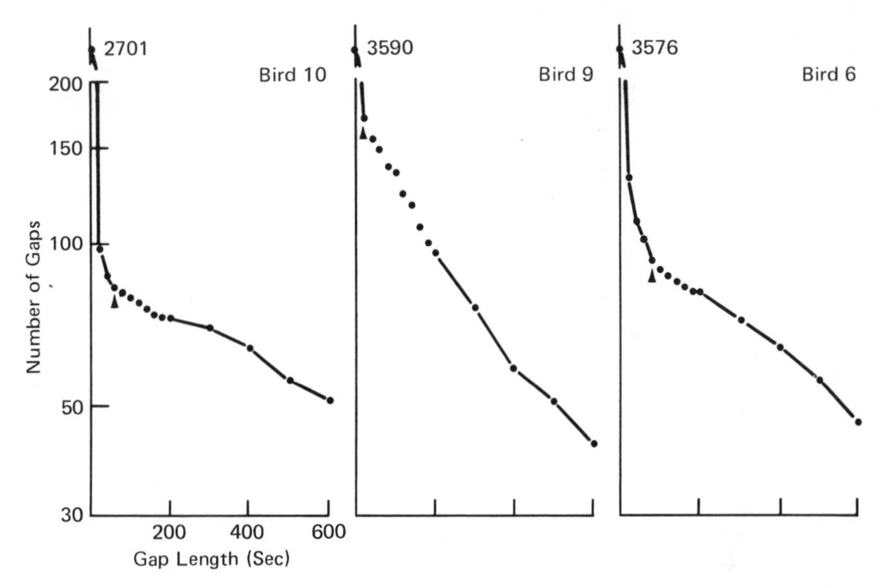

Figure 5-5 Log survivor functions for intervals between pecks at food for three birds. Each point gives the number of intervals longer than gap length shown on the abscissa. Arrows indicate the point chosen for bout definition for each bird. (From Slater, P.J.B., 1974, The temporal pattern of feeding in the zebra finch, *Anim. Behav.* 22(2):508, Fig. 1.)

The gradual portion of the curve can generally be fitted with a straight line, making it an exponential function. Machlis (1977:14) devised a procedure ". . . to determine the maximum number of long intervals which can be incorporated into the 'tail' of the curve but still have this tail fit reasonably well to an exponential function." This provides a very objective procedure for defining the bout criterion interval.

When long-term records of behavior are examined, bouts may be found clustered into "super-bouts" (Machlis 1977). The log survivorship curve may then be a composite of three types of bout intervals: 1) the initial steep portion of within-bout intervals, 2) the intermediate slope reflecting between-bout intervals, and 3) the very gradual slope representing between-cluster intervals (Fig. 5-6). The "super-bouts" may reflect diurnal rhythms, such as morning and evening feeding periods. We then have a hierarchy of behavioral units as in Figure 5-7.

4. Number of Behavioral Units to Measure

Selection of the number of behavioral units to measure will be affected by both the research question and logistics. The *minimum* number will be dictated by a determination of what units are impor-

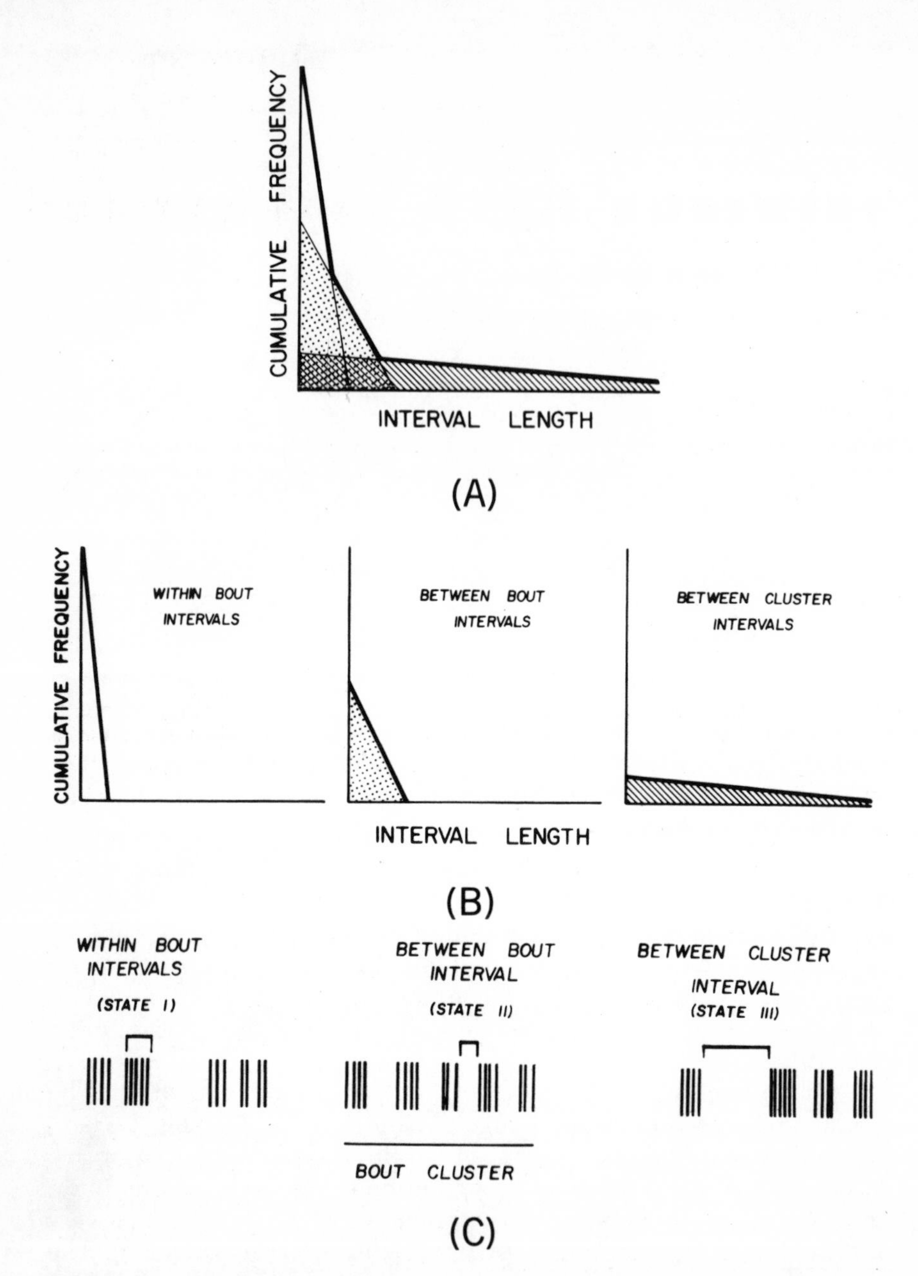

Figure 5-6 Model for the log survivorship curve based on the assumption that the intervals between pecks represent three different states within the chick (*C*) and that these intervals are Poisson generated. Such intervals will be exponentially distributed and will have log survivorship functions as shown in (*B*). The composite of these distributions is shown in (*A*). (From Machlis, L., 1977, An analysis of the temporal patterning of pecking in chicks, *Behaviour* 63(1–2):23, Fig. 7.)

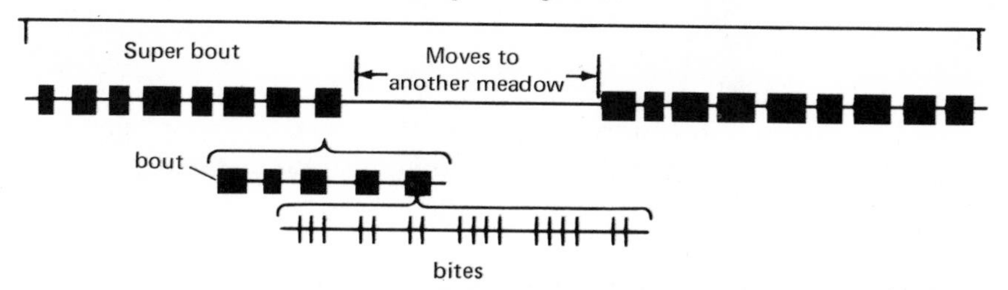

Figure 5-7 Hypothetical record of a portion of an elk's spring feeding pattern broken into a feeding period, super bouts, bouts and bites. Super bout = time in meadow spent primarily in feeding; bout = time with head bent to ground biting vegetation or seeking other clumps, separated by head-raised posture.

tant in answering the research question; that is, what is necessary for testing the hypothesis in a valid way. The *maximum* number will be determined by the experimental design, sampling method, data collection procedure (*e.g.*, equipment), and the ability and experience of the observer. Hutt and Hutt (1974) conclude that an observer is unable to cope reliably with more than 15 separate behaviors. However, many examples can be found in the literature where more than 15 behaviors have been recorded. Reliability (p. 129) may decrease as the number of recorded behavioral units increases.

5. Validity

Selection of the approach (naturalistic observation or experimental observation), variables, behavior units, and number of units to measure will have an effect on the validity of your research.

Validity refers to how well the behavior units you study and the methods you employ answer the research question. If you do not believe that the behavior units you have selected for study are appropriate to answer the research question, then your research is not valid, your objectives cannot be met, and you should go no further. All your methodology, including data collection, analysis, and interpretation (discussed later), will affect the validity of your study. At each step in the preparation of your research you must reflect on the validity of what you intend to do.

There are two types of validity:

Internal validity—how well does your research methodology answer your research question for your chosen sample?

External validity—how applicable are your results to other situations or populations?

If you are conducting *species-oriented* research you will be concerned primarily with internal validity. However, if your research is *concept oriented* you will also have to demonstrate external validity.

D. RESEARCHER-STATISTICIAN

Few researchers are both good ethologists and good statisticians. Consequently many maintain a basic understanding of statistics while developing their greatest skills in ethology. However, although this is the age of specialization and computers, the ethologist should have a basic knowledge of experimental designs and statistical procedures before setting out to collect data.

> . . . to write of the "experimenter" and the "statistician" as though they are separate persons is often convenient; the one is concerned with undertaking a piece of research comprehensively and accurately yet with reasonable economy of time and materials, the other is to provide technical advice and assistance on quantitative aspects both in planning and in interpretation . . . the statistician can produce good designs only if he understands something of the particular field of research, and the experimenter will receive better help if he knows the general principles of design and statistical analysis. Indeed, the two roles can be combined when an experimenter with a little mathematical knowledge is prepared to learn enough theory of design to be able to design his own experiments. [FINNEY 1960:3]

Just as the researcher relies on the statistician, the statistician relies on the researcher to provide the proper data.

> The statistician . . . reasons, never with complete certainty, from observations back to the causal relations presumed to have led to these observations, and in doing so makes inferences about any probabilities that may be intervening to produce loose coupling between cause and effect. [HOTELLING 1958:11]

In order to assure that the proper data are collected the researcher should consult a statistician during the planning stages.

The discussions of experimental designs and statistical analysis in this book are introductory and cursory. The material represents what I believe to be the minimum knowledge ethologists should have in order to conduct their research with insight and logic. It is meant to supplement, not substitute for, consultation with a biometrician.

E. BASIC EXPERIMENTAL DESIGNS

Experimental designs are protocols for manipulating and/or measuring independent variables (treatments) in such a way that their singular or combined effects on the behavior being studied (dependent variable) can be determined. Choice of the proper design is based on knowledge of the 1) objectives of the research, 2) hypothesis to be tested, 3) feasibility of gathering various types of data, 4) types of experimental designs available and their relative attributes of 5) precision 6) power, and 7) efficiency. More will be said about the evaluation of experimental designs later.

Four basic experimental designs—*completely randomized*, *randomized block*, *incomplete block*, and *Latin square*—will be considered here. Almost all other designs can be built by combining two or more of these designs. Each subsequent design discussed measures the effect of an additional independent variable and consequently reduces the effect due to the nuisance variables (error effect).

Each variable should contain a zero treatment level, or *control*. The controls should be exactly the same in every way except that no treatment is applied. Controls provide a baseline (zero treatment) with which to compare the effects of different levels of the treatment.

1. Completely Randomized Design

This is one of the simplest designs, but it can be used to compare any number of treatment levels of a single independent variable (variable a). Treatment levels are the qualitative or quantitative differences in the independent variable which you have hypothesized do (or do not) have a measurable effect on the behavior you are studying.

a. Tabular Form

	Treatment Levels		
	a_1	a_2	a_3
	x_{11}	x_{12}	x_{13}
	x_{21}	x_{22}	x_{23}
	x_{31}	x_{32}	x_{33}
	x_{41}	x_{42}	x_{43}
	x_{51}	x_{52}	x_{53}
Treatment means	$\bar{x}_{.1}$	$\bar{x}_{.2}$	$\bar{x}_{.3}$
		Grand mean $\bar{x}_{..}$	

Treatment levels: a_1, a_2, a_3

x_{ij} = dependent variable to be measured; ij designates the ith measure in treatment j.

$\bar{x}_{.j}$ = a bar over a variable indicates a mean for that variable; in this case this is the mean for treatment j; the dot in the subscript indicates the variable over which the summation occurred. Hence the grand mean = $\bar{x}_{..}$

For example, Wood-Gush (1972) measured the amount of prelaying pacing by domestic hens of two different strains. Six birds were selected and placed in individual pens equipped with trap-nests.

Strain			
White		Brown	
Bird No.		Bird No.	
P2370	x_{11}	1932	x_{12}
P2338	x_{21}	2912	x_{22}
P2376	x_{31}	3337	x_{32}
P2335	x_{41}	3305	x_{42}
P2346	x_{51}	3292	x_{52}
P2345	x_{61}	3306	x_{62}

x_{11} = amount of prelaying pacing by bird no. P2370

b. Linear Model

A simple equation can be used to show all the sources of variation that affect the individual measurements.

$$x_{ij} = \mu + \alpha_j + \epsilon_{ij}$$

The model (equation) states that the individual measurement (x_{ij}) is equal to the population mean (μ) plus the treatment effect (α_j) plus an error effect (ϵ_{ij}) which is unique for each individual subject.

As in most experiments, population parameters are not known. However, our sample is used to estimate those parameters. In this experiment μ, α_j, and ϵ_{ij} are unknown; but they are estimated respectively by the following:

$$\hat{\mu} = \bar{x}_{..} \text{ (estimates } \mu)$$

$$\hat{\alpha}_j = (\bar{x}_{.j} - \bar{x}_{..}) \text{ (estimates } \alpha_j)$$

$$\hat{\epsilon}_{ij} = (x_{ij} - \bar{x}_{.j}) \text{ (estimates } \hat{\epsilon}_{ij})$$

The error effect is the summed effect of all the uncontrolled *nuisance variables*. It is an estimate of all effects not attributable to a particular treatment. We can rearrange the linear model to show that the error effect ($\hat{\epsilon}_{ij}$) is what remains of an individual measurement (x_{ij}) after the treatment effect and grand mean are subtracted from it.

$$\hat{\epsilon}_{ij} = x_{ij} - \hat{\alpha}_j - \hat{\mu}$$

2. Randomized Block Design

This design attempts to control for additional variability (expressed in the error effect) by assigning subjects that are similar in one or more characteristics (or ways treatments are applied to them) to blocks. That is, subjects within each block (*e.g.*, same sex or age) should be more homogeneous than subjects between blocks. With this design we are measuring the effect of two independent variables (variables a and b).

a. Tabular Form

	Treatment Levels				
Blocks	a_1	a_2	a_3	Block Means	
b_1	x_{11}	x_{12}	x_{13}	$\bar{x}_{1.}$	
b_2	x_{21}	x_{22}	x_{23}	$\bar{x}_{2.}$	
b_3	x_{31}	x_{32}	x_{33}	$\bar{x}_{3.}$	
b_4	x_{41}	x_{42}	x_{43}	$\bar{x}_{4.}$	
b_5	x_{51}	x_{52}	x_{53}	$\bar{x}_{5.}$	
Treatment means =	$\bar{x}_{.1}$	$\bar{x}_{.2}$	$\bar{x}_{.3}$	$\bar{x}_{..}$	= Grand mean

As an example, Wood-Gush (1972) also wanted to determine the effect of type of pen on pacing by the two strains of domestic hens (see example for completely randomized design). He therefore used pen type (ordinary wire battery cages vs. cages with solid metal walls on two sides and the back) as a blocked treatment. These were termed open and enclosed, respectively.

	Strain	
Blocks	White	Brown
Open cage	8 hens	8 hens
Enclosed cage	8 hens	8 hens

b. Linear Model

The equation for this design includes β_i, the effect attributable to the ith block:

$$x_{ij} = \mu + \alpha_j + \beta_i + \epsilon_{ij}$$

Therefore, by assigning similar subjects to blocks we have partitioned out an additional source of variability. We can show this by rearranging the linear model

$$\epsilon_{ij} = x_{ij} - \hat{\alpha}_j - \hat{\beta}_i - \hat{\mu}$$

Estimates for μ and α_j remain $\bar{x}_{..}$ and $(\bar{x}_{.j} - \bar{x}_{..})$, respectively.

$$\hat{\epsilon}_{ij} = (x_{ij} - \bar{x}_{.j} - \bar{x}_{i.} + \bar{x}_{..}) \text{ (estimates } \epsilon_{ij})$$
$$\hat{\beta}_i = (\bar{x}_{i.} - \bar{x}_{..}) \text{ (estimates } \beta_i)$$

If the block effect (β_i) is appreciable then we will have been successful in reducing the error effect (ϵ_{ij}) by blocking. The relative efficiency of the statistical analysis we use will be increased by reducing the error effect as much as possible. Therefore, that should be one of the goals of selecting an experimental design.

3. Incomplete Block Design

This particular design is applicable when the number of subjects available for study is not large enough to measure each treatment effect for each block.

a. Tabular Form

	Treatment Levels			
Blocks	a_1	a_2	a_3	Block Means
b_1	x_{11}		x_{13}	$\bar{x}_{1.}$
b_2		x_{22}	x_{23}	$\bar{x}_{2.}$
b_3	x_{31}	x_{32}		$\bar{x}_{3.}$
Treatment means =	$\bar{x}_{.1}$	$\bar{x}_{.2}$	$\bar{x}_{.3}$	$\bar{x}_{..}$ = Grand mean

This design is *balanced*, which demands that each block contains the same number of subjects, each treatment level occurs the same number of times, and subjects are assigned to the treatment levels so that each possible pair of treatment levels occurs together within some block an equal number of times.

In *partially balanced* designs some pair of treatment levels occur together within the blocks more often than do other pairs.

For example, Kinsey (1976:181) tested his hypothesis that "... Allegheny woodrats *(Neotoma floridana magister)* would exhibit territorial behaviour when confined in relatively low-density populations in a large observation cage and at populations of higher density would exhibit increased agonistic interactions and a dominance hierarchy type of social organization." He placed wild-trapped male and female woodrats together in a 65-m² enclosure at densities varying from 2 to 14. However, not all densities were represented nor did each group contain both males and females. The densities (blocks) were considered in two larger blocks of low density (2–4) and high density (5–14). The table shows the densities and sexes that were represented in the groups used.

		Treatments		
Blocks (individuals per 65 m²)		All Male	All Female	Mixed
Low Density	2	X	X	X
↓	3			X
	4			X
High Density	5		X	
	6	X		X
	7			
	8			X
	9–13			
↓	14			X

At this point the question arises: If I have a limited number of subjects should I go to an incomplete block design or fall back on the completely randomized design? Although this will be discussed in greater detail later, *you should always strive to collect the data (dictated by the experimental design) in such a way as to reduce the error effect as much as possible.*

The equation for the incomplete block design is the same as that for the randomized block design:

$$x_{ij} = \mu + \alpha_j + \beta_i + \epsilon_{ij}$$

4. Latin Square Design

This design uses the blocking principle to reduce the error effect by studying the effect of three independent variables simultaneously. The levels of the nuisance variables are assigned to the rows and columns of a Latin square.

a. Tabular Form

	Blocks (a variable)			Block Means
Blocks (b variable)	a_1	a_2	a_3	(b variable)
b_1	c_1 x_{111}	c_2 x_{122}	c_3 x_{133}	$\bar{x}_{1..}$
b_2	c_2 x_{212}	c_3 x_{223}	c_1 x_{231}	$\bar{x}_{2..}$
b_3	c_3 x_{313}	c_1 x_{321}	c_2 x_{332}	$\bar{x}_{3..}$
Block means (c variable)	$\bar{x}_{.1.}$	$\bar{x}_{.2.}$	$\bar{x}_{.3.}$	
Treatment means	$\bar{x}_{..1}$	$\bar{x}_{..2}$	$\bar{x}_{..3}$	
			Grand Mean	$\bar{x}_{...}$

The treatments c_1, c_2, c_3 are assigned to the cells in the Latin square in a balanced fashion. That is, each treatment is equally represented in both rows and columns.

The three subscripts represent a particular block (b variable), a particular block (a variable), and treatment level (c variable) respectively. A Latin square must have the same number of rows, columns, and treatment levels.

For example, Huck and Price (1976) in their study of the effect of early experience on the climbing behavior of wild and domestic Norway rats had sufficient numbers of each sex to assign both males and females to treatment groups. Had they been limited they could have used a Latin square design as shown below, although this would have been less desirable.

	Blocks	
Blocks	Wild	Domestic
Enriched environment	males	females
Unenriched environment	females	males

b. Linear Model

The equation for this design includes variables for the row effect (b variable; β_i), column effect (a variable; α_j), and treatment effect (c variable; υ_k)

$$x_{ijk} = \mu + \alpha_j + \beta_i + v_k + \epsilon_{ijk}$$

Through this double blocking we have partitioned another source of contribution to the error effect.

$$\epsilon_{ijk} = x_{ijk} - \hat{\alpha}_j - \hat{\beta}_i - v_k - \hat{\mu}$$

This double blocking could reduce the error effect significantly enough to make this a more powerful design than either the completely randomized or randomized block designs.

The following show the assignment of treatments *A*, *B*, etc. to the cells of a 4 × 4 and a 6 × 6 design.

4 × 4

A	B	C	D
B	C	D	A
C	D	A	B
D	A	B	C

6 × 6

A	B	C	D	E	F
B	C	D	E	F	A
C	F	B	E	A	D
D	E	A	B	F	C
E	A	D	F	C	B
F	D	E	C	B	A

F. EFFICIENCY OF EXPERIMENTAL DESIGNS

A measure of efficiency should include the cost (time and money) of collecting the data balanced against the strength of those data. The relative efficiency of two designs is often assessed by comparing their respective error effects (experimental errors). *Experimental error* is the extraneous variation in the measurements due to all the nuisance variables. Its ultimate effect is to mask the effect due to the independent variable.

Federer (1955:13) has proposed the following formula to measure efficiency:

$$\text{efficiency} = \frac{\left(\dfrac{n_2 c_2}{\hat{\sigma}_1^2}\right)\left(\dfrac{df_1 + 1}{df_1 + 3}\right)}{\left(\dfrac{n_1 c_1}{\hat{\sigma}_2^2}\right)\left(\dfrac{df_2 + 1}{df_2 + 3}\right)}$$

$\hat{\sigma}^2$ = estimate of experimental error per observation
n = number of subjects
c = cost of collecting data per subject
df = experimental error degrees of freedom

The subscripts designate the two experimental designs. If the ratio is greater than one, then the first design is more efficient than the second.

Because of the limited control ethologists generally have in field studies, compromises are usually necessary.

G. DETERMINATION OF SAMPLE SIZE

In order to determine the sample size required for your particular study, you should have an estimate of the variability of the data. This can be determined by gathering some preliminary data (or referring to data gathered during reconnaissance observations) and calculating the standard deviation of the measurements (see p. 236). The following formula for determining required sample size (Snedecor 1950) can then be used.

$$n = \text{number of samples required} = \frac{s^2 t^2}{d^2}$$

where s = standard deviation
t = tabular "t" value (Table A1) at the selected confidence level (see p. 230), and for the degrees of freedom (see p. 247) in your sample
d = margin of error (mean × designated accuracy)

For example, let us say we want to determine the difference in mean durations of coyote howls given nocturnally and diurnally. We get out our field notes and find that we have measured the durations of six individual nocturnal howls as follows:

5.2 sec.	3.1	
6.1	7.2	mean = 5.4 sec.
4.3	6.7	

We calculate the standard deviation (s) = 1.5.

The tabular t value for 5 degrees of freedom (6–1) at the 0.95 confidence level = 2.015.

We decide to accept a 0.05 level of accuracy; then:

$$n = \frac{(1.5)^2 (2.015)^2}{(5.4 \times .05)^2} = \frac{(2.25)(4.06)}{(.07)} = 130$$

Therefore we must obtain 130 samples of nocturnal coyote howls in order to have a reasonable estimate of the true mean duration. We would have to make the same calculations based on a sample of diurnal howls.

If you are unable to obtain an estimate of the variation to be expected in the data, then use as large a sample size as possible and feasible. *Statistics estimate population parameters from sample measurements.* The larger the sample, the better the possibility that the sample statistics will closely approximate the population values.

6
Experimental Manipulation

In the previous chapter naturalistic observation and experimental manipulation were discussed relative to designing a research project. Naturalistic observation was described as the approach in which the observer studies the behavior of animals as it occurs naturally with as little intrusion as possible. In experimental manipulation variables are manipulated (or allowed to change naturally) in order to measure their effect on selected behavior.

If we make a further dichotomy (actually a procedural continuum) between field and laboratory investigations, then we can categorize ethological investigations in what is often (but not necessarily) an increasing order of artificiality as follows:

I. Field investigations
   A. Naturalistic observation
   B. Experimentation
      1. Natural variation
      2. Artificial manipulation
         a. Manipulation of the environment
         b. Manipulation of the animal

II. Laboratory investigations
   A. Naturalistic observation
   B. Experimentation
      1. Artificial manipulation
         a. Manipulation of the environment
         b. Manipulation of the animal

This chapter is organized according to this categorization. Selected experiments from the literature will be cited in order to illustrate examples of various procedures.

A. VARYING THE VARIABLES

The objective of experimental manipulation is generally to determine the factors that are important in affecting a given behavior. These factors may be long term (evolutionary and ontogenetic) or short term (proximal); refer back to the behavioral model on page 7.

To understand the variables that may affect an organism's behavior we must first recognize that an individual exists in time and space in a dynamic state, continually under the influence of its environment and the result of its evolutionay and ontogenetic history. Figure 6-1 is a diagram of the relationship between an organism and the *environment*. The environment has both *biotic* (biological) and *abiotic* (physical) features, some of which will affect the organism. Biotic features may include vegetation type in habitat selection, predator-prey relationships, and mate selection. Abiotic features are important in regulating certain animals' temporal activity patterns, such as diurnal-nocturnal patterns and seasonal cycles. Temperature, humidity, wind, etc., may all be important in determining the behavior patterns shown by an animal or group of animals.

The *organism* consists of a *genotype* (genetic material) that bears the mark of natural selection over many previous generations. The *phenotype* is the sum of the behavioral and physical characteristics of the individual. The organism can exert forces on both the biotic environment (*e.g.*, intra- and inter-specific social behavior) and the abiotic environment (*e.g.*, a badger burrowing into a hillside).

Since all the factors in Figure 6-1 are variables that affect all behavior to different degrees, the researcher must carefully select one (or a few) of the factors to study at one time. The complexity of the interacting variables must be recognized and dealt with as skillfully as possible within the limits available to the researcher.

The first step is to list all the variables that are known to affect the behavior in question or are suspected of having some effect. Some examples of the different types of variables are listed below:

I. Environment
   A. Biotic
      1. Members of social group
      2. Predator-prey relationships
      3. Vegetative characteristics of habitat
      4. . . .

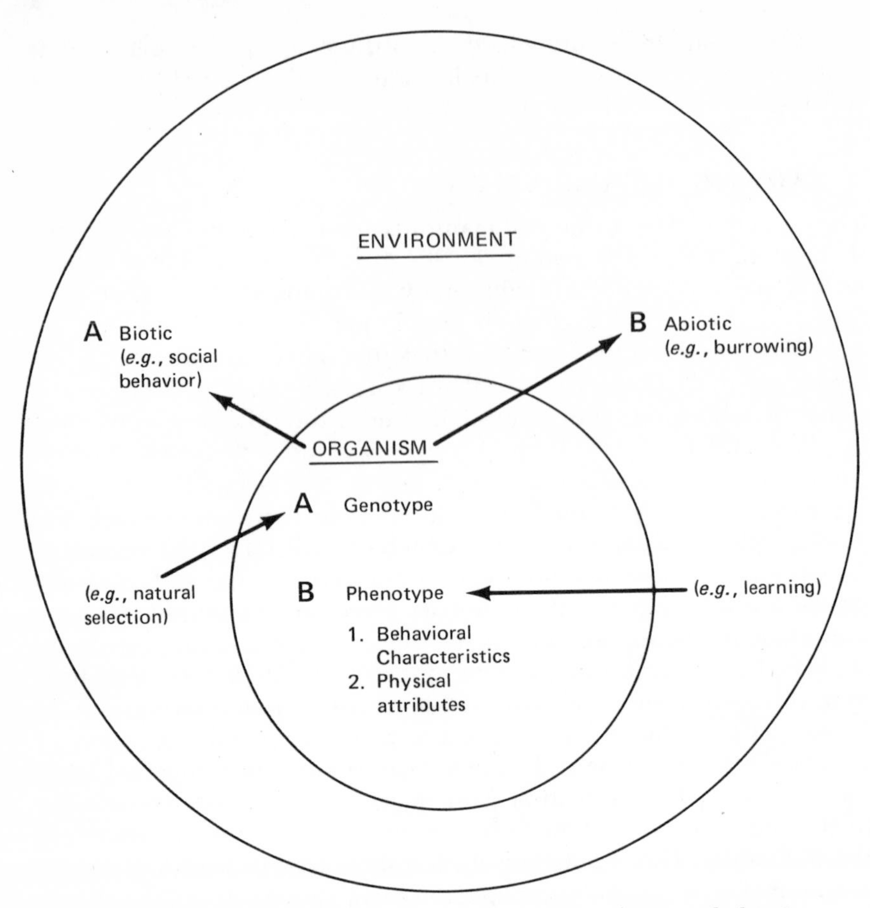

Figure 6-1 Interactions between an organism and its biotic and abiotic environment. Arrows indicate forces from the environment acting on the organism and *vice versa*. (See text for explanation.)

B. Abiotic
1. Temperature
2. Wind
3. Humidity
4. Cloud cover
5. Topography
6. Time—circadian and seasonal
7. ...

II. Organism
A. Genotype
1. Sex

2. Parent stock
3. . . .

B. Phenotype
1. Behavioral characteristics
   a. Description of behavior
   b. Frequency
   c. Rate
   d. Duration
   e. Temporal patterning—circadian and circannual
   f. Spatial characteristics
2. Physical attributes
   a. Morphological characteristics; *e.g.*, shape of feathers, color patterns
   b. Physiological characteristics
   c. . . .

This procedure is useful, whether you are attempting to provide perspective on the causation of a particular behavior in anticipation of designing a study or whether you have already decided on the variable(s) you want to manipulate and want to account for other potential sources of variation. The number of factors which could potentially affect a behavior is extremely large; so the researcher should be judicious in spending time compiling the list. In addition, you should enlist the help of your colleagues in determining the most important variables.

The usual procedure is to isolate the variable of interest (*e.g.*, light intensity) and systematically manipulate that variable artificially or follow it through natural changes. The *other factors must remain constant or vary randomly*, so that they can be considered to have no systematic effect on the behavior being studied. The factor being manipulated is the *independent variable* (*e.g.*, light intensity), and the behavior being measured is the *dependent variable* (*e.g.*, time spent walking/sample period). See Chapter 5, Section B for a further discussion of variables. Several variables may be manipulated and measured simultaneously in order to measure both individual effects and interactions. Selected analyses of this type are discussed under multivariate analyses in Chapter 11.

1. Natural Variation

Observation of behavior under conditions of naturally occurring changes in both the environment and the animal provides the first level of experimentation.

The observational work has to be followed up by experimental study. This can often be done in the field. The change from observation to experiment has to be a gradual one. The investigation of causal relationships has to begin with the utilization of "natural experiments." The conditions under which things occur in nature vary to such a degree that comparison of the circumstances in which a certain thing happens often has the value of an experiment, which has only to be refined in the crucial tests. [TINBERGEN 1953: 136]

Regular changes in the environment can be utilized to study their effects on the behavior of selected species. Pengelley and Asmundson (1971) showed that the yearly activities of golden-mantled ground squirrels *(Spermophillus lateralis)* fluctuated in synchrony with climatological variables in the environment. Foraging activity of the nocturnal bee *(Sphecodogastra texana)* was shown by Kerfoot (1967) to be based on the lunar cycle. Sunrise and sunset apparently trigger the onset and cessation of activity in cottontail rabbits *(Sylvilagus floridanus)* and snowshoe hares *(Lepus americanus)* (Mech et al. 1966).

Some environmental parameters fluctuate within seasonal ranges but vary somewhat irregularly from day to day. For example, decreasing light levels near sunset apparently trigger the initial departure towards the roost of foraging starlings *(Sturnus vulgaris)* (Davis and Lussenhop 1970). Nisbet and Drury (1968) compared measurements of density of songbird and waterbird migration to nineteen weather variables in the area of takeoff. They found that migration densities were significantly correlated with high and rising temperature, low and falling pressure, low but rising humidity, and the onshore component of wind velocity.

The response of animals to simultaneous variations in the environment can also be studied. Heinrich (1971) examined the feeding pattern of the caterpillar *(Manduca sexta)* and found that it was consistent for given leaf shapes and sizes. The use of simultaneous variation in the environment is the basis for field studies of habitat selection. MacArthur (1958) studied the distribution of five congeneric species of warblers while they fed on individual white-spruce trees. He divided the trees into 16 zones and measured the percent of total number of seconds of observation and percent of total observations for each species in each zone. He found that the five species distributed themselves on the trees utilizing different microhabitats (Fig. 6-2).

The age and experience of the animals under investigation can be allowed to advance naturally and their behavior observed at various sequential stages. J.P. Scott and Fuller (1965) observed the changes in behavior of several domestic dog breeds from birth to maturity and

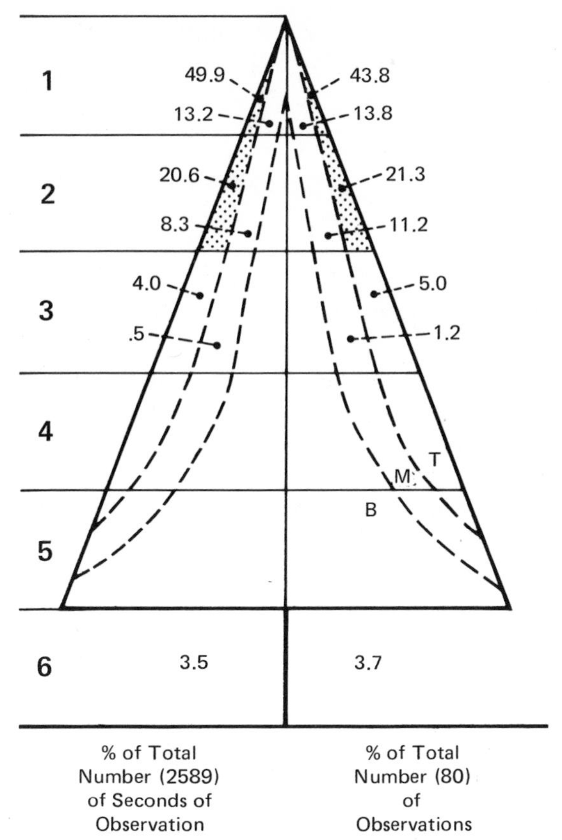

Figure 6-2 Cape May warbler feeding position. The zones of most concentrated activity are shaded until at least 50 percent of the activity is in the stippled zones. Each branch was divided into three zones: (*B*) bare of lichen-covered base, (*M*) old needles, and (*T*) new (less than 1.5 years old) needles and buds. (From MacArthur, R.H., 1958, Population ecology of some warblers of northeastern coniferous forests, *Ecology* 39:599–619, Fig. 2.)

were able to divide their development into four periods: neonatal, transition, socialization, and juvenile. Development of behavior in the song sparrow was divided into six similar stages by Nice (1943). Drori and Folman (1967) showed a marked effect of experience on the copulatory behavior of male rats, and Carlier and Noirot (1965) demonstrated that experience improved pup retrieval in female rats. Stefanski (1967) showed that the average territory size of black-capped chickadees varied during six stages of the breeding season: prenesting, nest building, egg laying, incubation, nestling, and fledgling.

The use of natural variation has limitations which are both qualitative and quantitative. Waiting for the proper conditions to arise and attempting to gather a sufficient number of observations often drives the ethologist to artificial manipulation:

> Systematic exploitation of such natural experiments—that is, systematic comparison of the situations which do and those which do not release a given response—can be almost as good as planned experiments; the important thing seems to me is not to miss the natural experiments and yet to know when it becomes necessary to continue by planned tests. [TINBERGEN 1958:289]

2. Artificial Manipulation

Another approach to the study of cause and effect of behavior is to take control of the variables and manipulate them in the field or the laboratory. Although the manipulation is artificial, every attempt should be made to approximate the natural situation (*i.e.*, the appropriate and important stimuli) as closely as possible.

Models (*i.e.*, dummies) have a long history of use in ethology and have the advantage of allowing the experimenter to vary stimuli (*e.g.*, visual, auditory, chemical, tactile) in a systematic way in order to measure the effect of qualitative and quantitative differences. Tinbergen and Perdeck (1950) presented models of an adult herring gull's head to herring-gull chicks. They found that the color of the spot on the bill (qualitative property) of the model had an effect on the number of pecks given by the chicks. The gaping response of blackbirds *(Turdus merula merula)* and thrushes *(Turdus ericetorum ericetorum)* is directed by the relative size of the head to the body. This was demonstrated by Tinbergen and Kuenen (1939) using simple models.

Models are often used in a context in which it is believed that they are a reasonable facsimile of the natural stimulus (see Curio 1975 for an excellent example of extensive and proper use of models). In other words, ". . . an underlying assumption of the method is that response to the model depends on much the same causal system as response to the natural stimulus" (Losey 1977:224). However, this assumption is rarely validated. In Losey's experiments on the response of host fish *(Chaeton aurign)* to a cleaner *(Labroides phthrirophagus)*, he demonstrated the validity of his use of a cleaner model through three indicators: pose duration, pose-to-inspect ratio, and approach behavior of the host fish to both live cleaners and models.

The use of models must not only be carefully planned, but the results of such experiments must be carefully interpreted. In another aspect of Tinbergen and Perdeck's (1950) experiments on the begging response in neonatal herring-gull chicks, they changed the position of the red spot from the model's bill to its forehead. The chicks delivered significantly more pecks to the model with the spot on the bill than they did to the model with the spot on the forehead. They concluded that the position of the red patch had a large effect on the chicks' responses. Hailman (1969) re-investigated this phenomenon by placing the models at different distances from the pivot point of the rod holding the model. Further, he adjusted the height of the chick so that it was always at eye level with the red spot. He had created three models: a "normal model" with the spot on the bill, a model with the spot on the forehead and the pivot point the same ("slow model"), and a model with the spot on the forehead but with the pivot point the same distance from the spot as on the bill-spot model ("fast model"). The fast forehead-spot model received more pecks than the slow forehead model, although fewer than the "normal model," revealing the effect of speed of movement on the chicks' responses. Therefore, Tinbergen and Perdeck (1950) were correct in concluding that position of the spot is important; but Hailman demonstrated that speed of movement is also a contributing factor.

Artificial manipulation is a powerful technique in ethology, but its use must be carefully monitored. Models may be too simple with the important stimuli absent or too complex with extraneous stimuli confounding the experiment. As with any powerful tool, however, in the hands of a skilled experimenter models can be an important means of artificial manipulation. See page 96 for a further discussion on the use of models and also of preference tests, another important artificial manipulation used by ethologists.

B. EXAMPLES OF EXPERIMENTAL MANIPULATION

1. In the Field

Many experiments arise from descriptive studies in the field and progress through experiments using natural variations to experimental manipulation of either the animal or its environment (see p. 105).

a. Manipulation of the Animal

The role of sensory receptors and physiological state can be studied by manipulation of the animal *per se*. Layne (1967) studied the role of vision in diurnal orientation of the bat *(Myotis austroriparius)* by releasing normal, earplugged, and blinded bats at various distances from the home cave. None of the eye-covered bats homed. Ehrenfeld and Carr (1967) measured the role of vision in the sea-finding behavior of female green turtles *(Chelonia mydas)* by blindfolding them or fitting them with spectacles containing different filters. Blindfolded

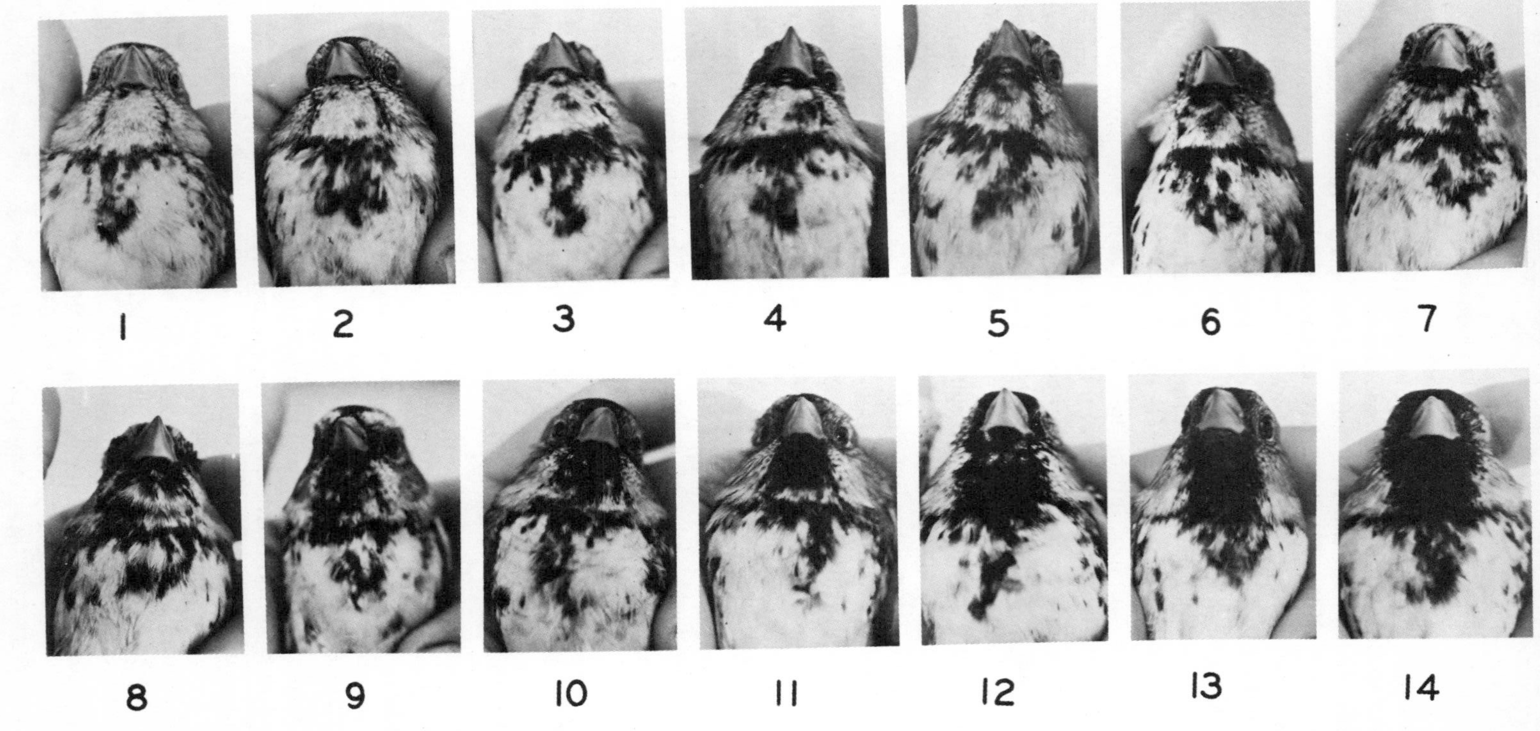

Figure 6-3 The 14 "studliness" categories (in ascending order of studliness) used by Rohwer in his study of male Harris' sparrows. (From S. Rohwer, 1977, Status signaling in Harris sparrows: some experiments in deception, *Behaviour* 61(1–2):107–129, Fig. 1, Pl. 4.)

turtles and those wearing red, blue, and 0.4 neutral density filters had significantly reduced orientation scores.

Morphological changes are often made on animals in the field and the effect on the animal's ability to obtain and/or retain a mate, social status, or a territory is then measured. In these studies it is the change in behavior of the other individuals that are engaged in interactions with the altered individual which is actually being measured; but the effect can be determined by observing the altered individual. Bouissou (1972) showed that dehorning and reduced weight decreased the ability of domestic cattle to obtain and maintain high social rank in the herd. Harris sparrows *(Zonotrichia quereula)* signal their dominance status by variations in the amount of black feathering on their crowns and throat. Rohwer (1977) ranked individuals into 14 "studliness" categories (Fig. 6-3) and then altered individuals to determine the effect of their status. Subordinates dyed to mimic the highest ranking birds were still persecuted by legitimate "studlies," and bleached birds eventually exerted their normally high-ranking dominance. The data suggested that "cheating" (*i.e.*, lower-ranking birds being elevated in status simply by having a darker crown and throat) is socially controlled. The role of the red epaulets of male red-winged blackbirds *(Agelaius phoeniceus)* was studied by D.G. Smith (1972) by dying the epaulets black on selected males. He found that the epaulets were important in maintenance of territories against rival males, but they had little effect on the males' ability to obtain mates. N.G. Smith (1967) changed the eye-ring color of one member of mated pairs of sympatric glaucous gulls *(Larus hyperboreus)*, Kumlien's gulls *(L. glaucoides)* and herring gulls *(L. argentatus)*. In all cases where the female's eye-ring color had been changed the pair broke up, but altering the male's eye-ring appeared to have no effect on the pair's behavior.

It is important in all research where animals are manipulated and the effects are studied in interactions with other individuals to *observe the effects on both the manipulated animal and others responding to it*. This is true in both intra- and inter-specific studies, such as the effects of altered males on selection by females and altered prey on selection by predators, respectively.

b. Manipulation of the Environment

Altering the environment in order to study its resultant effect on behavior ranges from gross-perturbation experiments to subtle changes in one or a few stimuli.

Stewart and Aldrich (1951) were able to get an indication of the extent of the surplus "floating" population of unmated male birds in the spruce-fir forests by drastically reducing (by shooting) the number of territorial holders on a 40-acre tract. From June 6 to June 14 they removed 148 territorial males, reducing the population to 19% of the original. They continued to shoot birds as they moved into the area, and by July 8 they had collected a total of 455 individuals. This is a rather drastic perturbation experiment, and as they admit "... the breeding territories were completely disrupted during the period when the original occupants were being removed and at the same time new adult males were constantly invading the area." On a smaller scale Krebs (1971) shot six pairs of great tits occupying territories and observed that residents expanded their territories and four new pairs took up occupancy. Tinbergen has been prone to concentrate on subtle environmental changes in order to study effects without greatly disturbing the normal activities of the animal.

> The trick is, to insert experiments now and then in the normal life of the animal so that this normal life is in no way interrupted; however exciting the result of a test may be for us, it must be a matter of daily routine to the animal. A man who lacks the feeling for this kind of work will inevitably commit offenses just as some people cannot help kicking and damaging delicate furniture in a room without even noticing it. [TINBERGEN 1953:138]

Manipulation of the environment can be conveniently divided into three types: *intraspecific*, *interspecific*, and *physical-environment* manipulations. The intraspecific facilitating effect of a female mallard on male courtship displays was demonstrated by Weidmann and Darley (1971) by introducing a strange female or male to resident groups of three males in the spring and autumn. Free (1967) manipulated the conditions in honeybee hives and showed that the amount of pollen collected increased with the amount of brood present and decreased in the absence of a queen. He went on to isolate some of the stimuli produced by the brood which are important in stimulating pollen collection.

Stout and Brass (1969) placed pairs of complete models or wooden-block models with adjustable stuffed heads (Fig. 6-4) in glaucous-gull territories and demonstrated that the head and neck are the parts of the body that release aggressive display in territorial behavior of this species.

Baerends and Kruijt (1973) measured the stimuli important in releasing egg retrieval in herring gulls by presenting them with three-dimensional dummies placed two at a time on the edge of a nest. The relative importance of the various configurations in releasing egg retrieval was 1) larger > smaller; 2) speckled > not speckled; 3) green > blue, red > grey; and 4) shape, other than roundness, relatively unimportant. The *titration method* used by Baerends and Kruijt is

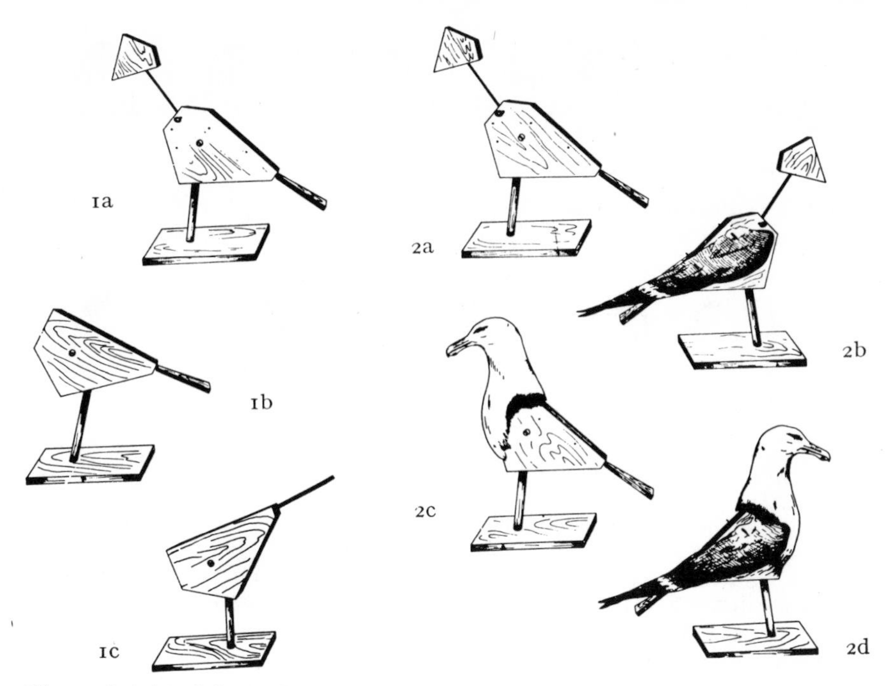

Figure 6-4 Models used by Stout and Brass in their study of glaucous-winged gulls. *1a,* upright threat-postured body; *1b,* trumpeting-postured body; *1c,* choking-postured body; *2a,* basic wooden control model; *2b,* upright threat posture; *2c,* control model without wings; *2d,* upright threat posture with wings. (From Stout, J.F., and M.E. Brass, 1969, Aggressive communication by *Larus Glaucesceus,* Pt. II, Visual Communication, *Behavior* 34(1–2):44, Figs. 1, 2.)

worthy of careful consideration for other studies using models. This method allowed them to rank the models on a relative basis between and within the four categories of features.

Our experiments with the size series showed position preference to be a quantitative phenomenon. A first choice for the smaller egg in the preferred position can always be overcome by increasing the size of the model in the non-preferred position. With our series of models gradually increasing in size it was possible to identify stepwise, in successive tests with the same bird [Fig. 6-5], the minimum size of a model required to overcome position preference, when in competition with a dummy of a smaller size in the preferred position. Thus, through this "titration," a model was found the value of which, in combination with that of the non-preferred site, could just outweigh the combined values of the smaller model and the preferred site. Empirically it turned out that the birds were acting in accordance with the ratios between the surfaces of the maximal projections (maximal shadows when turned around in a beam of parallel light) of the models. Different pairs of models,

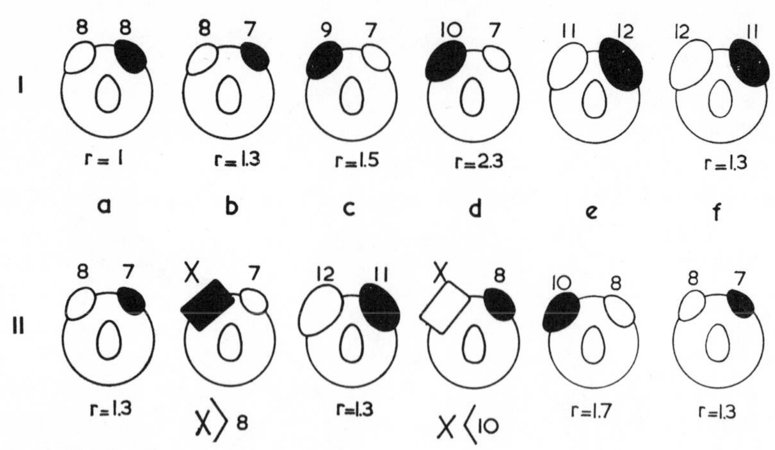

Figure 6-5 The "titration" method for determining the value of an egg dummy. The circle represents the nest with one egg in the nest bowl and two dummies on the rim. The code numbers 7, 8, 9, 10, 11, 12 refer to the dummies of the size series (*R*) shown in Figure 6-6. *X* is the model to be measured; *r* is the ratio between maximal projection surfaces of the dummies on the next rim. The black dummy is always the preferred one. I, determination of the value of the position preference. Ia shows that the right site is preferred. This preference remains when dummy 8 is replaced by the smaller dummy 7 (Ib), but can then be overcome by replacing 8 by 9 (Ic); this sequence shows that the value of the position preference lies between $r = 1.3$ and $r = 1.5$: this conclusion holds when another pair of dummies with the same ratio is used (If). Control test Ie shows that the size optimum for this gull exceeds size 11. II, determination of the value of model *X*. Tests IIa, IIc, and IIf show that the position preference has remained unchanged. Test IIb and IId indicate, in combination with the preceding and succeeding tests, that the value of *X* is between those of the models 8 and 10 of the reference size series. (From Baerends, G.P., and J.P. Kruijt, 1973, Stimulus selection, p. 31, Fig. 3 *in* Hinde, R.A. and J. Stevenson-Hinde, (eds.), Constraints on learning, Academic Press, London.)

matching each other with respect to other parameters tried (*e.g.*, volume), or equal with regard to the difference instead of the ratio in the parameters used, proved to be unequal in counteracting position preference. The ratio between sites often remained constant for a couple of hours, and within that period the relative value of dummies with any kind of stimulus combination could be measured and expressed with reference to the standard size series [Fig. 6-5]. [BAERENDS and KRUIJT 1973:30]

Their results (Fig. 6-6) show how the releasing value of a model egg with respect to size is affected by the other varying characteristics—for example, changing the egg shapes into a round-edged block, omitting the speckling on brown models, and adding speckling to green models.

Baerends and Kruijt caution that the exactitude of the method should not be overestimated. It was limited by the step sizes in the "titration" series, and there was considerable variability in the results of individual tests. However, it is clear that their "titration" procedure provided increased understanding of the role of the various stimuli in the egg retrieval behavior. Also note that they conducted over 10,000 tests in their experiment.

Manipulation of eggs, although considered here as an intraspecific manipulation, might be argued to be manipulation of the physical environment. The answer lies in the "eyes of the beholder," the gulls; hence we'll probably never know.

An *interspecific manipulation* was made by Littlejohn and Martin (1969) in their study of acoustic interaction between two sympatric

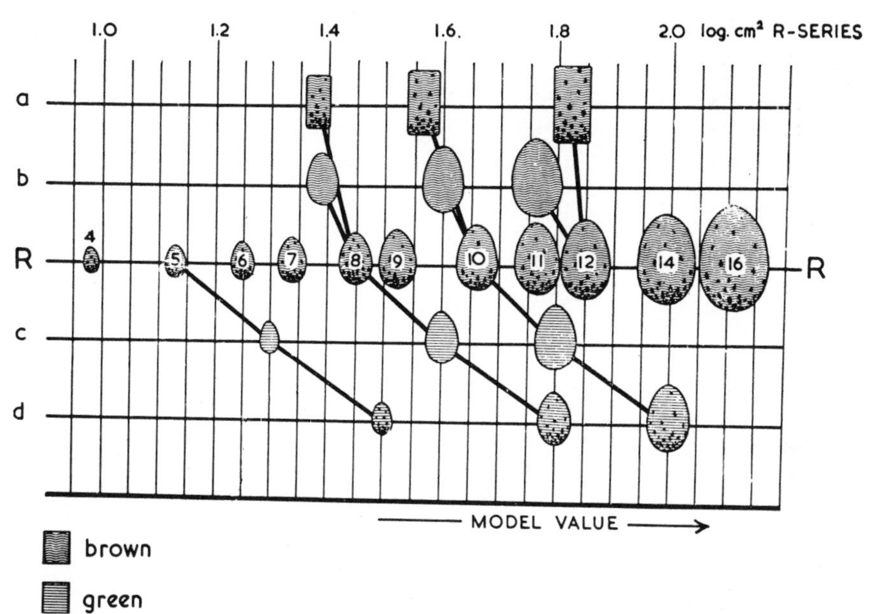

Figure 6-6 The average values found for various dummies with respect to the reference size series *R*. The position of different types of dummies (brown, speckled, block-shaped; brown, unspeckled, egg-shaped; green, unspeckled, egg-shaped; green, speckled, egg-shaped), each in different sizes, was determined with the method described in the legend of Figure 6-5. The code numbers 4 to 16 stand for, respectively, 4/8 to 16/8 of the linear dimensions of the normal egg size (8 = 3/8). The maximal projection surfaces of the eggs of the references series have been plotted (egg centers) along the logarithmic scale (cm²) of the abscissa. Equal distances between points on this scale imply equal ratio values. (From Baerends, G.P., and J.P. Kruijt, 1973, Stimulus selection, p. 32, Fig. 4 *in* Hinde, R.A., and J. Stevenson-Hinde (eds.), Constraints on learning, Academic Press, London.

species of frog, *Pseudophryne semimarmorata* and *Crinia victoriana*. They played a tape-recorded mating call of *C. victoriana* and synthetic signals to individual calling males of *P. semimarmorata*. The call of *C. victoriana*, if played above 80 dB, and synthetic pulsed signals with a carrier frequency of 1500 to 2500 Hz were all effective in inhibiting *P. semimarmorata* males from calling.

Manipulation of the physical environment can take many forms. Two basic procedures can be utilized: 1) change the environment in which the animal is presently located or 2) relocate the animal to another environment. These procedures are usually utilized to determine the effect of the physical environment; however, the confound-

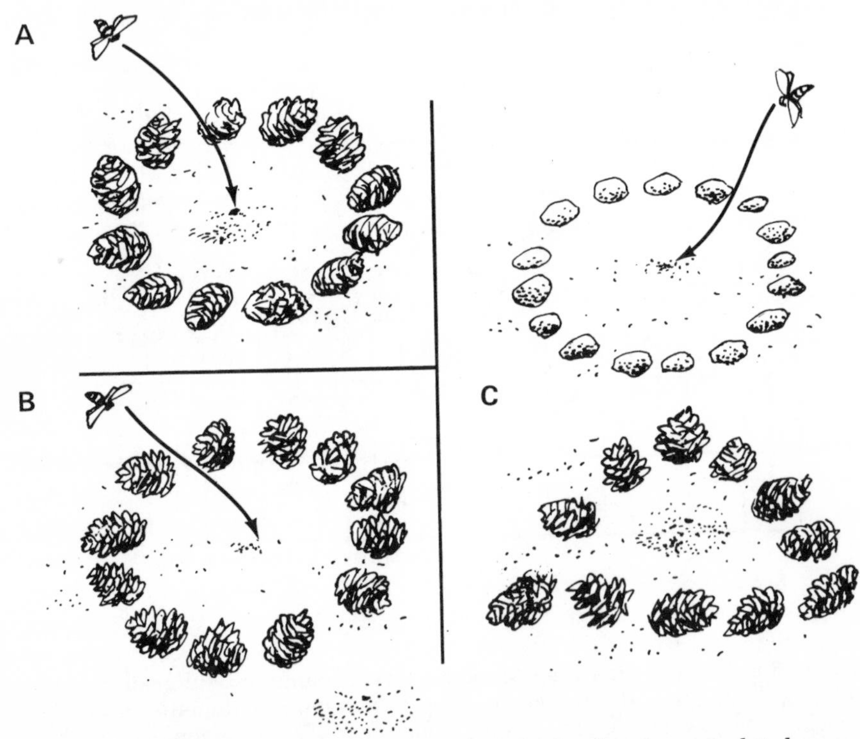

Figure 6-7 A digger wasp always memorizes the landmarks around its burrow so that it will be able to find its way back. In an experiment to test this, Dr. Tinbergen surrounded a burrow with a ring of pine cones (*A*), and the wasp immediately learned to recognize it. But when the ring was moved a foot or two (*B*), the wasp was unable to find its burrow just outside the ring. When the pine cones were arranged in a triangle and a decoy ring of pebbles was made (*C*), the wasp chose the pebbles, proving that it was their arrangement rather than the cones themselves that the insect was responding to (from Tinbergen 1965).

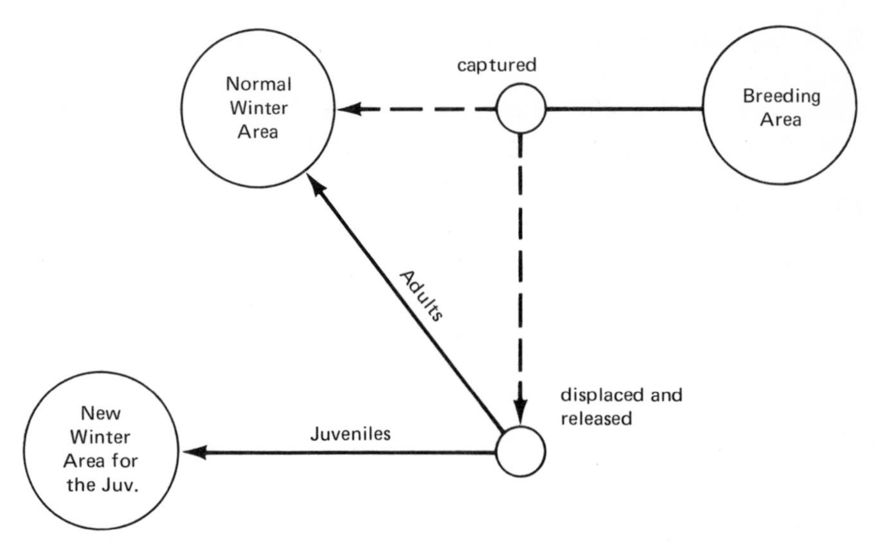

Figure 6-8 Starlings were captured by Perdeck on their autumn migration westward across Europe. They were displaced south and released. The adults were able to navigate to their normal winter area; but the juveniles continued to orient themselves westward and wintered in a new area (after Perdeck, 1958).

ing effects of intra- and inter-specific interactions are often difficult to eliminate.

Tinbergen and Kruyt (1938) investigated the role of landmarks in the ability of female digger wasps *(Philanthus triangulum)* to locate their burrows. They manipulated the type and geometric arrangement of objects around or near the burrow and recorded the response of the returning wasp (Fig. 6-7). This is typical of the simple, yet cogent, type of experimentation for which Tinbergen is famous. Remarking on Tinbergen's methodologies, Lorenz (1960b:xii) stated, "He knows exactly how to ask questions of nature in such a way that she is bound to give clear answers."

Perdeck (1958) relocated migrating starlings geographically to measure their ability to navigate to their normal winter areas. He captured adult and juvenile starlings in The Netherlands during their westward fall migration and transported them south to Switzerland where they were released. Adults were recovered northwest of the release sites in their normal winter areas along the coast of Western Europe. The juveniles, however, were recovered west of the release sites, indicating that they had continued to follow a westward orientation, not navigating northwest as did the adults to adjust for their southerly displacement (Fig. 6-8).

2. In the Laboratory

No one would argue that ethologists are found in the field; but some would suggest that they have crossed disciplinary lines when they work in the laboratory. That rationale is ridiculously primitive and shortsighted. Scientists are not judged by where they work but by what they do. More will be said about this in the next section.

a. Manipulation of the Animal

The animal under study can often be more easily and exactly manipulated and observed in the laboratory than in the field. For example, Buchler (1976) examined the wandering shrew's *(Sorex vagrans)* use of echolocation by training six shrews to echolocate the position of a platform in order to drop to it. The shrews preferentially directed ultrasonic transmissions toward the platform before dropping. When their ears were plugged their ultrasonic transmission rate increased, but their ability to locate the platform decreased significantly. When the ear plugs were replaced with hollow tubes they located the platform as well as when the ears were not plugged (unmanipulated = control).

The effects of hormones on behavior have been investigated in a large number of studies. For example, R.J.F. Smith and Hoar (1967) demonstrated that injections of prolactin failed to induce fanning behavior in male sticklebacks *(Gasterosteus aculeatus)*, castration reduced the behavior, and injections of testosterone restored it. Estrogen can stimulate nest-material preparation (cutting strips of paper) in peach-faced lovebirds *(Agapornis roseicollis)* at least two weeks before it would normally occur, but only after the female is at least 98 days old (Orcutt 1967). Lindzey et al. (1968) measured territorial-marking behavior in male mongolian gerbils *(Meriones unguiculatus)* which were either castrated or sham operated at 30 days of age. Marking did not develop in castrates, but when injected with testosterone they began to mark earlier and reached higher frequencies than did controls.

The effects of stimulation of various brain sites on general behavior patterns (*e.g.*, sitting, standing, eating, crowing) in domestic chickens were studied by von Holst and Saint Paul (1963). Dethier and Bodenstein (1958) were able to demonstrate that the recurrent nerve running from the foregut to the brain signals the brain of the blowfly when the foregut is distended and inhibits further feeding. By cutting the recurrent nerve they were able to show that the blowfly will continue to ingest until it bursts.

By studying age- or genotype-dependent behavior in the laboratory, one is essentially making use of natural variation. Fuller

(1967:470) focused on genotype effects and demonstrated that "albino [house] mice otherwise cogenic with strain C57BL/6J escaped more slowly from water, were less active in an open field and made more errors on a black-white discrimination task than their pigmented congeners." Van Abeelen (1966), also interested in genetic effects, used 30 behavioral components performed by individual and pairs of male house mice to measure differences between strains DBA/2J, C57BL/6J, their F_1 hybrids, and homozygous and heterozygous short-ear animals. Hybrids between peach-faced lovebirds *(Agapornis roseicollis)*, which carry nest material under feathers on their backs, and Fischer's lovebirds *(A. personata fischeri)*, which carry nest material in their bills, initially tried tucking nest material in their plumage as well as carrying it in their bills. Even though feather tucking was unsuccessful for these hybrids, it took two years before feather tucking diminished to any great extent and carrying in the bill was almost exclusive (Dilger 1962). This demonstrated the interaction between genotype and experience.

b. Manipulation of the Environment

Examples of environmental manipulation in the laboratory are widespread in the ethological literature. As in the field, manipulation of the environment in the laboratory can consist of altering intraspecific, interspecific, and physical-environment variables. Providing experience in an intraspecific flock of domestic hens can change the dominant-subordinate relationships seen in paired encounters (King 1965). Marsden (1968) artificially induced changes in rank in young rhesus monkeys by introducing a "strange" adult male at a time when the second-ranking female was in oestrus or by removing and reintroducing the currently top-ranking female.

Various interspecific manipulations have been made in laboratory investigations. The following three experiments with birds will serve as examples. Kalinoski (1975) observed agonistic behavior between house finches *(Carpodacus mexicanus)* and house sparrows *(Passer domesticus)* simply by maintaining mixed groups in laboratory cages. He systematically controlled the species and sex composition as follows: Group I—four male house finches and four male house sparrows; Group II—four male finches and four female sparrows; Group III—four female finches and four male sparrows; Group IV—four female finches and three female sparrows (one female sparrow died prior to the experiment).

Turner (1964) investigated social feeding in house sparrows and chaffinches *(Fringilla coelebs)* by allowing a caged "reactor" (either species) to simultaneously observe two individually caged "actors" (both of the same species, either chaffinch or sparrow), one which was feeding and the other not feeding. He found that individuals of each species were attracted to feeding and nonfeeding conspecifics. Also, chaffinches are attracted more by feeding than nonfeeding sparrows, but this was not true for sparrows observing chaffinches.

The responses of a caged chaffinch to a stuffed owl located at various distances were measured by Hinde (1954). He found that at distances closer than 17 feet the chaffinch moved away, while at greater distances it moved predominantly towards the stuffed owl.

As another example of manipulation of interspecific stimuli Wells and Lehner (1978) were able to differentially affect the ability of coyotes to find a rabbit by manipulating the sensory stimuli available to them. Visual, auditory, and olfactory stimuli were eliminated, respectively, by testing the coyotes in the dark, with dead rabbits and with an intense masking odor of rabbit feces and urine. Also, Metzgar (1967) exposed pairs of mice to a screech owl in a laboratory test area for 2–30 minutes. One was a "resident mouse" (had spent several days in the test area), and the other was a "transient mouse" (had no prior experience in the area). The owl took "transient mice" significantly more frequently.

Potash (1972) measured changes in the physical environment in his study of the effects of environmental noise on separation crowing in Japanese quail *(Coturnix coturnix japonica)*. He found that ambient noise increased the frequency of separation crowing and the number of crows per bout, both of which should increase the detectibility of the signal and the localizability of the transmitter. Bradbury and Nottebohm (1969) varied the amount of light in a flight chamber and measured the ability of auditorily impaired and untreated little brown bats *(Myotis lucifugus)* to negotiate a vertical string maze. Dim light and high contrast enhanced the bats' ability and was interpreted as evidence that they use pattern vision while in flight. Reynierse (1968) investigated the effect of refrigeration and nonrefrigeration during daily maintenance, the intertrial interval, and the experimental session on earthworm locomotion. He found that refrigeration before an experimental session at room temperature inhibited their locomotion, but had no effect if the session was also run under refrigeration.

The role of early experience on habitat selection in the chipping sparrow *(Spizella passerina)* was investigated by Klopfer (1963) in the laboratory. He released individuals into a room in which he had placed pine boughs on one side and oak branches and leaves on the other. He found that wild caught adults and hand-reared isolated adults preferred the pine, but that hand-reared individuals which had been previously exposed to oak preferred the oak. Emlen et al. (1976) tested the orientational capabilities of indigo buntings *(Passerina cyanea)* in a caged situation with minimal exposure to visual cues but with a geomagnetic field provided by Helmholtz coils surrounding the cage. When the horizontal component of the magnetic field was deflected clockwise by 120°, the orientation of the buntings shifted accordingly (clockwise to geographic east-southeast). Emlen (1971) also used a planetarium to demonstrate that the axis of celestial rotation was important in development of migratory orientation in indigo buntings.

The effect of different types of feedback (see model, p. 7) provided by differently treated-seeds was tested in black-capped chickadees *(Parus atricapillus)* by Alcock (1970). He presented the birds with striped seeds which were empty, filled with a mealworm to which salt had been added, or contained mealworms treated with quinine sulphate (an emetic). There was a rapid and stable avoidance of the empty and emetic seeds, but they continued to attack the salted mealworms, perhaps because the food reward outweighed the punishment (salty taste).

C. FIELD TO LABORATORY: A CONTINUUM

Field and laboratory studies represent the extremes along one of the conceptual dimensions of ethological research (Chap. 1, Fig. 1-1). However, in practice they compliment each other in a cyclical continuum called a "research cycle" by Kelly (1967, 1969).

At the two extremes we have the natural experiments in the field and the highly manipulative studies in the laboratory. The middle of the continuum is illustrated by studies conducted in enclosures in the field. For example, Wecker (1964) set up an instrumented enclosure that was half in an oak-hickory woodlot and half in a field, in order to investigate habitat selection in prairie deer mice *(Peromyscus maniculatus)*. Wells (1977) used a large outdoor enclosure to investigate the relative priority of the coyote's distance senses in predation on rabbits. At what point does the field become the laboratory and vice versa? Hinde and Spencer-Booth (1967:169), in their study of rhesus-monkey behavior, attempted to reach a compromise in an enclosure with ". . . a moderately complex environment under conditions which permit a moderate degree of experimental control and moderately precise recording."

It is often feasible to take laboratory instrumentation into the field (Hess 1972) and likewise to make rather unobtrusive observations in the laboratory (Brockway 1964). However, movement into the laboratory is not always easy. Tinbergen, reflecting on some of the shortcomings of his field studies on gulls, concluded:

> It would seem to be more efficient to try to improve the field methods than to try and keep a large colony of gulls under laboratory conditions. [TINBERGEN 1958:251]

Not only is it sometimes difficult to move field studies into the laboratory, but conditions in the field often make manipulations very difficult. In reviewing ethological and behavioral ecology studies of African ungulates, Leuthold (1977) concluded that

> Experiments have rarely been carried out so far, partly because much descriptive work was required at first, and partly because of the physical difficulties of manipulating wild ungulates in experimental situations. [LEUTHOLD 1977:13]

Hoffman and Ratner (1973:541) suggested ". . . that laboratory investigations complement and explain the frequently puzzling data obtained in a natural setting." Ideally, however, research should undergo the field-laboratory cycle several times, utilizing to best advantage the important attributes of each. Menzel (1969) considers this process analogous to "zooming in" and "zooming out" with a lens. Avian orientation studies provide a good example. Matthews (1951) found that homing pigeons released in unfamiliar territory under clear skies were able to fly off directly toward home; but if the sky was overcast they became disoriented. That the sun was a cue used in orientation was given further support by Kramer (1952) who placed starlings in a circular cage with six windows giving a view of the sky only. The starlings showed migratory restlessness *(Zugunrühe)*, fluttering in the proper migratory direction when the sky was clear but in random directions when it was overcast. Kramer altered the apparent position of the sun with mirrors and was able to reorient the starlings in a predictable manner. Schmidt-Koenig (1961) kept pigeons under artificial day-night conditions six hours out of phase with the normal day. When the pigeons were released they oriented 90° from the correct direction, showing that they were using a biological clock and the sun's position as a cue. Kramer trained starlings to find food in particular trays in a circular cage using only the sun as a cue to direction. When the cage was covered and they were presented with a stationary light, they used the light as if it were the moving sun and changed their direction at the rate of 15°/hour. Meyer (1964) used discrimination tests in the laboratory to show that pigeons could indeed detect movement of 15°/hour. Night-migrating warblers were tested in a planetarium by Sauer (1957) and were shown to use stellar cues in orientation. Emlen (1967) measured the nocturnal orientation of caged indigo buntings outdoors under the natural night sky and then took them into the planetarium, where they continued to orient themselves correctly when the planetarium sky was set for local conditions. They reversed themselves when the north-south axis of the planetarium sky was reversed. They were disoriented when the planetarium sky was diffusely illuminated or darkened. Later, Emlen (1970) used a planetarium to demonstrate that indigo buntings learn to associate stellar cues with the axis of celestial rotation.

Both field and laboratory studies have provided convincing evidence that geomagnetic fields are an orientation cue sometimes used by birds. Moore (1977) showed that nocturnal free-flying passerine migrants responded to natural fluctuations in the earth's magnetic field. Electromagnetic fields produced by large antennaes were shown to alter the path of free-flying migrants (Larkin and Sutherland 1977) and of gulls held in an orientation cage (Southern 1975). Homing pigeons become disoriented when released under an overcast sky with a bar magnet attached to their backs (Keeton 1974) or with Helmholz coils on their heads (Walcott and Green 1974). In carefully controlled laboratory investigations with a cage surrounded by Helmholz coils, use of the inclination of the axial direction of the magnetic field for orientation was demonstrated for the European robin *(Erithacus rubecula)* (Wiltschko and Wiltschko 1972) and indigo bunting (Emlen et al. 1976).

These are only a few of the projects which have been conducted with the use of natural experiments or various degrees of manipulation in both the laboratory and the field. For example, D.E. Davis (1964) reviewed the relative contribution of field and laboratory research to our understanding of aggression and the role of hormones in aggressive behavior. The keen researcher recognizes the value of both field and laboratory research and utilizes various approaches to test his hypotheses.

I see neither halos nor horns on either a *real experiment* or on *accurate observations*. Any method is a special case of human experience, and it cannot surpass the limitations of its human interpreters. [MENZEL 1969:80]

7
Data Collection Methods

We are now at the point where 1) the research question(s) has been asked, 2) the objectives have been formulated, 3) the research hypotheses stated, 4) the approach, naturalistic observation or experimental manipulation selected, 5) the behavioral units to be measured determined, and 6) the experimental design established. The statistical tests to be used in the analyses (see Chapter 11) should also be considered before data are collected. Now it is necessary to decide how the data will be collected—what procedures and equipment will be used.

A. RESEARCH DESIGN AND DATA COLLECTION

Research design and data collection are mutual dictators. The research design chosen will dictate the data to be collected. Likewise, a knowledge of the type and amount of data that can be collected will partially dictate the research design to be used. Research design and data collection are in harness together, and the pushing and pulling that each does to the other will depend on the individual study and the experience of the researcher. The experienced observer will be able to push for the best research design through a knowledge of the animal's behavior and types of data that he can expect to collect. On the other hand, neophyte researchers may allow a research design, selected for statistical attributes, to pull them around in the field, attempting to collect nearly impossible (and sometimes behaviorally meaningless) data.

Research design and data collection must complement each other for the study to be effective and the researcher efficient. Even seemingly well planned research can often benefit from redesign and additional (or modified) data collection. Do not be afraid to evaluate your research design carefully while you are collecting data. However, *do not redesign your research until you have carefully assessed your present and future losses in time and data.*

B. SCALES OF MEASUREMENT

Data collection involves the assignment of numbers to observations and observations to categories. This process is often referred to as *measurement.*

Scales of measurement are various levels of refinement (or precision) of measurement. The four scales, from Stevens (1946), represent points along a continuum of scaling. That is, some types of data will appear to fall between two scales and be difficult to categorize.

The four scales are listed in order from the lowest resolution (nominal) to the highest resolution (or most restricted measure, Stevens 1946):

Nominal scale—Observations are classified into predetermined mutually exclusive, qualitatively different categories (*e.g.*, behaviors *A*, *B* and *C*). For example, we can record the presence or absence of feeding, swimming, calling, etc.

Ordinal scale—Same as nominal scale with the addition that the categories are ordered with respect to each other (*e.g.*, behavior $A > B > C$).

The behaviors must have a common qualitative property by which they are ordered; the ordering must be stable, and it must hold throughout the entire scale.

A linear dominance hierarchy meets the criteria:

$$A \rightarrow B \rightarrow C \rightarrow D$$

where individual *A* is dominant over *B*, *B* over *C*, and *C* over *D*. However, if circular relationships develop, then the ordering stability is lost:

$$A \rightarrow B \begin{matrix} \nearrow & C \\ & \downarrow \\ \nwarrow & D \end{matrix}$$

The ordinal scale can be used to rank behavior into categories such as broodiness: Excellent > fair > poor; or incubation > intention movement > approach (Tinbergen 1960).

Interval scale—Same as the ordinal scale with the addition that the amount of the differences between respective categories is known; this necessitates a unit of measurement which permits additivity; the zero point is not known or arbitrarily defined for measurements on an interval scale.

EXAMPLE: The length of time that it took individuals to fly after an alarm call had been given.

$$A = 2 \text{ sec.}, B = 2.5 \text{ sec.}, C = 1.9 \text{ sec.}$$

These times can be compared relative to each other, but the zero point for flight is not really known. We use our hearing of the alarm call as an arbitrary zero point.

Ratio scale—Same as the interval scale except the zero point is known.

EXAMPLE: Songbirds mobbing an owl.

Individual *A* flies at 3 m from the owl.
Individual *B* flies at 5 m from the owl.
Individual *C* flies at 6 m from the owl.
Individual *D* flies at 8 m from the owl.

To determine the scale of measurement used in a study, examine the data as it was recorded (*i.e.*, the "raw data"), as shown below:

Behavior	Data Recorded (Coded)	Scale of Measurement
A occurred.	A	Nominal
A occurred at intensity level 2.	A2	Ordinal
A occurred at intensity level 2 for 5 sec.	A2/5	Interval
A occurred at intensity level 2 for 5 seconds at a distance of 8 m from the stimulus model.	A2/5–8	Ratio

Note that data with a scale of measurement of less resolution can be extracted from those with higher resolution (*e.g.*, A2 from A2/5 or A2/5–8). This is sometimes done for data analysis with nonparametric tests. Also, in some cases nominal data may be summarized and analyzed as ordinal data (see p. 135). However, it is best to take the data with as high a resolution scale of measurement as is feasible (*e.g.*, ratio) and then sacrifice the resolution later if you find it was not necessary in that particular study. (You should, in fact, know this before you begin to collect data.) Remember to keep the original data, for you may want to come back to it later, and at that time you may need the higher-resolution scale of measurement.

The scale of measurement chosen for collecting data will in part determine the research design as well as the type of statistical analysis (Chap. 11) selected. Only nonparametric statistical tests can be used with nominal and ordinal data, whereas either parametric or nonparametric tests can be used on interval and ratio data.

C. SAMPLING METHODS

The discussion of sampling methods which follows is based almost entirely on J. Altmann's (1974) excellent review (see also Sackett 1978). The sampling method you select for your research will be based on 1) your research question and research objectives, 2) the behavioral units you have selected to measure, 3) your experimental design, and 4) a multitude of practical considerations, such as availability of equipment, animals, visibility, etc.

In the discussion to follow, an example of how each method is used will be based on Figure 7-1, a hypothetical behavior record for six mule deer *(Odocoileus hemionus)* showing only two events (standing-up and lying-down) and one state (feeding) directly. In order to be sure that you understand the diagram, confirm the following statements:

1. All animals fed during the 90-minute observation period.
2. All animals fed during at least two of the three 30-minute segments.
3. None of the animals fed while lying down (state).
4. Only animal IV definitely did not feed every time it was standing.

1. Random vs. Haphazard Samples

Mention of random samples is frequently made in the following discussion. Some observers use the term *random* to refer to the way in which their samples were collected, when really what they collected were *haphazard* samples. For example:

> This study is based on facts gleaned for the most part from *random* [italics mine] observations. [LORENZ 1935:90]

Random Sample—A sample drawn from a population in such a way that all possible samples have the same probability of being selected (*e.g.*, prescribed samples assigned numbers and then drawn from a hat).

Haphazard Sample—A sample taken on some arbitrary basis, generally convenience (*e.g.*, samples taken before lunch and after supper) or when the animals are thought to be most active.

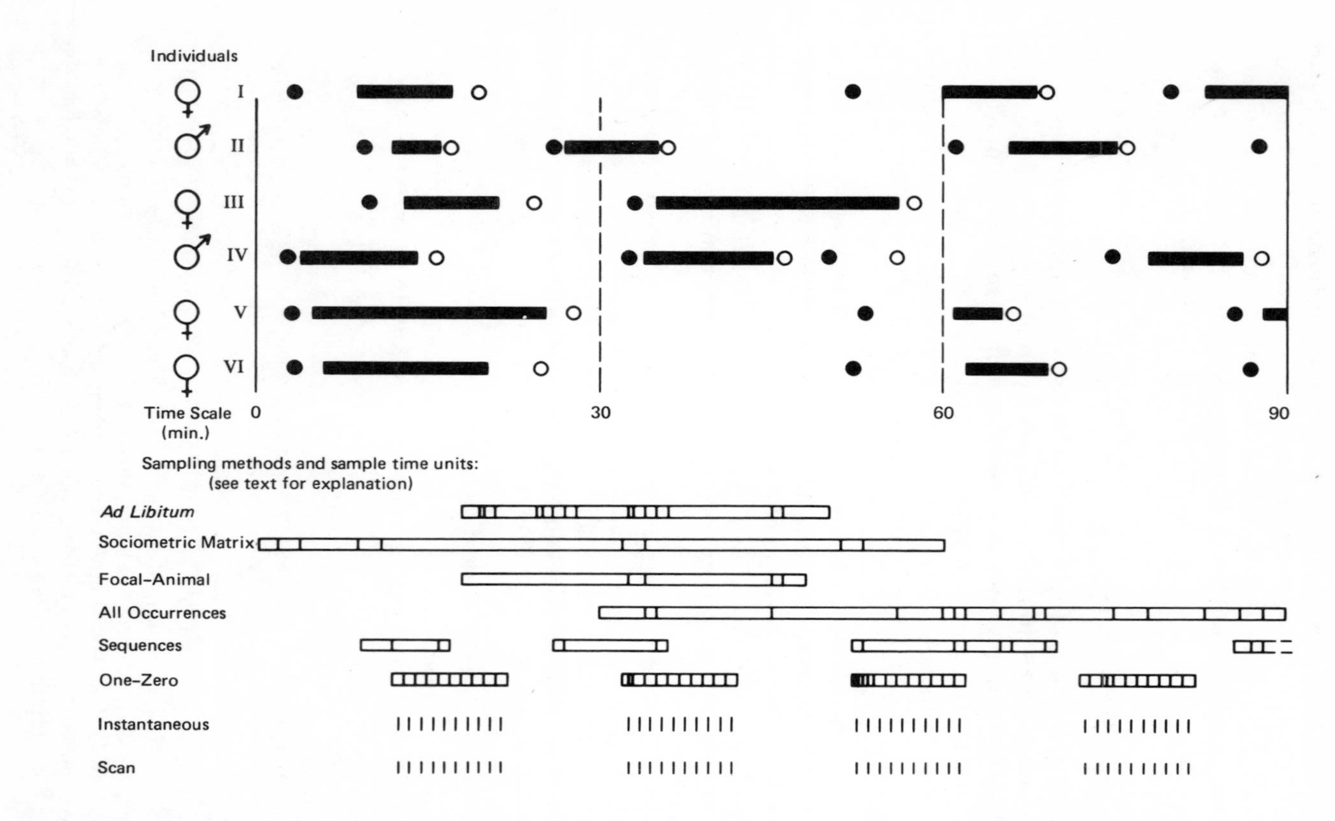

Figure 7-1 Hypothetical record of occurrences of three behaviors in six penned adult male deer: ● = standing up (event); ○ = lying down (event); ■ = feeding (state).

Even Tinbergen admitted that in his earlier studies with grayling butterflies they had presented their models in a haphazard rather than random sequence:

> We varied the sequence of the models irregularly (not being sophisticated enough to vary them in a random way). [TINBERGEN 1958:182]

True random sampling is often difficult in the field situation; but the researcher must consciously attempt to attain randomization through unbiased sampling efforts. S. Altmann (1964:494) ". . . tried to sample at random from among the monkeys. However, no systematic randomizing technique was used." When this situation occurs it is best to attempt to *equalize* the number and temporal distribution of observations among individuals.

2. *Ad Libitum* Sampling

As *ad libitum* implies, no restraints are employed in sampling behavior with this method. What is recorded are generally the behaviors of those individuals (or groups) that are most easily observed. It is the type that results in "typical field notes." J. Altmann (1974:235) states that "In field studies of behavior [*ad libitum* sampling is] perhaps the most common form of behavior record"—an unfortunate and regrettable fact. That is not to say that "typical field notes" are not important (see reconnaissance observation, p. 24), but rather that the relative abundance of observations made in this way so greatly outnumbers observations using some form of systematic or random (not haphazard) sampling.

Tinbergen has reflected on the importance of unplanned observations:

> Scientific examination naturally requires concentration, a narrowing of interest, and the knowledge we gained through this has meant a great deal to us. But it has become increasingly clear to me how equally valuable have been the long periods of relaxed, unspecified, uncommitted interest . . . an extremely valuable store of factual knowledge is picked up by a young naturalist during his seemingly aimless wanderings in the fields.
>
> Nor are the preliminary, unplanned observations one does while relaxed and uncommitted without value to the strict experimental analysis. [TINBERGEN 1958:287]

Ad libitum sampling is most often used when an ethologist is merely recording as much as he can during 1) an unplanned encounter with a species or 2) during reconnaissance observations for a later study.

However, when the observations are treated as data and quantitative comparisons made, the following assumption generally must also be made: The true probability of observing the different sexes, age classes and behaviors is reflected in your notes. Comparisons cannot be made across time unless the number of samples is large and the samples are taken randomly. For example, we might encounter the six deer represented in Figure 7-1 and observe them for the 32 minutes indicated. During that period we would record that 1) all but female I fed, 2) all individuals lay down, and 3) three of the six individuals stood up. That is, we could have recorded those behaviors or we could have been temporarily focusing on other behaviors and missed portions of the complete record. Let us say that when female VI laid down at minute 25, she laid close to female III, and they began head-butting intentions and making threats to each other. We focused our attention on them for the next fifteen minutes and missed seeing male II stand up, feed, and lie down again. That is, unless we had decided previously to focus on the behaviors diagrammed in Figure 7-1, it is unlikely that we could have duplicated that portion of the diagram in our 32-minute sample from our field notes.

Since it is rare that all individuals are equally visible, some researchers have attempted to measure individual observability (Chalmers 1968; Sade 1966). At some regularly scheduled time period (*e.g.*, half-hour intervals) censuses were taken of the individuals which were visible—so-called *observability* samples (see also p. 116). J. Altmann (1974:239) suggests that these adjustments are of limited compensatory value: "Observability samples provide an accurate correction only to the extent that the probability of a behavior being recorded if any individual performs that behavior is directly proportional to the percent of time that the individual is visible." Altmann goes on to point out that there are at least the following three potential sources of failure to obtain consistently proportional samples:

1. Individual or class-specific differences in observability may vary with different behaviors.
2. A specific behavior may affect observability.
3. An observer's preferences (decisions) in sampling specific behaviors introduce biases, as well as his attempts to compensate for these biases.

Ad libitum sampling provides ideas for future research and often reveals rare, but significant, behavioral events.

3. Sociometric Matrix

A *sociometric matrix* is really an experimental design or a way of tabulating data. Collection of data for a sociometric matrix can be considered a special type of *all-occurrences* sampling in which the observer searches for interactions between pairs of individuals (*e.g.*, transmitter-receiver, groomer-groomee) or records interactions by an individual *(focal animal)* during a specified sampling period.

> **EXAMPLE:** We might want to measure the interactions involved in standing up in the group of six mule deer in Figure 7-1. We suspect that the behavior of certain individuals when standing up stimulates others to stand up, so we construct the following sociometric matrix. We record the initiator and follower, if the follower stands up in less than 60 seconds after the initiator has stood up. The data in the matrix below are from the 60-minute sample period in Figure 7-1.

Initiator \ Follower	I	II	III	IV	V	VI
I					1	
II			1			
III						
IV	1		1		1	1
V	1					1
VI					1	

The above data suggest that individual IV is more of a leader than a follower; but we would, of course, need to record a large number of these interactions in order to demonstrate significant correlations.

In most instances the researcher uses a sociometric matrix to test for one-sidedness in dyadic interactions. Therefore, an attempt is made to record as many interactions as possible without regard to random or systematic sampling. Hence the data cannot be compared between cells, and the matrix cannot be treated as a true contingency table, but more a form for tabulating data.

4. Focal-Animal Sampling

With this method, one individual is the focus of observations during a particular sample period. That is, a particular individual receives highest priority for recording its behavior, but it does not necessarily restrict us to *only* that individual. Where social behavior is recorded, a focal-animal sample on an individual provides a record of all acts in which that animal is either the actor or receiver (J. Altmann 1974). In some instances, observation of focal subgroups may be appropriate. It is necessary to record the length of each sample period and the amount of time that the focal animal is in view during that period. The problem of an *animal under observation temporarily disappearing from view* (an error of apprehending, p. 128) has not been successfully resolved. There is no valid procedure for determining (predicting) what behavior(s) occurred while that animal was out of sight. However, there is a general relationship among predictability of what behavior occurred, the duration of behaviors most commonly observed, and the time the animal is out of sight. Four methods for dealing with this problem are discussed below relative to the duration of behaviors and time out of sight. Although *none of these methods is valid*, the hazards in using them can be reduced by following the guidelines:

For out-of-sight periods of long duration and when the durations of common behaviors are short (relative to the out-of-sight periods) do the following:

1. Delete the time out of sight from the sample; duration of the sample period is reduced accordingly.
2. Delete the time out of sight from the sample, but increase observation time until the time the animal was actually observed equals the time required for the predetermined sample period.

For out-of-sight periods of short duration and when the durations of common behaviors are long (relative to the out-of-sight periods) do this:

1. Assign the behavior seen when the animal goes out of sight to the out-of-sight period.
2. Assign the behavior seen when the animal comes back into view to the out-of-sight period.

Behaviors occupying the largest percentage of the animal's time budget are those that are most likely to be interrupted. Two factors which should, perhaps, override or dictate use of the above methods are experience and common sense. Probably no one knows the animal better than you do; therefore follow the course of action which you consider to be the most appropriate. Also, it is often wise to deal with data using two or more methods for comparison.

For our example of focal-animal sampling, we begin our observations of the six mule deer in Figure 7-1 at the same time that we began our *ad libitum* sampling; but we terminate our focal-animal sample at exactly 30 minutes. We focus on male IV and note that he is lying down when we begin our sample period. He stands up during the 15th minute, begins feeding 1 minute later, feeds for 11 minutes, lies down during the next minute, and is still lying down when our sample period ends. In this very simplified example, it would probably have been possible to record behavior of other individuals (*e.g.*, the other males).

Focal animals may be chosen selectively because of a research objective or experimental design restraint. Likewise, the design might call for a random selection of an individual within some class or treatment group.

Focal-animal sampling may push an observer's ability to record data, especially in highly social species (*e.g.*, monkeys). It does provide for a rigorous examination of the behavior of individuals or groups, and *it is this type of sampling which will pay the biggest dividends in the future*.

J. Altmann (1974) concludes that with the proper choice of behavior units, sample periods, and focal individuals, this method will generally be the best to use.

5. All Occurrences

It may be desirable to focus on one or a limited number of *behaviors* and record all occurrences (called "event-sampling" by Hutt and Hutt [1974] and "complete record" by Slater [1978]). This contrasts with focal-animal sampling where the focus is on the individual. All occurrences of selected behaviors is possible if the following factors exist:

1. Observational conditions are adequate.
2. The behaviors have been carefully defined so that they are easily recognized.
3. The behaviors do not occur more often (or more rapidly) than the observer can record them.

This method of sampling can provide the following types of information:

1. Rate of occurrence (and temporal changes in rate) of the selected behavior(s); see p. 266 for discussion of the analysis of rates of behavior.
2. Restricted sequencing (see example below).
3. Behavioral synchrony (see example below).

EXAMPLE: Let us say we are interested in the sequence of initiation and synchrony of feeding in the six mule deer in Figure 7-1. We selected the

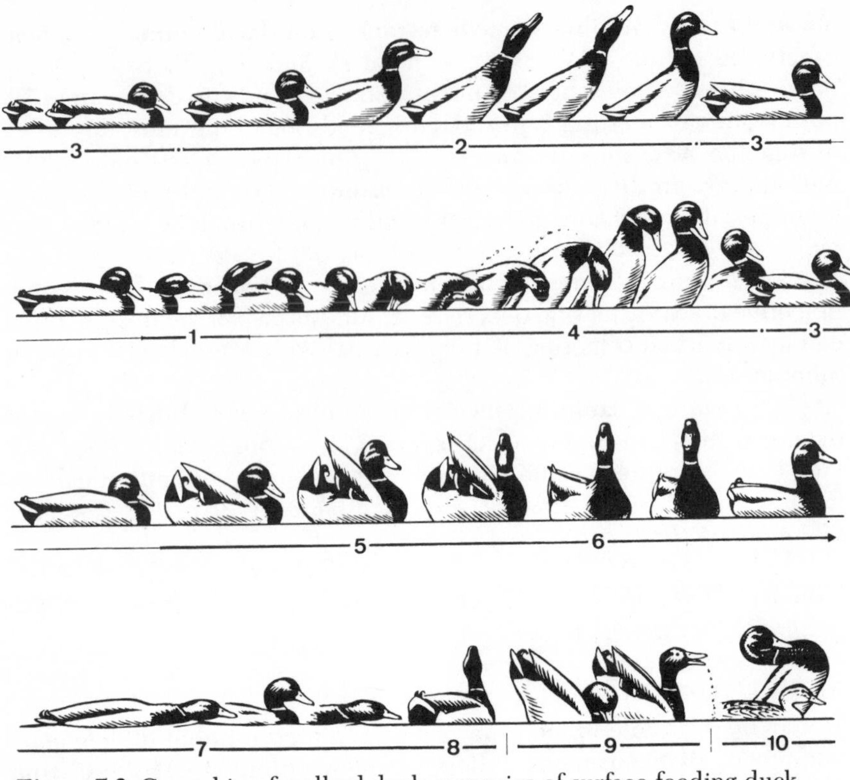

Figure 7-2 Courtship of mallard duck, a species of surface feeding duck, involves the following sequence of behaviors: (3) tail-shake, (2) stretch-shake, (3) tail-shake, initial posture, (1) head-flick, (4) grunt-whistle, (3) tail-shake, initial posture, (5) head-up-tail-up, (6) looking towards the female, (7) nod-swimming, and (8) showing the back of the head. Also shown are down-up (9) and bridling (10). Bridling is a postcopulatory display. (Drawing by Hermann Kacher in collaboration with Konrad Lorenz; permission granted by Hermann Kacher.)

60-minute sample period shown in the diagram and record the initiation and termination of feeding by each individual (events indicated in the diagram). We can then examine the data for the sequence of initiation of feeding for the six individuals, as well as how many and which individuals fed together. In this case we can gain information about a state (feeding) by recording all occurrences of two events (initiation and termination of feeding).

6. Sequence Sampling

In sequence sampling the focus is on a chain of behaviors. These may be performed by a single individual (*e.g.*, courtship displays in male ducks, Figure 7-2) or they may be behaviors alternating between two (or more) individuals (*e.g.*, courtship in the queen butterfly, Figure 7-3).

The initiation of a sample period is usually determined by the beginning of a sequence. An experienced observer can often anticipate the initiation of a sequence in an individual and an impending interaction between two or more individuals. The sample period terminates when the observed sequence terminates.

There may be difficulty in specifying the beginning and end of a sequence, as well as choosing individual sequences or social interactions at random. J. Altmann (1974:250) discussed the sampling bias caused by differing lengths of individual sequences (or social interactions):

> If . . . the observer always begins sampling at the onset of a sequence and chooses the next sequence to sample at random among sequence onsets or in any other way that samples sequences of each length in proportion to their *frequency of occurrence*, the resulting data will be unbiased with respect to sequence length: the total time spent with sequences of, say, duration d_i, will be proportional to d_i times f_i, where f_i is the frequency of sequences of length d_i. Then the time spent with sequences of different lengths, not the probability of choosing such sequences, will be in proportion to the total time taken up by sequences of that length.

Sequence sampling of social interactions has some other potential problems. Interactions may branch or converge; that is, an interacting group may break up (branch) into subgroups or other individuals may join an interacting group which is under observation (converge).

It is up to the observer to record as clearly as possible the branching and converging of interactions. These are recognizable events that are often important parts of social interactions among large groups. They may force the observer to develop additional observational skills, including peripheral vision; but they should be included in the observational record or they should be eliminated from the experimental design. For example, Hazlett and Bossert (1965) restricted their observations to interactions between two individual crabs (dyads).

As an example, we might be interested in the sequence of behaviors that mule deer go through from the time they stand up to feed until they lie down. In Figure 7-1, the first two bars (sample periods) after sequence sampling show our observations of male II. We begin our sampling as he stands up and terminate it when he has lain down. We might, of course, be interested in additional behaviors that are not included in the diagram, such as limb movements associated with lying down and standing up. The third and fourth bars illustrate a

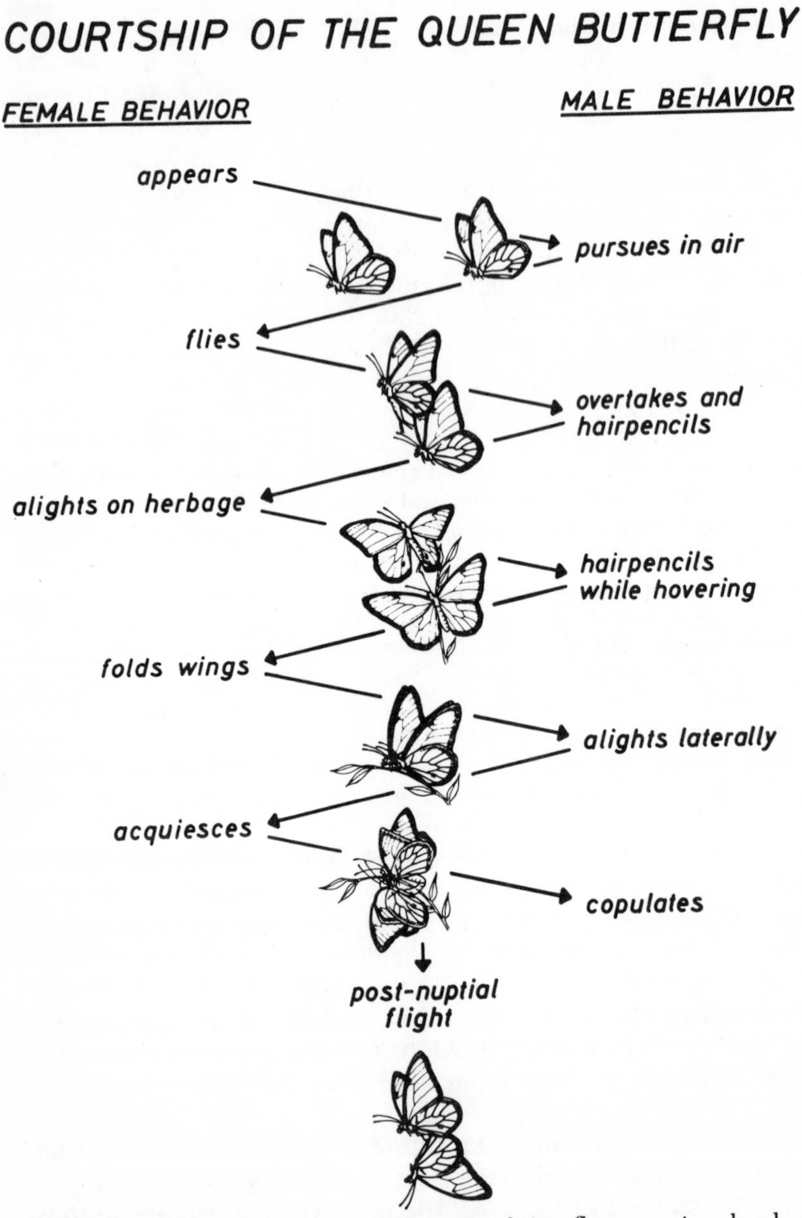

COURTSHIP OF THE QUEEN BUTTERFLY

FEMALE BEHAVIOR

MALE BEHAVIOR

appears

pursues in air

flies

overtakes and
hairpencils

alights on herbage

hairpencils
while hovering

folds wings

alights laterally

acquiesces

copulates

post-nuptial
flight

Figure 7-3 Courtship behavior of the queen butterfly, a species closely related to the monarch. Note that the female permits copulation to occur only after the male completes a series of courtship actions, including the release of several different pheromones. (From Brower *et al.*, 1965, Courtship behavior of the queen butterfly, *Danaus gilippus berenice* (Cramer), *Zoologica* 50:18, Fig. 9.)

sample of interactions in which we are looking at a relationship between the behavior of females V and VI. We record the events which occur for each of them, beginning with the first one to stand up and terminating when the other has lain down. The next sample period begins when one of them stands up, etc. The data would then be analyzed for correlations between the two individuals' behavior; if significant correlations existed then we would design additional experiments to test for cause and effect.

Note that in the previous example we used all-occurrences sampling to determine the sequence of initiation of feeding in the six deer.

7. One-Zero Sampling

One-zero sampling is a method in which the observer scores whether a behavior occurs (one) or not (zero) during a short interval of time (sample period). It is suitable for recording states and/or events. This method is often referred to in the literature as "time-sampling" (Hutt and Hutt 1974) or the "Hansen system" (Fienberg 1972).

This method has the following features:

1. In each sample period the occurrence or nonoccurrence (not frequency of occurrence) is scored.
2. Behaviors of one or more individuals are recorded in each sample period.
3. Occurrence refers to either an event or a state (ongoing at some point during the sample period).
4. The sample periods are generally short (*e.g.*, 15 seconds), and several (*e.g.*, 20) are used in succession.

Although the method states that a behavior is scored only once for a sample period regardless of the number of times it occurs, some researchers have recorded all occurrences but analyzed the data as one-zero scores (*e.g.*, Kummer 1968). Since the results are generally presented as frequencies (number of sample periods in which it occurred/total number of sample periods), the number of sample periods should be relatively large. J. Altmann (1974:253) points out a common fallacy in interpreting one-zero data: "It is too easy for both author and reader to forget that a one-zero score is not the frequency of *behavior* but is the frequency of *intervals* that included any amount of time spent in that behavior."

Another potential problem arises when recording a state (ongoing behaviors) that continues through several sample periods and is scored for each one. In this case there is no close relationship between the number of scores (*i.e.*, intervals in which the behavior occurred) and actual frequency of occurrence (Dunbar 1976). However, if the sample period is sufficiently short relative to the behavior's duration and the interval between successive occurrences, then the observer can obtain (with reasonable accuracy) both frequency and duration through careful data analysis. For this to be fairly accurate, the probability of both a termination and an onset occurring in one sample must be negligible. In addition, Adams and Markley (1978), using computer-generated "behaviors" with a variety of known mean durations and frequencies, found instantaneous sampling (see next section) to be superior to one-zero sampling for estimating duration of behaviors. Their results support the conclusions of several similar studies.

Caution should be used when converting one-zero scores to percent of time spent in a behavior (Simpson and Simpson 1977). This would be correct only if the behavior lasted for the complete sample periods in which it was scored. If the observer desires to use one-zero data for "time-spent" estimates, he must determine how closely his data approximate the above condition. S. Altmann and Wagner (1970) have described a method of estimating the mean rate of occurrence of events if the events approximate a Poisson distribution; that is, ". . . that the behavior occurs randomly at a constant rate, that the chance of two or more simultaneous occurrences of the behavior is negligible, and that the chance that a particular behavior will occur during an interval is independent of the time that has elapsed since the last occurrence of that behavior." Slater (1978) suggests that one-zero data may be used as a first-approximation look at associations between behaviors, by determining how frequently they occur in the same time unit as each other.

The major disadvantages of this sampling method is that a large amount of information about frequency and duration is lost. The researcher has to weigh this disadvantage against the case of scoring and high interobserver reliability which this method provides. J. Altmann (1974:258) concludes that "In short, neither ease of use nor observer agreement *per se* provides an adequate justification for the use of this technique."

EXAMPLE: In Figure 7-1 you can see that there are four groups of 10 sample periods each. The sample periods indicated are of one minute duration, although generally they would be much shorter. If we are interested in one-zero scoring of the event of standing up, then the first set of 10 sample periods would all contain "0" scores, the second set two "1" scores, the third set three "1" scores, and the fourth set two "1" scores.

8. Instantaneous and Scan Sampling

Instantaneous sampling is a special type of one-zero sampling in which the observer scores an animal's behavior at predetermined "points" in time (called "time-sampling" by Hutt and Hutt 1974). This method has also been termed "point" sampling by Dunbar (1976) and "on-the-dot" sampling by Slater (1978). It is used to sample states since the probability of scoring events with this method is remote. This method can be used to obtain data on the time distribution of behavioral states in an individual.

For example, suppose we wanted to record the time distribution of feeding in female III in Figure 7-1. We set out sampling points in a manner which is comparable to the periods used for one-zero sampling. Sample points 2 through 9 would be scored ones in the first set, 4 through 10 in the second set, 1 through 4 in the third set, and none in the fourth set. The frequency of scores decreased as our sample periods progressed. You can see how closely this approximates the real situation in the diagram.

Scan sampling is simply a form of instantaneous sampling in which several individuals are "scanned" at predetermined points in time and their behavioral states are scored. The same sample points can be used as in instantaneous sampling (Fig. 7-1). The observer should attempt to be as instantaneous as possible, for the longer he lingers on one individual, the more the sample approximates a series of short focal-animal samples of unknown durations. Behavior categories chosen for study should be clearly delineated to assist in quick scoring. Estimates of time spent scanning individuals, as well as groups, should be made.

One important use of instantaneous and scan sampling is to estimate the percentage of time that individuals spend in various activities (*e.g.*, time budgets). One-zero sampling is equally effective in determining time spent in some activity under limited conditions only (see Simpson and Simpson 1977), hence it is not recommended. Caution should be used when estimating rates and relative frequencies for instantaneous and scan samples, since they tell us nothing about the frequencies of behaviors beginning and ending. The caution is the same as that expressed for one-zero samples. "In the special case where the interval between instantaneous samples is short enough that no more than one transition can occur between consecutive samples, the resulting data are essentially equivalent to that of focal-animal sampling for rate and relative frequency estimates, but have a greater margin of error for duration estimates" (J. Altmann 1974:261).

9. Summary

Each sampling method described above has recommended uses. By combining more than one method, a researcher is often able to maximize efficiency of data collection and insure that the proper data are collected for testing the research hypothesis.

A researcher can make a first approximation of the type of sampling he might use by deciding whether his research question is more behavior-oriented, individual-oriented, or a combination of both (Fig. 7-4).

J. Altmann (1974) has also provided a table to assist in the selection of the proper sampling method (Table 7-1).

Different sampling methods produce different types of data. Hence, the validity of the research will be affected by the sampling method used.

Dunbar (1976) presented an excellent demonstration that the use of various behavioral parameters and sampling methods *to answer the same question* often leads to different conclusions. His demonstration challenged the internal validity of several methodologies. He estimated "social relationships" among individual gelada baboons *(Theropithecus gelada)* divided into eleven age-sex classes, using three different behavioral units that showed the extent to which individuals 1) *interacted* with each other, 2) *groomed* each other, and 3) were in a particular *spatial relationship*.

Data were collected in the following seven combinations of behavioral units and sampling methods:

1. The number of social contacts between individuals (*i.e.*, the number of times they interacted, irrespective of the number of social acts exchanged in each interaction) (focal-animal and all-occurrences sampling).
2. The number of social acts exchanged between individuals (all-occurrences sampling).
3. The number of 30-second time intervals in which individuals interacted (one-zero sampling).

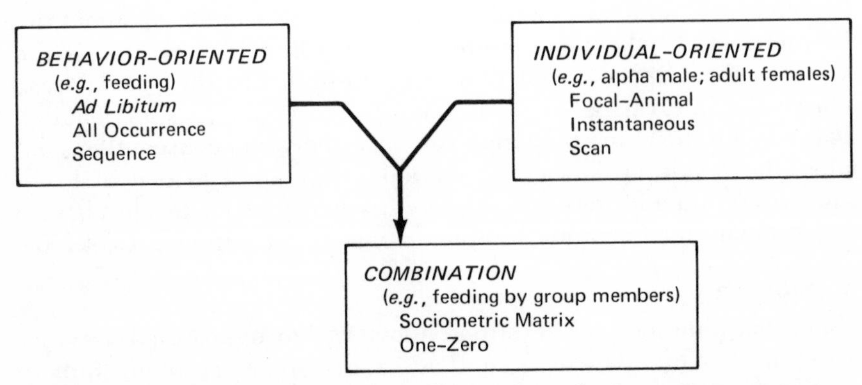

Figure 7-4 Sampling methods divided into groups for first approximation selection.

Table 7-1. Sampling methods and recommended uses (J. Altmann 1974).

Sampling Method	State or Event Sampling	Recommended Uses
1. *Ad libitum*	Either	Primarily of heuristic value; suggestive; records of rare but significant events.
2. Sociometric matrix completion	Event	Asymmetry within dyads.
3. Focal-animal	Either	Sequential constraints; percentage of time; rates; durations; nearest neighbor relationships.
4. All occurrences of selected behaviors	Usually event	Synchrony; rates.
5. Sequence	Either	Sequential constraints.
6. One-zero	Usually state	None
7. Instantaneous and scan	State	Percentage of time; synchrony; subgroups.

4. The number of 60-second instantaneous or point samples in which individuals were interacting.
5. The number of grooming bouts exchanged between individuals (all-occurrences sampling).
6. The number of 60-second instantaneous or point samples in which individuals were grooming.
7. The number of 30-second time intervals in which individuals were within arm's reach of each other (one-zero sampling).

Dunbar (1976) described his method as follows:

The recording method was so designed that data were simultaneously obtained on each of the seven different measures. In each case, a single subject animal was selected at random from the herd, and his behaviour was recorded on a checksheet graduated into sixty 30-sec. intervals. Everything the subject did (or had done to him) and the age-sex class of his interactee was recorded in the interval in which it occurred. Persistent behaviours such as grooming were recorded in each interval in which they occurred, the end of a bout being marked by a vertical line behind the last entered symbol. At 60-sec. intervals, an instantaneous sample of what the subject was doing (and, where appropriate, with whom) was made. In addition, the exact times of onset and termination of all grooming bouts were noted to the nearest sec. (with a possible error of 2-3 secs.) and the age-sex classes of all animals within arm's reach of the subject during any part of each 30-sec. interval were recorded.

The data collected by the seven methods are shown in Table 7-2. There are obvious differences in the data among the seven sam-

Table 7-2. Frequencies with which subjects "interacted" with the various age-sex classes as determined in different ways (all subjects combined) (from Dunbar 1976).*

	Age-Sex Class of Interactee											
Estimator	Adult Male	6-yr. Male	5-yr. Male	4-yr. Male	Adult Female	3-yr. Male	3-yr. Female	2-yr. Male	2-yr. Female	Yearling	Infant	Total
1. Number of contacts	3	3	1	2	8	11	3	3	2	10	1	47
2. Number of acts	19	10	3	7	47	34	8	15	8	17	1	169
3. One-zero interacting	34	33	3	17	99	49	11	39	9	10	1	305
4. Point interacting	15	15	1	7	44	19	3	16	5	1	—	126
5. Grooming bouts	16	9	1	5	38	18	7	15	2	3	—	114
6. Point grooming	15	15	1	7	44	17	3	16	4	—	—	122
7. Spatial measure	45	37	4	19	111	51	17	40	21	67	3	415

**Note:* these data were pooled and treated as though they came from the same individual or sex-age class for the purpose of comparing methodologies. No conclusions about gelada-baboon behavior should be drawn from the data.

Table 7-3. Relative strength of relationship with each age-sex class as determined by the different estimators given as a probability (based on data in Table 7-2) (from Dunbar 1976).

	Age-Sex Class of Interactee										
Estimator	Adult Male	6-yr. Male	5-yr. Male	4-yr. Male	Adult Female	3-yr. Male	3-yr. Female	2-yr. Male	2-yr. Female	Yearling	Infant
1. Number of contacts	0.064	0.064	0.021	0.043	0.170	0.234	0.064	0.064	0.043	0.213	0.021
2. Number of acts	0.112	0.059	0.018	0.041	0.278	0.210	0.047	0.089	0.047	0.101	0.006
3. One-zero interacting	0.111	0.108	0.010	0.056	0.325	0.161	0.036	0.128	0.030	0.033	0.003
4. Point interacting	0.119	0.119	0.008	0.056	0.349	0.151	0.024	0.127	0.040	0.008	0.000
5. Grooming bouts	0.140	0.079	0.009	0.044	0.333	0.158	0.061	0.132	0.018	0.026	0.000
6. Point grooming	0.123	0.123	0.008	0.057	0.361	0.139	0.025	0.131	0.033	0.000	0.000
7. Spatial measure	0.108	0.089	0.010	0.046	0.267	0.123	0.041	0.096	0.051	0.161	0.007

pling method-behavioral unit measurement techniques. When each pair of measurements is compared to determine if they rank the interactees in the same order, the correlations are quite high. Therefore, any of the methods would appear to be suitable if we were interested in only an ordinal scaling of the interactees (*e.g.*, adult females interacted more than three-year males, who interacted more than adult males).

However, if we try to make direct comparisons among relative probabilities of interaction (interval scaling) (Table 7-3), we find larger discrepancies are due in part to the following:

1. By using seven different measurement techniques a "social interaction" has been operationally defined seven different ways.
2. Social interactions between different sex-age classes are often expressed in different ways.
3. Each type of sampling method has inherent biases including the problems of frequency and duration (discussed previously; see also Dunbar 1976).

Dunbar's exercise exemplifies the need to state clearly the hypothesis to be tested and to carefully define the behaviors to be measured. Make certain that the sampling method you choose is a valid measure of the behavior you want to measure. This is essentially a subjective decision dictated by your definition of the behavior you want to measure; but it is strengthened by a sound knowledge of the biases inherent in the various sampling methods. If possible, compare different sampling methods on the same data, as did Dunbar.

D. OBSERVER EFFECTS

As methods have biases, so do observers. Bias is only one factor that contributes to observer error (Rosenthal 1976). Observer errors can contribute to a decrease in both reliability and validity; therefore the results are only as good as the observer.

Good data mean that they are an *accurate* measurement of the true situation. An observer may be *precise*; that is, his data will not vary greatly, but because of biases they might not be accurate. For example, consider two riflemen firing at two targets at a rifle range (Fig. 7-5). Rifleman *A* groups his five shots closely (good precision) and in the bull's-eye (good accuracy). Rifleman *B* groups his five shots closely (good precision) but to the right of the bull's-eye (poor accuracy). Rifleman *B*'s accuracy is biased by his always pulling his rifle slightly to the right while squeezing the trigger.

To carry the analogy to completion, we may say that each rifle-

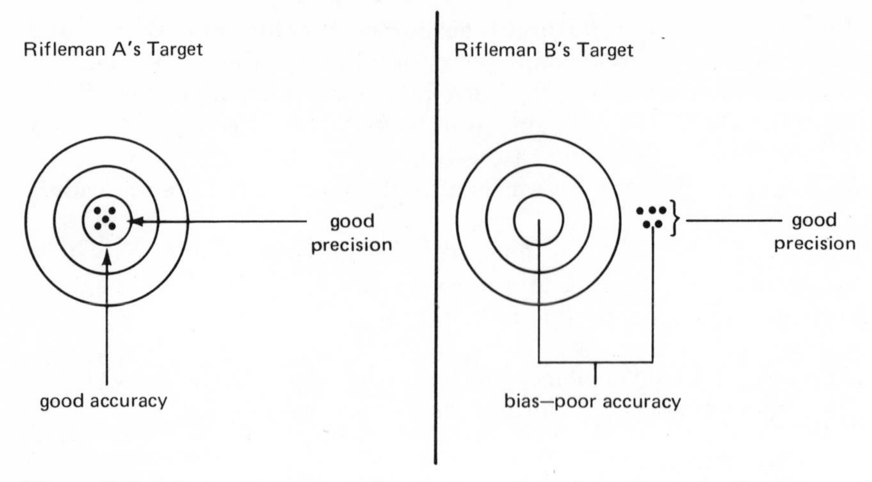

Figure 7-5 Relative accuracy and precision of two hypothetical riflemen.

man's shots were measurements of the location of the bull's-eye. Rifleman *A*'s measurements (shot holes in target) were both reliable (precise) and valid (accurate). Rifleman *B*'s measurements were reliable but inaccurate.

Rosenthal (1976) has discussed experimenter effects in research; several of the potential errors, which may have severe effects on the results of ethological research, are depicted in Figure 7-6. *Observer effect* is due to the presence of the observer or stimuli (*e.g.*, odor) from the observer and results in a change in the animal's behavior.

> Of course, it is sheer metaphysical conceit to claim that we ever do achieve a strictly naturalistic picture of behavior. To do so would require access to phenomena "as they are." The best we can do is to try for samples of events that are representative and valid, and hope that our analysis of what happened is reasonably accurate. [MENZEL 1969:80]

Error of apprehending is due to the physical arrangement of both the animal and observer making it difficult to observe the behavior. *Observer error* is caused by many factors, including inexperience and poorly defined behavioral units. Observer error also includes observer "drift" and "decay" defined by Hollenbeck (1978:96) as follows: "Drift refers to the *movement of an observer* in time from some base point either in a positive or negative direction. Decay, on the other hand, implies *that the instrument*, which includes the observer and the scoring categories, is drifting beyond the bounds of acceptable measurement error." *Observer bias* is principally related to expectancies of the observer. Biases may be consistent or inconsistent and either high or low. Geist recognized this inherent problem in his study of mountain sheep.

> How could I, the same person, separate the process of data gathering from any thought I might have about those data? As the lone investigator I had to record observations and simultaneously think about the observed events in order to recognize relationships. This is a very important problem, for if data gathering is not clearly separated out, it can be biased by some subconscious preference. [GEIST 1971:xii]

In truth, research hypotheses are really personal prophesies, and to strengthen confidence in our ethological insight most of us would like to see them come true. This is a basic human trait which we must guard against if we are to reduce observer bias to an acceptable level. *Error of recording* may be due to poor techniques and equipment, mental lapses in the observer, and inexperience. *Computational error* is usually due to an inappropriate choice of a statistical test. All of these factors can have a severe effect on the results of the research. The results may not only be invalid, but they may be the basis for another ethologist's research and cause compounded problems.

E. RELIABILITY

Selection of the proper behavior unit, scale of measurement, states or events, and sampling method will all affect your study's reliability and validity (p. 75).

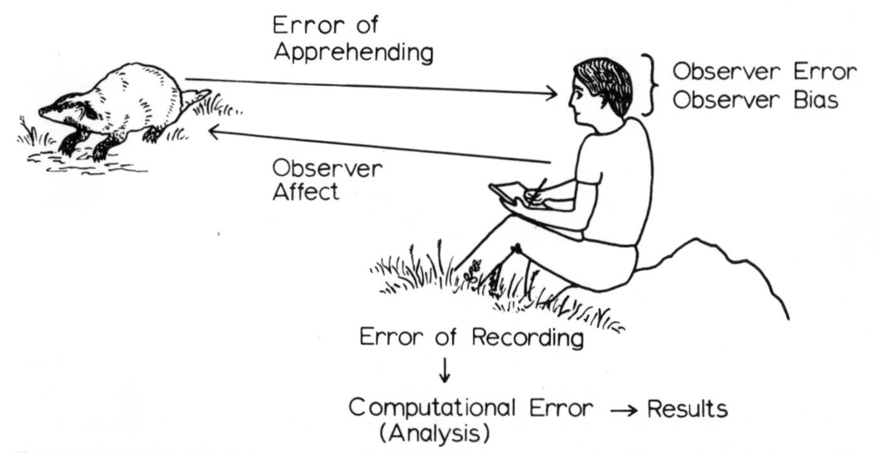

Figure 7-6 Types of observer effects encountered in ethological research. (Drawing by Dan Thompson.)

Reliability refers to the reproducibility of the measurements. Can we rely on our own ability to obtain very similar data again; that is, how good is our *intraobserver reliability*? Other observers should also be able to replicate our measurements which means, in part, that we should have good *interobserver reliability*. This is, of course, often difficult, since skill in observation develops through practice. Accurate and precise observation is not only a general skill (good ethologist), but also a specific skill (*e.g.*, good butterfly ethologist or good acoustical-communication ethologist).

Hollenbeck (1978) concluded that reliability consists of both stability and accuracy. However, this is true only of interobserver reliability. As I discussed earlier (p. 127), an observer may be *reliable* but still have poor *accuracy* as long as *precision* (stability) is maintained. Therefore, intra-observer reliability is solely a measure of stability (or precision), whereas accuracy affects validity (p. 128). However, accuracy will almost certainly affect *inter-observer* reliability since few observers are likely to have the same biases. An accuracy criterion can be established by using an "expert" observer or the consensus of several observers.

1. Intraobserver or Self-Reliability

I know of no observer who sees and hears all he would like to during each observation. Many times in the field, after having observed an unusual behavior, I have muttered "I wish I could see that again." We rarely have that opportunity. Only if we have captured it on film or videotape are we afforded the privilege of seeing it again (then only in two dimensions) and few observers are able to film all their observations. Therefore, we must be as efficient and accurate as possible in recording behavior as it occurs.

Experience is the most important factor in increasing efficiency and accuracy and is necessary with the following:

1. Observing animals
2. The species being studied
3. The behavioral units being recorded
4. The sampling methods
5. The techniques and equipment being used

Intraobserver reliability can be measured by filming or taping a segment of the animal's behavior which is under investigation and then observing the same segment several different times at varying intervals. Your perception of the behavior, and the data you record, should not differ whether you see the segment three times in succession or at two-day intervals.

We all question our own observations and are prone to attempt improvement. However, strive for objectivity; do not redefine behavior units as you proceed. Be careful not to overlay interpretations on your observations as you record data. This is often very difficult; but every observer must check the tendency to see the behavior of animals in light of their own experiences (see Chapter 3). Continually assess the validity of your data and the behavioral units you have selected. However, *do not redesign your research in midstream without careful consideration of the consequences*. Search for and acknowledge weaknesses in your study, such as an inability to identify individuals.

Many types of studies, particularly of social behavior, necessitate the *recognition of individual animals*. Any longitudinal study of a group of animals is strengthened by recognition of individuals. This can be assisted by marking individuals (see p. 141) or by relying on individual variation in morphology and/or behavior (*e.g.*, limp). D.K. Scott's reliability in identifying individuals in her study of the social behavior of Bewick's swans (*Cygnus columbianus bewickii*) was measured by Bateson (1977). More than 100 individual swans were identified by Scott as they were photographed in color by Bateson. The 30 clearest slides were selected and shown to Scott 14 days later, at which time she correctly identified 29 of the 30. She then correctly identified 23 swans from 30 inferior photographs. This test demonstrated that Scott could reliably identify a large number of individual swans and provided increased credibility to her claim that she could identify some 450 individual swans.

2. Interobserver Reliability

> . . . the greatest source of errors in using observations made by others is that no two people who look at the same thing see the same thing. [LORENZ 1935:93]

Lorenz was referring to basic differences in observers' perceptions of the same behavior that are a result of both *inherited* and *learned* traits. For example, poor eyesight and hearing may be inherited, but training will provide an observer with the ability to use his senses with increased effectiveness.

Several factors mentioned above as contributing to observer effects also affect interobserver reliability:

Error of apprehending—can each observer see and hear the animal(s) equally well?
Observer error—are the behavior units clearly defined and are they mutually exclusive? How good is the intra-observer reliability?
Observer bias—do one or all observers have preconceptions about what the animal is likely to do?
Error of recording—is each observer recording data in the same manner?

Poor interobserver reliability is a frequent problem in field studies, where errors of apprehending are difficult to overcome. The other factors can be reduced to a minimum through training and experience.

Interobserver reliability should be measured periodically in order to insure reasonable reliability in the study as a whole. The simplest measurement of interobserver reliability is to take a sample of observations made by all observers and determine the percentage which are the same; this also allows you to detect observer effects as a source of aberrant data.

For example, four observers are alternating in collecting instantaneous samples (every 20 seconds) of behavior of a zebra stallion (focal animal) living in a small family group in Ngorongoro Crater, Tanzania. They are interested in developing a time budget for the following behaviors: Feeding *(F)*, resting *(R)*, grooming *(G)*, walking *(W)*, agonistic (intragroup A_1, intergroup A_2).

In order to measure interobserver reliability they all decide to collect data for the same 10-minute period (30 samples). The hypothetical results are presented in Table 7-4.

The four observers recorded the same behaviors only 60% ($^{18}/_{30}$) of the time. However, observers 1, 2 and 4 recorded the same behaviors 93% ($^{28}/_{30}$) of the time, and observers 1 and 4 had 100% reliability. Therefore, the team as a whole was not very reliable, observer 3 being the most unreliable. Observer 3 should not be allowed to collect data, and efforts should be made to increase his reliability to >95%. The reliability of 93% of observers 1, 2 and 4 is marginal, and attempts should be made to improve it. However, reliability >90% may be the most we can hope for in field studies, due primarily to error of apprehending.

Both Sackett et al. (1978) and Hollenbeck (1978) have cautioned that this measurement, called "percentage agreement," does not give a complete picture of reliability until that agreement which can be accounted for by chance alone is calculated. Hollenbeck went on to list four additional weaknesses of "percentage agreement" provided by Hartmann (1972) and concluded that it should not be used at all.

Table 7-4. Hypothetical results of four observers collecting instantaneous samples of feeding (*F*), resting (*R*), grooming (*G*), walking (*W*), and agonistic behavior (intragroup A_1, intergroup A_2) from a single zebra stallion (focal animal).

Instantaneous Sample	Observers				Same Observations Observers		
	1	2	3	4	1,2,3,4,	1,2,4	1,4
1	*F*	*F*	*F*	*F*	*x*	*x*	*x*
2	*F*	*F*	*F*	*F*	*x*	*x*	*x*
3	*F*	*F*	W	*F*		*x*	*x*
4	*G*	*G*	*G*	*G*	*x*	*x*	*x*
5	*G*	*G*	*G*	*G*	*x*	*x*	*x*
6	*G*	*G*	*R*	*G*		*x*	*x*
7	*F*	*F*	*F*	*F*	*x*	*x*	*x*
8	*F*	*F*	W	*F*		*x*	*x*
9	*F*	W	W	*F*			*x*
10	*F*	*F*	W	*F*		*x*	*x*
11	*G*	*G*	*G*	*G*	*x*	*x*	*x*
12	*G*	*G*	*G*	*G*	*x*	*x*	*x*
13	W	W	W	W	*x*	*x*	*x*
14	A_1	A_2	A_2	A_1			*x*
15	A_1	A_1	A_2	A_1		*x*	*x*
16	A_1	A_1	A_2	A_1		*x*	*x*
17	*G*	*G*	*G*	*G*	*x*	*x*	*x*
18	A_1	A_1	A_1	A_1	*x*	*x*	*x*
19	*F*	*F*	*F*	*F*	*x*	*x*	*x*
20	W	W	*F*	W		*x*	*x*
21	*F*	*F*	*F*	*F*	*x*	*x*	*x*
22	*G*	*G*	*G*	*G*	*x*	*x*	*x*
23	*F*	*F*	W	*F*		*x*	*x*
24	*F*	*F*	*F*	*F*	*x*	*x*	*x*
25	*F*	*F*	*R*	*F*		*x*	*x*
26	*R*	*R*	*R*	*R*	*x*	*x*	*x*
27	*R*	*R*	*F*	*R*		*x*	*x*
28	*R*	*R*	*R*	*R*	*x*	*x*	*x*
29	*F*	*F*	*F*	*F*	*x*	*x*	*x*
30	*F*	*F*	*F*	*F*	*x*	*x*	*x*
		Totals			18	28	30

Instead, Hollenbeck (1978) suggested several alternative measures of reliability that take chance agreement into account. Two of these are Kappa and Kendall's coefficient of concordance.

For nominal data:
Kappa, a statistic developed by Cohen (1960)

$$\text{Kappa} = \frac{(P_o - P_c)}{(1 - P_c)}$$

where: P_o = observed proportion of agreements
P_c = chance proportion of agreements

Kappa is derived from an agreement matrix, such as Table 7-5, which shows agreement on the diagonal. As an example I have used the data from observers 1 and 2 in Table 7-4.

TABLE 7-5. An agreement matrix for observers 1 and 2 in Table 7-4.

	Behavior Codes	Observer 1 *F*	*R*	*G*	*W*	A_1	A_2	Proportion of Total for Observer 1 (P_1)
Observer 2	*F*	13	0	0	0	0	0	13/30 = 0.43
	R	0	3	0	0	0	0	3/30 = 0.10
	G	0	0	7	0	0	0	7/30 = 0.23
	W	1	0	0	2	0	0	3/30 = 0.10
	A_1	0	0	0	0	3	0	3/30 = 0.10
	A_2	0	0	0	0	1	0	1/30 = 0.03
Proportion of Total for Observer 2 (P_2)		0.47	0.10	0.23	0.07	0.13	0.00	

30

$$P_o = \frac{\text{sum of diagonal entries}}{\text{total number of entries}} = \frac{28}{30} = 0.93$$

$$\begin{aligned} P_c &= \Sigma(P_1 \times P_2) \\ &= (0.43 \times 0.47) + (0.10 \times 0.10) + (0.23 \times 0.23) + \\ &\quad (0.10 \times 0.07) + (0.10 \times 0.13) + (0.03 \times 0.00) \\ &= (0.20) + (0.01) + (0.05) + (0.01) + (0.01) + (0.00) \\ &= 0.28 \end{aligned}$$

$$\text{Kappa} = \frac{(0.93 - 0.28)}{(1 - 0.28)} = \frac{0.65}{0.72} = 0.90$$

The kappa value of 0.90 is very close to the simple "percentage agreement" which is 0.93; however, note that by taking chance agreement into account kappa is lower.

For ordinal data:
Kendall's coefficient of concordance (W) developed by Kendall (1948)

$$W = \frac{\Sigma(R_j - \bar{R})^2}{(\frac{1}{12})K^2(N^3 - N)}$$

where: $\bar{R}$ = mean rank = $\frac{\Sigma R_j}{N}$

R_j = sum of the ranks for each behavior category across observers

N = number of behavior categories

K = number of observers

As an example, data from all four observers in Table 7-4 will be used. The data is treated in ordinal format by ranking the behaviors in order of frequency of occurrence as recorded by each observer.

TABLE 7-6. Data from Table 7-4 organized in an ordinal format for the calculation of Kendall's *W* statistic

	Behavior Codes					
Observers	*F*	*R*	*G*	*W*	A_1	A_2
1	1	4	2	5	3	6
2	1	4*	2	4	4	6
3	1	4	2.5	2.5	6	5
4	1	4	2	5	3	6
R_j =	4	16	8.5	16.5	16	23

*The value 4 is given to behaviors *R*, *W* and A_1 for Observer 2, since he observed each three times. This is the average for ranks 3, 4 and 5.

$$N = 6$$

$$\bar{R} = \frac{\sum R_j}{N} = \frac{84}{6} = 14$$

$$\begin{aligned}\sum (R_j - \bar{R})^2 &= (-10)^2 + (2)^2 + (-5.5)^2 + (2.5)^2 + (2)^2 + (9)^2 \\ &= (100) + (4) + (30.2) + (6.25) + (4) + (81) \\ &= 225.7\end{aligned}$$

$$W = \frac{225.7}{(\frac{1}{12})\,(4)^2\,(216 - 6)} = \frac{225.7}{280} = 0.80$$

The calculated coefficient ($W = 0.80$) indicates a relatively low level of agreement between the four observers when their data is rank ordered according to frequency of occurrence.

Hailman (1971) tested the effect of stimulus orientation on the begging response of newly hatched laughing-gull *(Larus atricilla)* chicks. Most data were collected simultaneously by two observers, and interobserver reliability was checked by using "percentage agreement." In another experiment, Hailman tested all the chicks, and then another observer tested the same chicks while Hailman was not present. The results showed good interobserver reliability with regard to stimulus orientation; but the repeated testing had suppressed the response rates (Fig. 7-7).

Successive tests of interobserver reliability, such as Hailman's above, are more a test of methods than of interobserver reliability. To measure interobserver reliability the observers must see the *same* behavior, either simultaneously or by videotape or movies. If

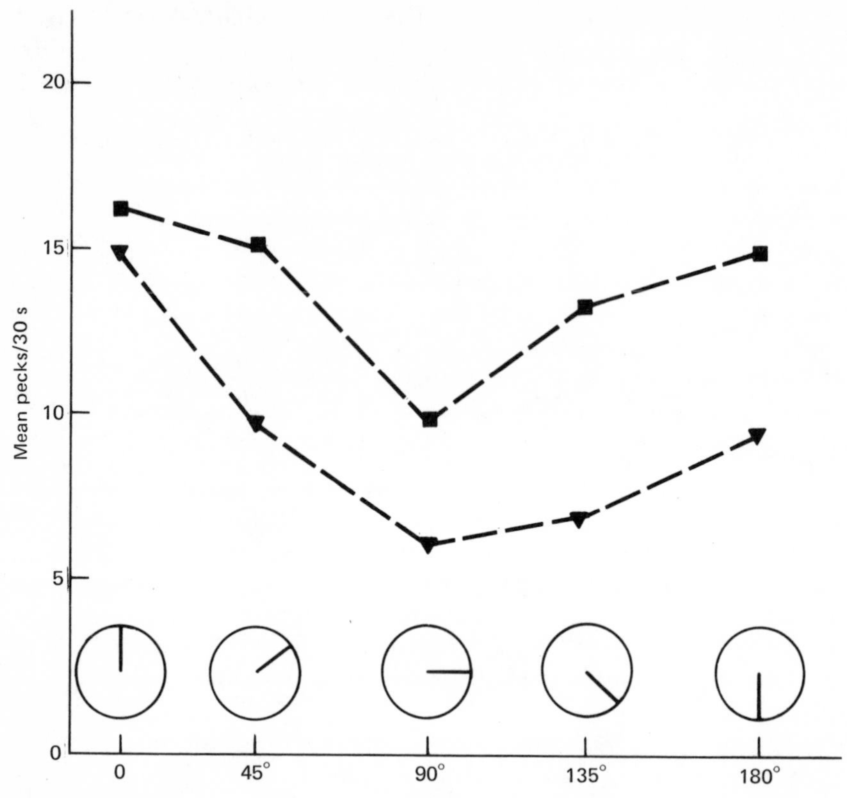

Figure 7-7 Mean responsiveness of newly-hatched chicks to two-dimensional stripes of various orientations. The upper curve (squares) was determined first by one observer, and then the same subjects were tested again by a different observer (lower curve, triangles). (From Hailman, J.P., 1971, The role of stimulus-orientation in eliciting the begging response from newly-hatched chicks of the laughing gull *(Larus atricilla). Anim. Behav.* 19(2):332, Fig. 3.)

videotapes or movies are used, observers viewing the behavior in these media should not be tested against those viewing the live behavior. Simply put, interobserver reliability can be measured only if everything is the same except the observer.

Reliability (intra- and interobserver) can be measured for behavior units, frequencies, durations, and sequences, in a manner similar to those described above. Remember that your major objectives are to answer the research questions as validly as possible and transmit that information to others as accurately as possible. These objectives can be met only if your data are reliable.

F. IDENTIFICATION AND NAMING OF INDIVIDUALS

A necessary prerequisite for many ethological studies, beyond the initial reconnaissance observation stage, is that the observer be able to recognize individuals. In particular, longitudinal studies (*i.e.*, studies of individuals or groups over long periods of time) make individual recognition necessary. As more is learned about the behavior of animals, it becomes increasingly clear that generalizations are difficult to make. Early naturalists talked about the behavior of species. We now know that there are often major differences between populations, social units, and individuals. Hence, ethological research is building from knowledge gained from studies of various individuals and groups and attempting to produce limited generalizations.

1. Natural Marks

The best situation an ethologist can have with regard to individual identification is to be studying a species in which morphological differences are sufficiently great to provide easy identification.

In some cases differences are obvious to even the casual observer. These can be the result of natural mutilations or mutations (*e.g.*, broken horns, melanistic ground squirrels). However, these types of markings are generally not sufficiently widespread in a population to allow large numbers of individuals to be recognized. Many species provide sufficient individual variation for easy recognition, *e.g.*, facial patterns in oryx (Saiz 1975); stripe patterns in zebras (Klingel 1965); spots on each side of the body of bonnethead sharks (Myrberg and Gruber 1974). Variations in the dorsal fins of bottlenose porpoises (Fig. 7-8) allowed Wursig and Wursig (1977) to identify 53 individuals during their 21-month study of group composition.

It is often necessary for the ethologist to spend considerable time in order to solve the problem of individual identification. Rudnai (1973) developed a method of individual identification for African lions *(Panthera leo)* based on variations in the pattern of vibrissa spots, which lie in four to five parallel rows between the upper lip and nose (Fig. 7-9A). This same method has been used by Patty Moehlman (pers. comm.) as an aid in identifying individual black-backed jackals *(Canis mesomelas)*.

The individual vibrissa-spot patterns were recorded in the field on schematic profile sheets. Figure 7-9*B* shows the schematic profile of an individual with its pattern of vibrissa spots and nick in the left ear. Note also that the vibrissa-spot patterns are not the same on both sides. Photos of each side of the lion's head were blown up and the

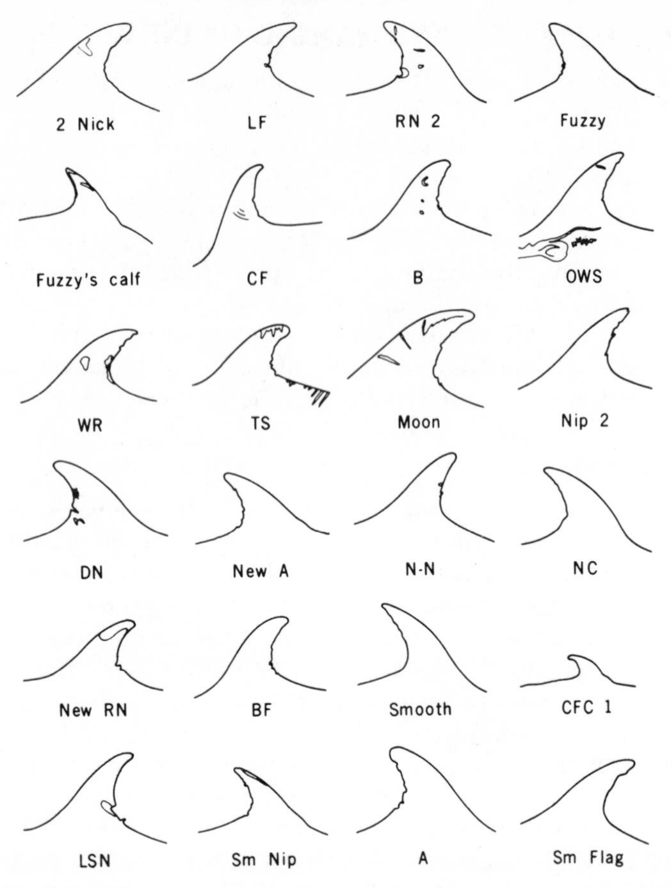

Figure 7-8 *Above:* A sample of 24 fine variations found within the population of bottlenose porpoises. Lines within the fin boundaries represent light pigment spots or scar tissue. *Opposite:* Three individual porpoises followed photographically through time. Compare these fins with the corresponding line drawings on the left. The fin shape and trailing edge nicks appear to be relatively stable. (From Würsig, B., and M. Würsig, 1977, The photographic determination of group size, composition, and stability of coastal porpoises *(Tursiops truncatus), Science* 198:755–756, Figs. 1, 2; copyright 1977 by the American Association for the Advancement of Science.)

vibrissa-spot patterns were checked against the schematic (Fig. 7-9*B*) drawn in the field.

The basis for individual identification was the number and position of the vibrissa spots in Rows *A* and *B* (Fig. 7-9*A*). Decisions were made about whether the spots in Row *A* were above or between spots in Row *B*. This pattern was then transferred to a schematic version

(Fig. 7-9*C*). The spots in Row *B* were numbered consecutively from anterior to posterior. All the possible positions for spots in Row *A* (above and between spots in Row *B*) were provided in the diagram. With continued practice Rudnai was able to use this method of individual identification with ease and confidence.

Pennycuick and Rudnai (1970) also developed a method to test

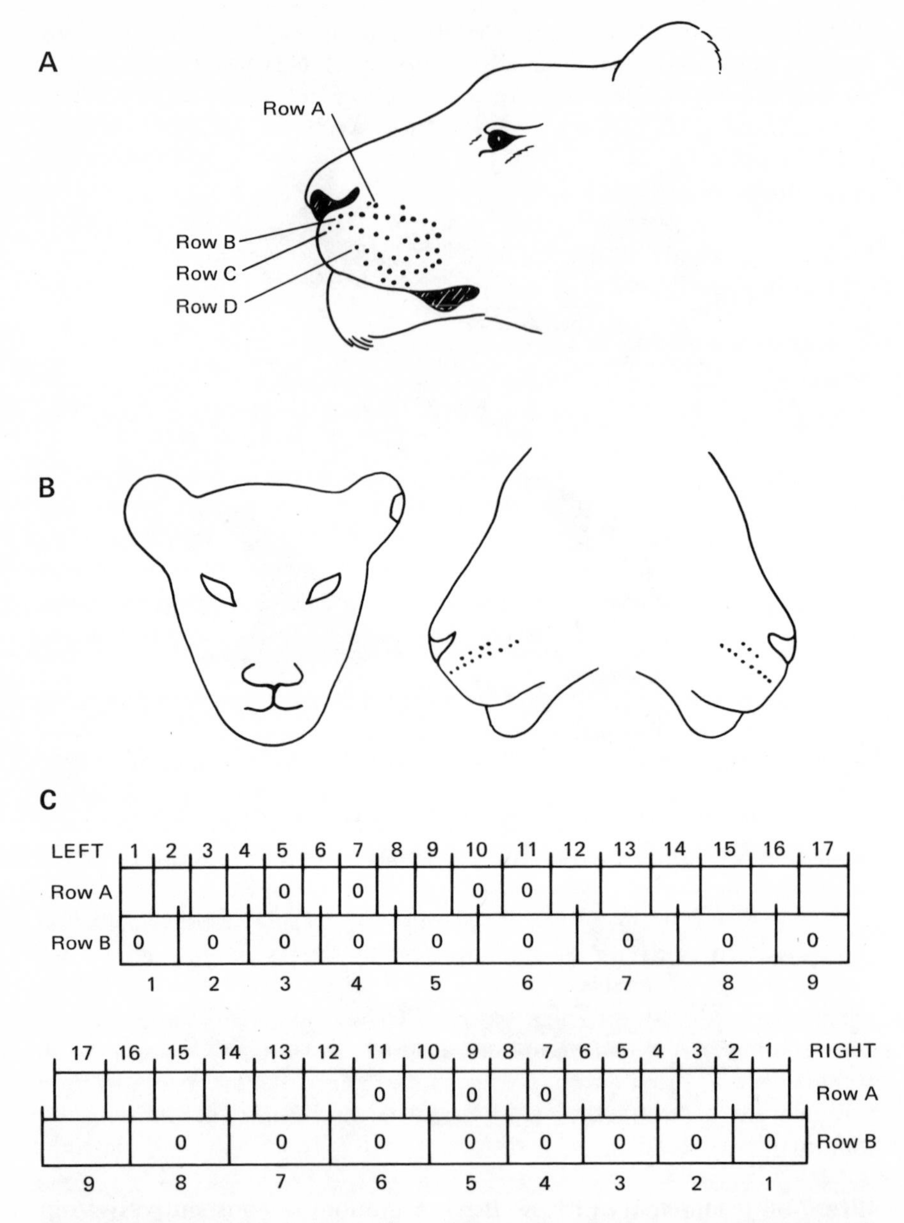

Figure 7-9 *A*, profile of a lion's face, showing the vibrissa spot pattern. Row *B* is used as a scale for recording the positions of the spots in Row *A*. (Adapted from Pennycuick, C.J., and J. Rudnai, 1970, A method of identifying individual lions *Panthera leo* with an analysis of the reliability of identification, *J. Zool. Lond.* 160:497–508, Fig. 1.) *B*, schematic profiles, on the form used for recording in the field. The full-face outline is used to record

the reliability of identifying individual lions in this manner. Their method was based on an estimate of the probability that two individuals would have markings that could not be distinguished. They began by determining the probability of occurrence of the different patterns in the population studied. Spots in Row *A* have never been observed in positions other than 1 through 13.

Regardless of the physical characteristics selected for use in identifying individuals, photographs of each individual are valuable as a continuing "field guide" (Fig. 7-10).

2. Capture and Marking

When natural markings are not available or when individuals are observed only rarely (*e.g.*, homing and migration studies), the researcher must mark the animals in some way. This generally necessitates capture of the animal, although some techniques have been developed for marking at a distance (*e.g.*, dye darts) and for self-marking by the animal. The various capture techniques are too numerous to discuss here, but an excellent review can be found in *Wildlife Management Techniques* (Giles 1969) and Young (1975). The use of drugs in the capture and restraint of animals has been reviewed by Harthoorn (1976).

Various types of markers for animals have been developed for different species and purposes (*e.g.*, dyes, leg bands, ear tags, collars, and radio transmitters, see p. 199). An annotated bibliography of bird marking techniques has been prepared by Marion and Shamis (1977), and Buckley and Hancock (1968) have devised a computer program for generating individual combinations of color and aluminum bands for birds.

A marker that allows the observer to follow visually the nocturnal movements of individual small rodents is the Betalight. These are sealed glass capsules coated internally with phosphor and filled with tritium gas, which emits low-energy beta particles causing the phosphor to glow. They can be obtained in disc or tube shapes ranging in length from approximately 6.5 to 76 mm (Fig. 7-11*A*). Available in a wide variety of colors, they are being used to a limited extent on small mammals, especially in conjunction with binoculars or starlight

---

Figure 7-9 (continued)
such imperfections as nicks in the ears. The lion's sex, approximate age, and pride (if any) are recorded on the same sheet. (*Ibidem*, Fig. 2.) *C*, schematic representation of the spot pattern shown in *A*. The positions in Row *B* are by definition consecutively filled, starting with No. 1, although the total number is variable. Spots in Row *A* have so far been observed as far back as position 13. (*Ibidem*, Fig. 3.)

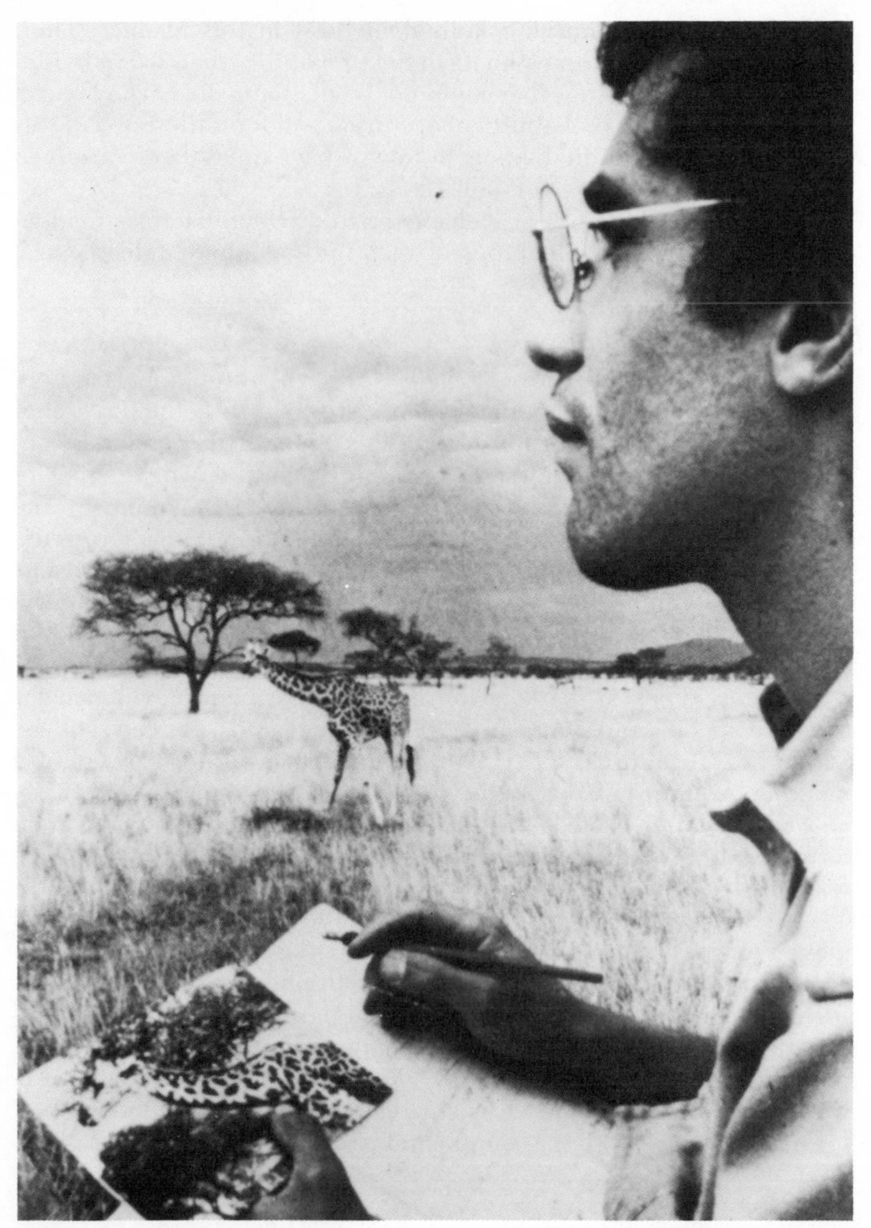

Figure 7-10 Carlos Mejia with photos of individual giraffes for identification in the field. (From Moss, C., 1975, Portraits in the wild, Houghton Mifflin Co., Boston.)

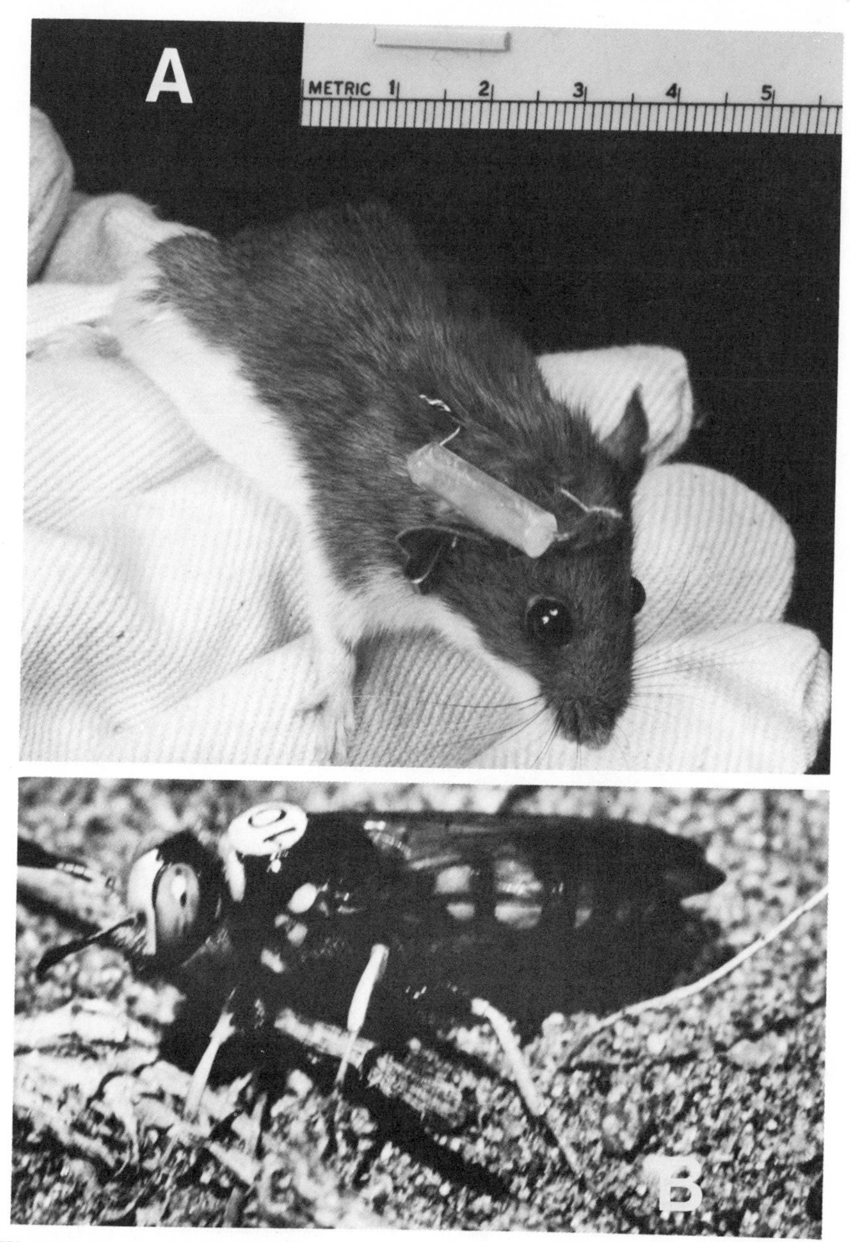

Figure 7-11 *A*, Betalight attached to mouse *(Peromyscus leucopus)*. (Photo by Mark Stromberg.) *B*, numbered-tag 10 attached to the back of a wasp *(Philanthus bicinetus)*. (Photo by Darryl Gwynne.)

scopes (O.J. Reichman, pers. comm.). They are available through Saunders-Roe, Ltd., Middlesex, United Kingdom. *They can only be used under a permit from the Nuclear Regulatory Commission* in Washington, D.C.

The type of marking to be used should be determined after considering the following factors:

1. Number of individuals to be identified
2. Distance over which identification is necessary
3. Length of time identification is necessary
4. Effects the marker might have on resultant behavior
5. Effect capture might have on the animal
6. Ease of capture

A volume on animal marking, edited by Stonehouse (1978), includes sections on methods of capture, tagging, marking by tissue removal and modification, recognition without marking, and radioactive and radio-tracking techniques. The *Wildlife Management Techniques* (Giles 1969) should also be consulted for types of markers for different species. In addition, technical reports of research on particular species often contain methods of marking developed for that particular species.

The ethologist needs to weigh the relative advantages and disadvantages of utilizing natural marks or artificial markers before launching a program. Aspects to be considered are shown in Table 7-7.

Table 7-7. Relative advantages and disadvantages of natural marks and artificial markers.

Type of Mark	Advantages	Disadvantages
Natural	Unnecessary to capture or handle the animal	Possible ambiguity (lack of reliability) in the markings: 1. Change of markings over time 2. Often less inter-observer reliability
Artificial	Positive identification	Markers being lost (Royall et al. 1974) Capturing, handling, or marking affecting behavior Markers themselves affecting behavior

3. Assignment of Numbers and Names

There are often inherent problems in assigning numbers or names to individual animals. We all would like our subjects to become well known like Jane Goodall's chimps "Fifi," "Leakey," and "Graybeard" and the Gardners' "Washoe." Certainly the individual animal given a name will be remembered longer than the animal which is assigned a number. The way a researcher assigns an identifier to an animal can affect the way he views that animal later in the research.

The naming of a wild animal should not be done casually, for a name colors one's thinking about it forever afterwards. [SCHALLER 1973]

I do not mean that the ethologist must necessarily coldly assign numbers to his research animals and remove all romanticism from his observations and interactions. Rather, the ethologist should be aware of the potential biases which arise when naming animals and guard against letting them influence his interpretation of results.

Names carry certain connotations or overtones for each of us. This is obvious in the naming of our children and our hearing a name and

Table 7-8. Potential biases associated with assigning identifiers to animals.

Type of Identifier	Example	Potential Biases
A. Number or Letter	No. 117; RG	Only those biases we superstitiously associate with numbers like 13, our birth date, higher numbers being "better" or "worse", or our initials
B. Named for object or physical characteristic	"White Face" "Gibraltar"	Only those characteristics reminding us of other individuals towards whom we are biased
C. Named for behavioral characteristic	"Limpy"	Researcher unconsciously carrying over behavior attributes which actually change
D. Named for person	"Leakey"	Naming in honor of a person or because animal possesses particular characteristics similar to that person's; the researcher might not want to see those characteristics change or that honor tarnished (*e.g.*, alpha individual becoming most subordinate)

forming a mental image of the person. Tinbergen has been acutely aware of these potential biases, but appreciates the intimacy one gains through assigning names to individuals under study. He reflects on a female kittiwake that was in the colony studied by Dr. Esther Cullen:

> . . . another bird, a female, was too shy to mate; although she kept visiting males through season after season, she was always too nervous to stay with any of them. (This bird was inadvertently named Cleopatra before her character was known.) [TINBERGEN 1958:210]

The biases inherent in the types of identifiers applied to animals are listed in increasing order in Table 7-8.

8
Data Collection Equipment

Data-collection equipment varies from the simple notebook and pencil to the complex on-line computer terminal. Most of us are mesmerized by automation, so we strive for the most sophisticated electronic methods for collecting data; but there are inherent problems. For increased speed and ability to handle quantities of data there is generally a decrease in reliability. You get immediate feedback from what you have written in a notebook or on a data form, but without a printout (or electronic readout) you are not sure whether your computer keyboard is working properly or not. One argument for the computer keyboard is that when it is working it is so much more efficient that it more than makes up for down time. This rationale depends on your sampling method; in some studies no gaps in data can be tolerated. Table 8-1 lists some points to consider when different data collection methods are weighed. The reader will undoubtedly be able to add several of his own (see also Table 3 in Holm, 1978).

Several data-collection methods are described below in ascending order of complexity. The reader is advised to consider the characteristics in Table 8-1 in deciding on a method to use. It is always wise to have a simpler method available as a backup. Remember Murphy's Law: If anything can go wrong, it will.

At the first level of data collection are the *ad lib*, notes taken in a notebook or spoken into a tape recorder. This method is recommended only for reconnaissance-type observations. A tape recorder can be substituted for some of the data forms discussed below; but caution should be taken to be sure that the recorder is always in working order. Many researchers will admit to losing valuable data by speaking into a dead recorder or letting the tape run out.

Table 8-1. Characteristics of two data collection methods.

Characteristic	Notebook	Computer Keyboard
Reliability		
Equipment	High	?
Interobserver	Less training	More training
Intra-observer	High	High
Speed	Low	High
Quantity of data	Medium	High
Data feedback	Can be reviewed in the field	Must be computer-crunched; remote computer access possible
Power	Limits of the observer	Battery restricted when in field
Altering data-collection format	Easy, which is not always a good characteristic	More difficult

Trotter (1959) added a time base to his note-taking by devising a motorized device in which a strip of paper was unwound from a roll and passed behind a slit in small steps. The steps were of equal size and occurred at a set time interval. This method can be viewed as either *ad libitum*, sampling with an automatic time base, or a type of event recorder with flexibility in the type of behavioral observation to be recorded at each interval. The size of the slit must be designed for the amount of data to be recorded, and the speed at which the paper steps must be in concert with the data being collected. Too small a slit and steps occurring too rapidly could be very frustrating.

A. DATA FORMS

Data forms are the next level of organization above *ad libitum* data collection and Trotter's method. They are the minimum data-collection method for any well-designed study. Hinde (1973:393) cautions that ". . . every check sheet [data form] must be designed with an eye both to the problem in hand and to the idiosyncrasies of the observer."

Kleiman (1974) described two types of check sheets (data forms) used at the National Zoological Park. One is a "time-sample check sheet" which is used for focal-animal samples of 4–5 minutes taken every hour. The second form is used for all-occurrence sampling, in which the animal(s) is observed for 30–60 minutes at a prescribed time each day and the occurrence and duration of selected behaviors recorded.

The steps discussed below for designing data forms are the same as those that should be used before employing one of the more complex data-collection methods. Those methods are merely more sophisticated ways of collecting the same data you would get from a data form. A computer does not do anything (indeed it cannot) that cannot be done by hand.

1. Characteristics of Behavior Units

An essential part of the design of research is defining the units of behavior to be measured (see p. 64). The units selected will not only affect the validity of the study but will also determine the efficiency with which the data are collected.

> The degree of selection [of behavioral units] depends in part on the problem, but also on the fact that there is an inverse relation between how much is recorded and the precision with which it is taken. For quantitative treatment, it is as important to know that an item of behavior did not occur as to know that it did, so items selected for recording must not be missed. [HINDE 1973:396]

Be aware of whether your behavioral units are *mutually exclusive, partially overlapping*, or *completely redundant*. The more exclusion there is between behavioral units, the more easily the data can be recorded.

2. Columns and Rows

Most data forms are two-dimensional representations of three (or four) dimensional data. Figure 8-1, for example, illustrates three variables (individual, behavior, and time) of the four commonly recorded in behavior studies. The fourth, spatial relationships, can be treated separately or incorporated on a standard data form. It could be incorporated on the form in Figure 8-1 by merely including the code of NW with the 0812 to indicate that it occurred in the northwest quadrat of the enclosure.

Columns should be *grouped* and clearly delineated, when possible, to reduce the time it takes the observer to find the appropriate column. Often behaviors can be grouped according to larger categories or animals can be grouped according to social units. The emphasis, however, should be on ease of recording with an eye to organization for data transfer later. For example, if the data are to be keypunched on computer cards, it is helpful to have the data form organized so that the keypunch operator can read across rows without jumping over columns. This, however, should be of lower priority than ease of getting data onto the forms.

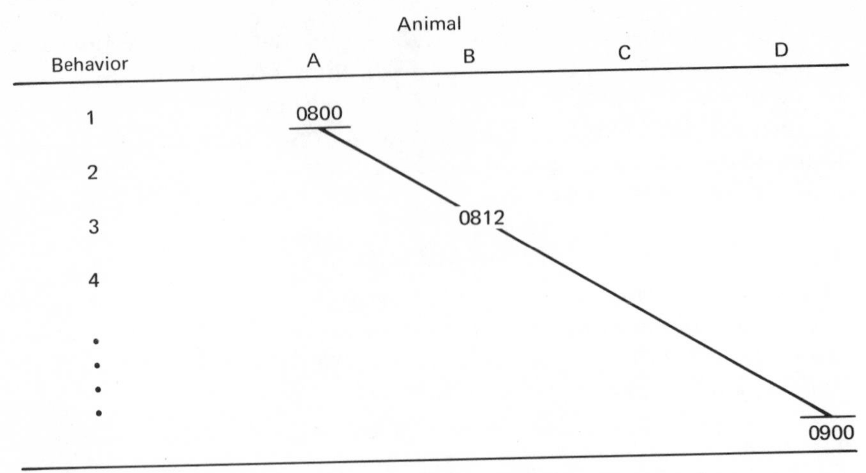

Figure 8-1 Example of a two-dimensional data form representing a three-dimensional format. Animal *B* engaged in Behavior 3 at Time 0812 during a sample period which ran from 0800 to 0900.

Leave both *blank columns and rows* and a large column on the right for *comments*. Notes taken in the comments column are sometimes the key to later interpretation of enigmatic results. Blank columns and rows provide flexibility in adding important behavioral units or animals to your data collection. These columns are probably used more often than not.

3. Coding

Coding data increases speed of recording. The shorter the code the more rapidly it can be recorded and the smaller the space that is necessary for entering the data. However, accuracy and reliability should not be sacrificed for speed; so every effort should be made to devise a code that is simple and easy to recall. Some examples are given below:

Behavior	Preening	P
	Flying	F
	Preening	Pr
	Pecking	Pe
Animal	Edward	E
	Mary	M
	Flash	F
	Red	R
	Green	G
	Blue	B

	Left blue	LB	(colored leg bands,
	Right blue	RB	ear tags, or marks)
	Left green	LG	
	Right green	RG	
Time	First sample period: first minute	1/1	
	Third sample period: fifth minute	3/5	
Spatial	quadrat 7	7	
	southeast quadrat	SE	
	in pond	P	
	on hillside	H	
	next to Edward	/E	

Coding should make data collection easier, not more difficult. It should be obvious, not cryptic.

4. Data-Form Examples

The following data-form examples are simplified to serve as a basis on which individual researchers can construct their own forms to meet their individual needs. The forms are designed to correspond to the sampling methods discussed.

a. Sociometric Matrix

Social-relationship data are generally collected as *all-occurrences* (*e.g.*, agonistic) or in *instantaneous samples* (*e.g.*, nearest neighbor at a particular time). However, other methods can also be used (Dunbar 1976).

The following forms can be used in the sampling of *all occurrences*:

Supplanter \ Supplantee	A	B	C	D
A		√ √ √	√	√ √ √ √ √
B			√ √ √	√ √ √ √
C	√ √ √ √	√ √ √		√ √ √ √ √ √
D			√	

A check is made in the appropriate box every time one individual supplants another.

Supplanter	Supplantee										
A	B	B	C	B	C	C	D	C	C	B	D
B	C	C	C	C	D	C	D	A	D	C	C
C	D	D	D	D	B	D					
D	C	C									

The code for the individual supplanted is entered sequentially in the boxes to the right of the supplanter; note that *sequential* data are obtained.

The following forms may be used when collecting sociometric data involving *instantaneous samples*:

Sample Period	Nearest Neighbor A	B	C	D
1	B	A	D	C
2	D	C	B	C
3	B	C	D	C
4	B	A	D	C

The code for the nearest neighbor is entered under each individual at each instantaneous sample; note that reciprocal relationships (*e.g.*, sample period 1) are not obligatory.

We can add zones of concentric circles around individuals and record the individuals in each zone.

Sample Period	Individuals and Zones A			B			C			D		
	1	2	3	1	2	3	1	2	3	1	2	3
1	C		B			A	A		D			C
2		B			A				D			C
3		B/C			C/A			A/B				

Note that here reciprocal relationships are obligatory, so that we can collect data on only half the animals.

Additional data concerning the exact location of individuals can be gained by using a data sheet with the spatial features represented directly. For example, we might be using a combination of focal-animal and instantaneous *sampling* to determine the social relationships between animal *A* and the other three members of the group. Our data form has a series of groups of concentric circles, each one centered on animal *A*. Hinde (1973) refers to these as "bull's-eye" check sheets (see also Fig. 12-9, p. 317). We use one for each instantaneous sample:

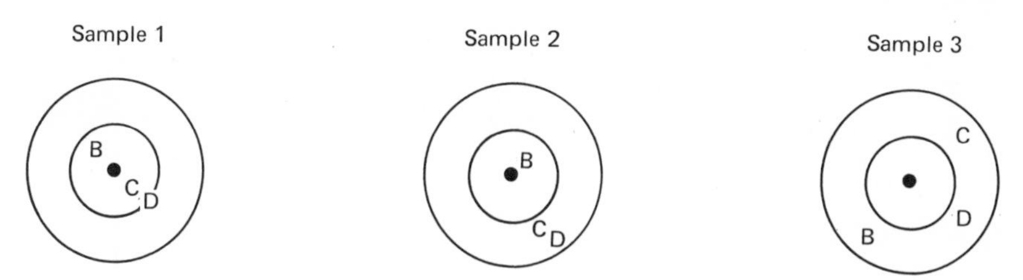

Eventually, sociometric data are generally cast into some form of a matrix in order to illustrate the relationships.

b. Focal-Animal Sampling

We would head each data form with the animal's designation (*e.g.*, Mary) and concentrate on two or more of the variables, *e.g.*, behavior, time, spatial relationships of other individuals, or the environment.

Animal: ____________________

Behavior	Time of Occurrence				
Feeding	0915	0918	0935	0937_1	0943
Grooming	0917	0937_2	0956		
Resting	0856	0900	0951		
Agonistic	1002				

Note that we also obtain frequency of occurrence and gross sequences; finer sequences can be obtained by showing the order of occurrence of events indicated at simultaneous times (*e.g.*, *A* 0937_1, *B* 0937_2).

c. All-Occurrences Sampling

We are interested in the who, when, and where of a particular behavior (or behaviors). A typical data form would be as follows:

Behavior: ____________________

Animal	When and Where the Behavior Occurred				
A	0830/SE	0900/SE			
B	0840/NW				

Behavior: ____________________

Animal	When and Where the Behavior Occurred				
C	0702/NW	0715/SW			
D	0702/NW	0717/SW			

If we are using quadrats *(NW, NE, SE, SW)* as indicated above, we can use a data form that resembles our quadrat layout. For example:

Animal A

	0830 0900

B

0840	

C

0702	
0715	

D

0702	
0717	

The times of occurrences are then recorded in the appropriate quadrats.

d. Instantaneous and Scan Sampling

In contrast to all occurrences sampling, here we are interested in the who, what, and where of behavior at particular points in time.

For example:

Animals, Behavior and Location

Sample Times	A		B		C		D	
	beh.	loc.	beh.	loc.	beh.	loc.	beh.	loc.
0800	Pr	SW	R	NW	R	NW	R	NW
0801	F	SW	R	NW	F	SE	F	SE
0802								
0803								
0804								

e. One-Zero Sampling

In this type of sampling we might ask: During sequential one-minute samples, what individuals are engaged in a particular behavior?

Behavior: ______________

Sample Times	Animals			
	A	B	C	D
0800	x	x		
0801		x		
0802	x		x	
0803				
0804		x		

We merely make a check mark if an individual animal is performing the behavior during each sample period.

f. Sequence Sampling

With this type of sampling we are interested in either intra- or inter-individual sequences of behavior. For example, we may be interested in all grooming sequences of male mallard ducks that are initiated by a tail-shake. We decide to discount the individual performing the behavior but begin sampling whenever a tail-shake *(TS)* occurs, then recording head-shakes *(HS)* and wing-flaps *(WF)* as they occur:

Sample	Sequence initiated by TS			
1	TS	HS	WF	HS
2	TS	HS	WF	
3	TS	HS	WF	HS
4	TS	HS	WF	TS
5	TS	HS	TS	

We could also list the possible movements and number them according to the order of occurrence.

Sample	Movements		
	HS	WF	TS
1	2/4	3	
2	2	3	
3	2/4	3	
4	2	3	4
5	2	3	

A *TS* always occurs first and may also occur at the end of a sequence.

BABY Whisky DATE 6/5/71 NO. 1 TEMP. 11°C WIND 15-20 CLOUD 7/10Se

Time	On Mother		Off Mother			Grooming			Leaves		Appr		Play	Remarks
	Eyes shut	On nip	Off nip	Under 60cm	Over 60cm	Other monkey	by *M*	by *B*	*M*	*B*	*M*	*B*		
00	✓	✓												
	✓	✓												
01	✓	✓												
	✓	✓												
02		✓	✓											
			✓	✓	✓					✓				
03					✓									
			R_3	✓	✓					✓✓				
04					✓									
					✓									
05					✓									
				✓	✓				✓			✓		
06					✓									

Time	On Mother			Off Mother		Grooming			Leaves		Appr		Play	Remarks
	Eyes shut	On nip	Off nip	Under 60cm	Over 60cm	Other monkey	by *M*	by *B*	*M*	*B*	*M*	*B*		
27					✓									
				✓	✓					✓	✓		2 Whoos	
28					✓									
					✓									
29		A_1✓	✓	✓	✓							✓		
		✓			✓									
Total	4	7	3	14	53				5	6	1	10		

Missed ____ Total on & off ____ Total <60 cms only ____ Total >60 cms only ____ Total off ____

*R*1 (mother moves) ____
*R*2 (mother rejects passively) ____
*R*3 (mother pushes, etc.) ____

*A*1 (mother puts arm round) ____
*A*2 (mother accepts passively) ____
M (mother's initiative) ____
Whoos (on) ____
Whoos (off) ____

Figure 8-2 Check sheet for mother-infant relations in captive rhesus monkeys. Each row represents one half minute. Ticks are placed in the columns if the activity in question occurred during the half minute except that "leaving" and "approaching" (that is, distance between mother and infant increases from less than 60 cm to more, or vice versa) are recorded each time they happen. $R_{1,2,}$ and $_3$ categories of rejection of the infant by the mother, $A_{1,2,}$ and *M* categories of acceptance. *C*, *R*, and *T* means that animal *C* initiated rough-and-tumble play with Whisky. "<60 cms only" refers to number of half mins. in which infant was off mother and not >60 cms from her: see also Section 6 infra (Hinde & Spencer-Booth, 1967) (from Hinde 1973).

We might also be interested in the sequence of occurrences of events in a group of individuals engaged in some type of behavior (*e.g.*, encountering predator or prey or leaving a sleeping site). Here we can record the times of occurrence of significant events.

	Individuals and Behavior			
Time	A	B	C	D
0520	St	Y	St	S
0521	W	Y	Y	S
0526	W	W	W	St
0528	D	D	W	W
0531	L	L	L	L

This is essentially a scan sample with the sample period being designated by the occurrence of significant changes in events.

Hinde (1973) also suggests the use of a data sheet with a precalibrated time scale (Fig. 8-2). This form is essentially the same as that provided by an event recorder, which is highly recommended if available.

B. CLOCKS AND COUNTERS

Pushbutton switches connected to electromechanical clocks and counters were used by Mitchell and Clark (1968) in their study of rhesus monkeys. In addition to the clocks and counters this equipment necessitates a 28-volt DC power supply and pulse formers, all of which are common in most experimental-psychology laboratories. Pressing a switch advances a counter and activates the respective clock which runs until the switch is released. The result is summary data in the form of number of occurrences and total duration for each behavior for the sample period. This equipment is not readily adaptable to field work since it is rather bulky, and the 28-volt DC power supply is normally inverted and transformed from 110-volt AC. The same functions can be obtained in the field using a handheld mechanical counter and stopwatch (p. 197).

C. EVENT RECORDERS

Event recorders are used to record the occurrence and duration of events and states. Switch closures activate solenoids that move a pen to one side as it is tracking along a piece of moving chart paper. Microswitches can be assembled into a keyboard (Fig. 8-3) so that the observer can record the occurrence of a behavior by depressing the assigned microswitch and holding it down for as long as the behavior

Figure 8-3 A 20-key microswitch keyboard and a 20-channel Esterline-Angus event recorder.

continues (Staddon 1972). Eisenberg (1963) used switches that locked closed when lifted up, could be depressed into the neutral open position, or pushed down and remained closed as long as they were depressed. Switches can also be triggered by the animal (*e.g.*, Wecker 1964), timers, etc., in the experimenter's absence.

The output from an event recorder consists of a series of tracings which travel at a set rate of speed. The tracings are displaced when a switch is closed, indicating the occurrence of a behavior (Fig. 8-3). These traces then provide a continuous record of the frequency, duration, and temporal patterning of behaviors (Mason 1960).

Hutt and Hutt (1974) describe the 60-channel Peissler event recorder with a built-in keyboard (Fig. 8-4). It was originally designed for use in studies of the social behavior of squirrel monkeys (at the Max Planck Institut für Psychiatrie in Munich) and hence has some limitations for widespread adoption. For example, depression of Keys 4–15 (actor channels), 16–48 (behavior channels), and 49–60 (recipient channels) merely programs the machine and recording does not

Figure 8-4 Peissler 60-channel event recorder. (From Hutt, S.J., and C. Hutt, 1974, Direct observation and measurement of behavior, courtesy of Charles C. Thomas, Springfield, Ill., p. 90, Fig. 24.)

begin until the BEGIN button is depressed. To end a recording the END button must be depressed after the appropriate actor-behavior-recipient buttons have been depressed. Hutt and Hutt note that this introduces delay into the recording system. However, the large number of channels does provide for the recording of many behaviors. This recorder can also be used to punch its output onto paper tape for computer storage and analysis.

A *cumulative recorder* (Fig. 8-5) is a special type of event recorder. Two pens can be separately operated by microswitch closures. One pen resets to its original position when the switch is opened. The other pen steps at each switch closure, making either 4 or 8 steps/mm or approximately 500 or 1000 steps across the paper before resetting automatically. The paper can be set to feed out at 15, 30 or 60 cm/hour. By selecting the proper paper-feed rate, the stepper pen can then be used to provide a visual record of the rate of occurrence of a particular

Figure 8-5 A cumulative recorder.

behavior. Wolach et al. (1975) devised a cassette tape recorder to record responses, converted the output into a relay closure, and then played the tape back to operate a cumulative recorder.

D. STENOGRAPH

Heimstra and Davis (1962) described their use of a stenograph machine for recording behavior simultaneously from two animals. However, the stenograph is adaptable to several additional formats of recording data. It prints several letters as well as numbers from 0 to 9. One or more numbers and/or letters can be depressed simultaneously. The record is printed on a 2.5-inch-wide paper tape which steps one line after a key (or keys) is depressed (Fig. 8-6). The stenograph is relatively compact, light, and very quiet.

Heimstra and Davis (1962) studied the effects of various drugs on the behavior of rats paired in a small wooden box. They separated the

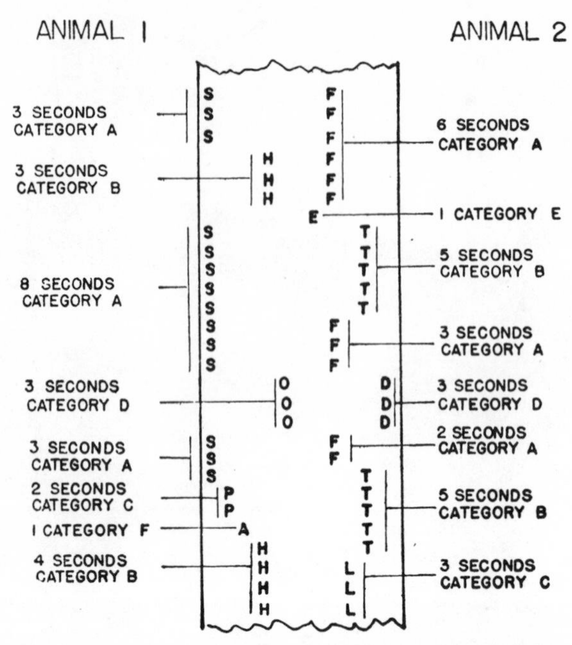

Figure 8-6 Sample printout from a stenograph machine used to record behavior from two rats simultaneously. (From Heimstra, N.W., and R.T. Davis, 1962, A simple recording system for the direct observation technique, *Anim. Behav.* 10(3–4):209, Fig. 1.)

behavior into nine categories *(A–I)* and recorded the occurrence of the behaviors of the two rats simultaneously. They recorded durations of behavioral states by depressing the same key at one-second intervals, cued by an electric metronome (see p. 198).

Data recorded on a stenograph are limited to the number of different keys available, but conversely are expanded by the user's capacity to depress any number of keys simultaneously. The stenograph could probably be modified to step one line automatically at set intervals through a motor drive rather than through the depression of the keys. Data recorded on a stenograph can provide information on the 1) frequency and rate of occurrences, 2) duration, and 3) sequences.

E. COMPUTER-COMPATIBLE DATA LOGGERS

Computers can be used directly to collect, organize, store, and analyze data through direct or indirect input of data. Figure 8-7 illustrates the use of a teletype to transfer behavioral records directly from videotape to a minicomputer.

Figure 8-7 Data collection with a teletype on-line to a PDP 8/e mini-computer.

In 1962 Tobach and her colleagues reported on a system that used the earlier-developed Aromson keyboard, with keys arranged in two semicircles to accommodate two-hand fingertip operation. This system (ASTL) was the forerunner of the modern data loggers used by ethologists. It linked the keyboard to a computer-compatible paper-tape punch and Esterline Angus event recorder through a converter developed by Laupheimer. More recently developed were two portable systems which encode observations and time onto magnetic tape for playback through a teletype or into a keypunch (R. Dawkins 1971; White 1971). Both Dawkins and White describe the operation of their remote keyboard and cassette recording systems, as well as the interface that is necessary for later playback into the teletype. White also compares the two systems.

Later, Stephenson et al. (1975) developed their SSR-VI system that encodes the incidence, duration, coincidence, and sequence of entries in real time onto magnetic tape for later playback into a com-

Table 8-2. Comparison of three field-operable encoding systems (from Gass 1977).

Features	R. Dawkins (1971)	Stephenson et al. (1975)	Digitorg
Capacity (number of codes)	100	48	256
Encodes durations directly	No	Yes	No
Encodes simultaneous events	No	Yes	No
Temporal resolution (seconds)	0.02	0.05	0.1
Designed to encode spatial information	No	No	Yes
Redundant recording	No	Yes	No
Encoding method	Analog (frequency)	Time-multiplexed digital	Serialized digital

puter. In 1977, Gass described his Digitorg system for recording both behavioral and spatial information in real time on magnetic tape. Table 8-2 compares the R. Dawkins (1971), SSR-VI, and Digitorg systems.

All data-logging systems provide a means for encoding, storing and inputting into a computer for reduction, storage, and analysis. Figure 8-8 illustrates these processes for the Digitorg system.

More recently several new or improved data-logging systems have been developed by investigators: ELOG (Fitzpatrick 1977), DCR-II (Celhoffer et al. 1977), SSR-VII (Stephenson and Roberts 1977), and Datamyte (Torgerson 1977). These and other systems are compared in Table 8-3.

The Datamyte 900 (Fig. 8-9) is a commercially available system developed by Electro/General Corporation, Minnetonka, Minnesota. The small unit (12 × 10 × 2 inches) weighing 4 lb. directly encodes input into a solid-state memory and automatically records real time. There is a built-in rechargeable battery that permits eight hours of use between charges. The keyboard has 10 numerical (0–9), 4 alpha *(C, F, H, *)* keys and four selectable input modes:

Mode 1 (ENTER with Time). Depressing a combination of the 14 character keys causes the characters to be displayed. Depressing ENTER records the displayed characters plus elapsed time.

Mode 2 (One-Character Autotime). Depressing any of the 14 character keys instantly records the character plus elapsed time.

Mode 3 (Two-Character Autotime). Depressing any of the 14 character keys displays the character. Depressing a second key instantly records the two characters plus elapsed time.

Mode 4 (ENTER). Depressing a combination of the 14 character keys causes the characters to be displayed. Depressing ENTER records the displayed characters.

The Datamyte system has been used in various studies of human behavior: parent-infant interactions (Sawin et al. 1977), infant visual behavior (K.G. Scott and Masi 1977), and group behavior (Conger and McLeod 1977). The SSR system has also been used in various ethological studies (see Stephenson et al. 1975; Stephenson and Roberts 1977).

Microprocessor Operated Recording Equipment (MORE) Co., Seattle, Washington builds and markets a microprocessor which is

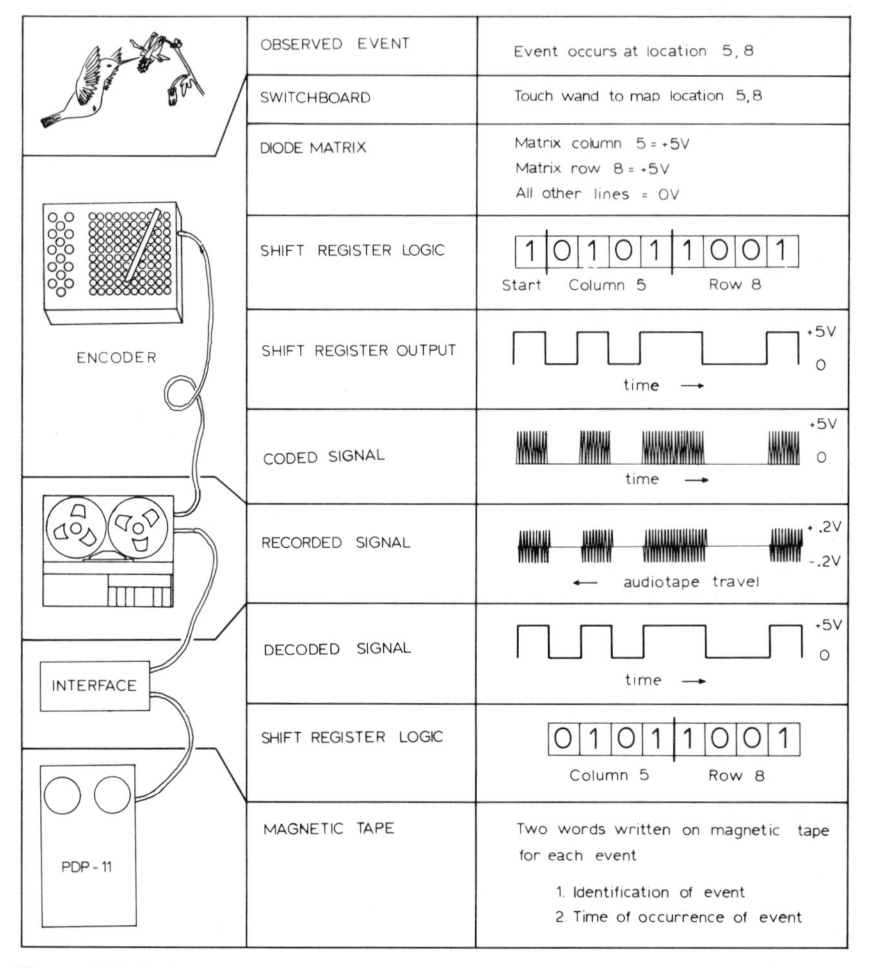

Figure 8-8 Information processing by Digitorg. Stages of information processing are shown from top to bottom. (From Gass, C.L., 1977, A digital encoder for field recording of behavioral, temporal, and spatial information in directly computer-accessible form, *Behav. Res. Meth. and Instrum.* 9(1):6, Fig. 1.)

Table 8-3. Comparison of features found in various input hardware (from Fitzpatrick 1977).

Feature	Standard Computer Data Terminals	Data Logging Input Devices							
		Azurdata	BOSS	ELOG	Epic Data	Esterline Angus	Datamyte	SSR6	SSR7
Timekeeping			X	X	X	X	X	X	X
Entry mode									
Serial	X	X	X		X	X	X	X	X
Parallel				X		X		X	X
Paper copy	Most					X			
Portability									
Timesharing accessible	X	X		X	X	X	X		X
Lightweight		X	X	X	X		X	X	X
Battery operation		X	X	X			X	X	X
Synchronization				X	*	*			X
Data transmission rate									
High (>300 bps)	X	X			X	X	X		
Medium (≈300 bps)	X	X		X	X	X	X		
Low (<300 bps)	X		X					X	X
Data storage	**								
Electronic memory		X			X	X	X		
Computer tape						X			
Audio cassette			***	X				***	***
Punched paper						X			

*Device has the capacity if an interface is built.

**Most Teletypes have paper-tape punches, and some data terminals contain tape or electronic memory.

***Reel-to-reel or cassette audio tapes run continuously and provide event timing; limited time capacity.

Figure 8-9 The Datamyte 900 data logger. (Courtesy of Electro/General Corporation, Minnetonka, Minnesota.)

designed for collection of observational data, but is capable of being programmed in BASIC language for both control and data collection in the laboratory. The observer selects the trial, code length, and recording mode for each trial. In the *timed event mode*, behaviors are timed in seconds from the entry of one behavior to the entry of the next behavior. In the *modified frequency mode* the sample period length (1–99 seconds) is selected and behaviors are then recorded in sequential sample periods. In either mode four characters of a code are displayed on the large, easily readable display, until the first character of the next behavior code is entered. It is a compact, lightweight unit.

Most authors discuss their data logger-to-computer interfaces and reference computer programs that have been used in conjunction with their systems (Fitzpatrick 1977; Stephenson and Roberts 1977; Celhoffer et al. 1977; Torgerson 1977; Gass 1977; Hollenbeck et al. 1977).

Fernald and Heinecke (1974) used punched *paper tape* to record events and later used the tape as input to a computer. Tape punching could be performed in four different input formats: 1) direct keyboard input, 2) keyboard input with explicit time punches, 3) keyboard with implicit time punches, and 4) external-event counter with implicit time mark.

1. Selecting a Data Logger

In selecting a data-logger system the researcher must take into account its several abilities:

Capability—What are your objectives?
Will it do what you want?
Will it do more than you need?
How efficient is it in logging the data you need?

Reliability—Is it a proven system?
What is its on-the-job record?
Do you or your organization have the capabilities for repair?

Compatibility—Is the system compatible with your present data collection and computing system?
How difficult will it be to integrate it?

Portability—Can it be easily carried into the field?
What is the battery life?
How much trade-off has there been in capability for portability?

Accountability—Is it really worth the cost?
Can it be quickly utilized or will large amounts of time be lost in adapting to the system?

The basic Datamyte 900 is priced from about $2500 to $3000 (twice the memory). The SSR-VII keyboard is available from Semeiotic Systems Corporation, Madison, Wisconsin. Fitzpatrick (1977) lists sources for other data-logging devices found in Table 8-3. Other systems can often be purchased from the investigators who developed them or built them, using schematics provided in the references. I caution against replowing old ground in developing your own devices, although our technician, Jim Starkey, designed and built an inexpensive but useful keyboard-cassette system for use in our research.

Remember that data loggers are only a faster and more efficient way of collecting data. They will not substitute for a poorly designed study by magically changing useless data into useful data. They are designed to assist researchers, not replace them.

F. AUDIO TAPE RECORDERS

The ethologist uses audio tape recorders for three different purposes: 1) to record observations verbally described by the observer, 2) to record sounds produced by animals under study, and 3) to store events in a manner compatible for later playback into a computer (see previous section) or cumulative recorder (p. 160).

1. Note-taking on Audio Tape Recorders

This method of note-taking has several advantages and disadvantages. The most noteworthy advantages are 1) being able to observe continuously while recording data and 2) flexibility of input; additional observations and comments can be easily recorded. The greatest disadvantages are 1) recorders stopping and tape running out, 2) real time being difficult to keep, 3) transfer of data from the tape often being difficult since most observers do not adhere to a strict format when recording the data, and 4) recording perhaps disturbing the animals being observed. If the behavior is complex or occurs rapidly, some type of coding should be used. The code must be clearly defined and the sounds of the code words must be easily discriminable for future transcription. Data transfer from audio tapes can take many forms. You may want to go directly to a data check sheet or to a format that can then be key-punched onto computer cards. A complete sequential transcript may be advisable, such as that used by Hutt and Hutt (1974) in their study of "free field" behavior in children (Table 8-4).

Audio tape recorders available for note-taking vary in size from the small pocket-sized minicassettes (short tape time) to the large decks (long tape time). There is no need for high fidelity in this type of use, but an on-off switch on the microphone is almost a necessity. Also, a rechargeable unit is especially useful to the researcher who makes daily forays into the field. Keep the recorders clean and in good working order; check from time to time while you are recording (VU meter) to be sure your voice is being recorded.

Examples of the use of tape recorders for note-taking are provided by Brockway (1964), Eisenberg (1967), Kaufman and Rosenblum (1966), and Rosenblum et al. (1964).

2. Recording Animal Sounds

When recording animal sounds on audio tape you will find it important to identify the recordist(s), animal(s), habitat, and behavioral context, geographic location, date, time, and climatological conditions, if possible. This may be given as commentary on the tape immediately after the recording or simultaneously on another track of the tape (see

Table 8-4. Transcript of the tape-recorded commentary made during a three-minute session of observation on a child's "free field" behavior. Numbers designate location in the room; strokes designate termination of an activity; numbers above the strokes are the duration of the activity (from Hutt and Hutt 1974).

9½ 2½
Standing 4 looking bricks, holding wire / / walks 7/8 twirling / / looking
2½ 4 3½
bricks / / walks to screen 15 / / turns to 0's call, walks 10/11 / / twirling,
7 2 2
looking bricks twirling 11 / / walks 10/11 to 2 / / bangs wall / / looking
4 2
bricks, twirling 5/6 / / walks 7 picks brick / / runs screen 15, puts brick
6 3
in mouth and bangs on screen / / rubbing screen walks 13 to 9 / / puts brick
2½ 2
window / / standing 5 bangs brick on window / / turns throws brick 15 and goes
4 4 6
after it / / picks up brick 15 throws at 0 / / walks 8, climbs on chair / /
3½ 5½
sitting chair, looks screen to 0 to window / / looks at brick / / gets off chair
2½ 4½ 3
walks 7 / / picks up brick throws at screen / / banging screen walks 13/14 / /
2 4½
throws brick at 0 / / turns, runs 8, climbs chair / / stands arm of chair
8 3½
holding door frame / / jumps on seat, turns / / bangs window holding on to chair
9 3½
/ / turns to 0's signal, reaches for 0's brick / / sits chair looking over
13 2
side at floor / / looks window / / looks ceiling, leaning over back of chair,
5 2 4 3
hand in mouth / / looks door / / gets off chair looking 12 / / walks 13 / /
8
stands 13 biting jumper looking corner / / turns, walks 13 to 9 to 5, biting
10 3 3½
jumper while walking to 6 to 7 / / throws brick to 16 / / twirls / / walks 16
4½ 2½ 5
to 15 / / picks brick throws it to 1 / / walks to 8 climbs on to chair / /

below). Additional written records of the recordings should also be kept in your field notebook.

In contrast to recorders used for note-taking, these recorders should be of the highest quality and fidelity. The following characteristics should be checked when you are selecting a recorder:

Frequency response—range from highest to lowest in hertz (Hz = cycles/sec.). The number of decibels (dB) from a flat curve is also usually indicated (*e.g.*, ±2dB).

Signal-to-noise ratio—the ratio of background noise from the recorder to the signal put on tape. A good s/n ratio should be about 55–60 dB.

Tape speed—represents a trade off between quality of recording (high speed) and economy (low speed). Good recorders should have the capacity for 38 cm/sec.

Most ethologists have chosen to use reel-to-reel recorders since they generally make better recordings than cassette recorders. However, suitable recordings can be made with the best cassette models (Bradley 1977). The two portable reel-to-reel recorders most often used by ethologists for recording avian vocalizations are the Uher (Fig. 8-12) and Nagra (Fig. 8-10).

Monophonic recorders come in models to record on the *full track* or *half track*. The full-track recordings are probably of the highest quality; but the half-track models allow you to turn the tape over and record on the second side. There is some loss of quality since only one half of the ¼-inch tape is being recorded on each side.

Stereophonic recorders split the tracks and allow you to record from two sources simultaneously. That is, you can record animal sounds on one track and a verbal description on the other track simultaneously (two-track). Also, two observers, each responsible for one animal, can simultaneously record their observations, one on each track (Grant and Mackintosh 1963). Four track recorders allow the recording of two simultaneous tracks on each side.

Audio tapes have three important characteristics: 1) *thickness*, 2) *backing*, and 3) *signal-to-noise ratio*.

Tapes are generally available in three thicknesses: 0.5, 1.0, and 1.5 mil. Trade-offs are involved when you choose the thickness of tape for your recordings. The thinner the tape, the more stretching that may occur, the more tape you will get per reel, and the more chance for print-through, *i.e.*, the tendency for a recorded signal to magnetize the adjacent wound tape (Bradley 1977). A 1.0-mil tape is usually a reasonable compromise.

Backings are generally either the newer polyester plastic (Mylar) or the older cellulose acetate. Polyester plastic is preferred, since cellulose acetate tends to be weaker and is more prone to stretching, warping, and wrinkling.

Low-noise tapes are superior to normal tapes in their ability to improve the signal-to-noise ratio. Some recorders have separate settings for low-noise tapes.

Microphones (mikes) come in two basic designs: dynamic (moving-coil) or condenser. The condenser mike may be superior, but it is more complicated and often requires additional repair. For field recording, dynamic mikes are probably the best choice. Look carefully at the *impedance* and *output*. The microphone's impedance must match the recorder and the output should be approximately −57 to −53 dB (Bradley 1977).

Differential directional characteristics are provided by *omnidirectional, cardoid*, and *supercardoid mikes*. Omnidirectional mikes are essentially sensitive to sounds in all directions, while cardoid mikes are most sensitive to sounds in front of them. Supercardoid mikes (shotgun mikes) are highly directional and increase the relative intensity of the sound at which they are directed (within a small angle in front of the microphone) by remaining insensitive to unwanted sounds (noise) outside of that angle. In essence, they increase the signal-to-noise ratio.

Cardoid microphones may be used with a *parabolic reflector* that focuses sound received over the width of the reflector onto the microphone, which is set at the focal point of the parabola (Fig. 8-10). Parabolic reflectors should be wider than the wavelength of sound that you are recording. For example, many songbird vocalizations are in the frequency range of 2–8 kHz, with wavelengths of 0.03 m to 0.15 m. Therefore a parabolic reflector with a width of 0.46 m is sufficient. However, coyote vocalizations are often around 500 Hz, with a wavelength of 0.61 m; therefore the parabolic reflector should be at least 0.61 m in diameter in order to make high-quality sounds. Parabolic reflectors are more effective than shotgun microphones when recording over distances that exceed 10–25 m.

3. Ultrasonic Detectors

There are two types of commercially available detectors which have been used to detect the ultrasonic sounds of animals (Fenton et al. 1973). These devices have been used in research on bats (Fenton 1970; Kunz and Brock 1975) and insects (Klein 1955). See Sales and Pye (1974) for a review.

One ultrasonic detector is manufactured by Holgates of Totton, Southampton, United Kingdom. It uses a capacitance microphone capable of responding to frequencies between 10 and 180 kHz as well as electronic tuning to limit the input band width.

Another that uses a crystal microphone adjusted for maximum sensitivity at 40 kHz is manufactured by Alton Electronics Co., Gainesville, Florida.

4. Playback of Sounds

Sounds are played to animals in an attempt to determine their effectiveness in stimulating or inhibiting behavior. In this way the function of animal-generated sounds can sometimes be determined.

Figure 8-10 Parabolic reflector covered with camouflage netting and wind-shielded microphone wired to a Nagra IV audiotape recorder.

Vocalizations are often played to conspecifics and their resultant effect observed. Emlen (1972) modified the recorded songs of indigo buntings *(Passerina cyanea)* and through playback demonstrated that 1) species recognition is coded in the note structure, internote interval, and note length; 2) individual recognition is coded in the details of note structure; and 3) motivation cues are reflected in song length and singing rate. Waser (1975*b*) demonstrated that playback of the gray-checked mangabey's *(Cercocebus albigena)* "whoopgobble" vocalization mediated intergroup avoidance. Lehner (1976) used the playback equipment in Figure 8-11 in his study of coyote vocalizations. Also, see page 180 for a discussion of pattern playback equipment. Using a unique approach, Simmons (1971) picked up bat cries in two condenser microphones and played them back to the bat with different delay times, simulating "phantom targets" at different distances.

Playback can be used to reveal the significance of interspecific sounds. For example, Cade (1975) showed that female parasitoid flies *(Euphasiopteryx ochracea)* containing living larvae were attracted to dead crickets attached to speakers, through which cricket songs were played.

Also, Larkin (1977) showed that tape recordings of thunder, birdcalls, and artificial sounds played to migrating birds through a loudspeaker system slaved to a tracking radar often caused the birds to turn away from the source of the sound. Heppner (1965) found that high-intensity white noise had no effect on the ability of captive robins *(Turdus migratorius)* to find earthworms, further supporting the hypothesis that the robins were primarily using visual cues.

G. SOUND SPECTROGRAPH

The sound spectrograph (Sonagraph is the trade name used by Kay Electronics; see Figure 8-12) is an electronic device for converting sound to a visual display. It will produce several types of displays, all of which are useful in the analysis of vocalizations or mechanical sounds (*e.g.*, grasshopper songs).

1. Sound-Spectrogram Types

The *normal display* of a sound spectrogram is shown in Figure 8-13A. Time is represented on the horizontal axis (1 sec/mark), frequency (pitch) is represented on the vertical axis (1 kHz/mark), and amplitude (intensity) is represented by the blackness of the mark. Sound spectrograms are produced on paper 14.5 cm × 32.4 cm, only a portion of which is shown in Figure 8-13.

From the sonagrams above we can see that the coyote's howl began with three bursts of energy over several frequencies, with none of the frequencies being clearly defined; these are essentially introductory "barks." The howl portion began at a relatively low frequency and then rose to approximately 1.6 kHz, where it remained for approx-

Figure 8-11 *A*, audio playback equipment: *a*, Uher 4000 Report-L recorder; *b*, Realistic MPA-20 amplifier; *c*, Realistic PA-12 trumpet speaker. *B*, equipment in *A* assembled into carrying case.

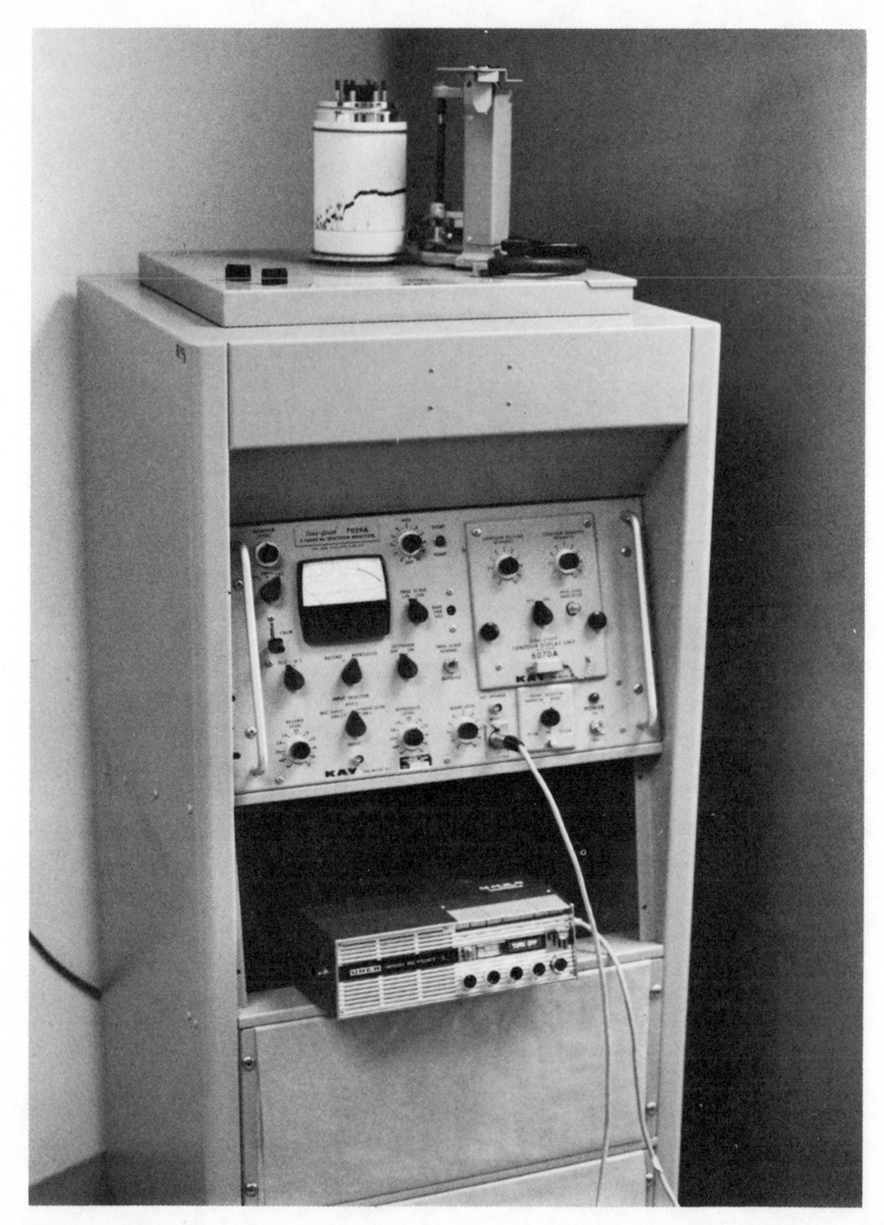

Figure 8-12 A Kay Electronics sound spectrograph and Uher audiotape recorder.

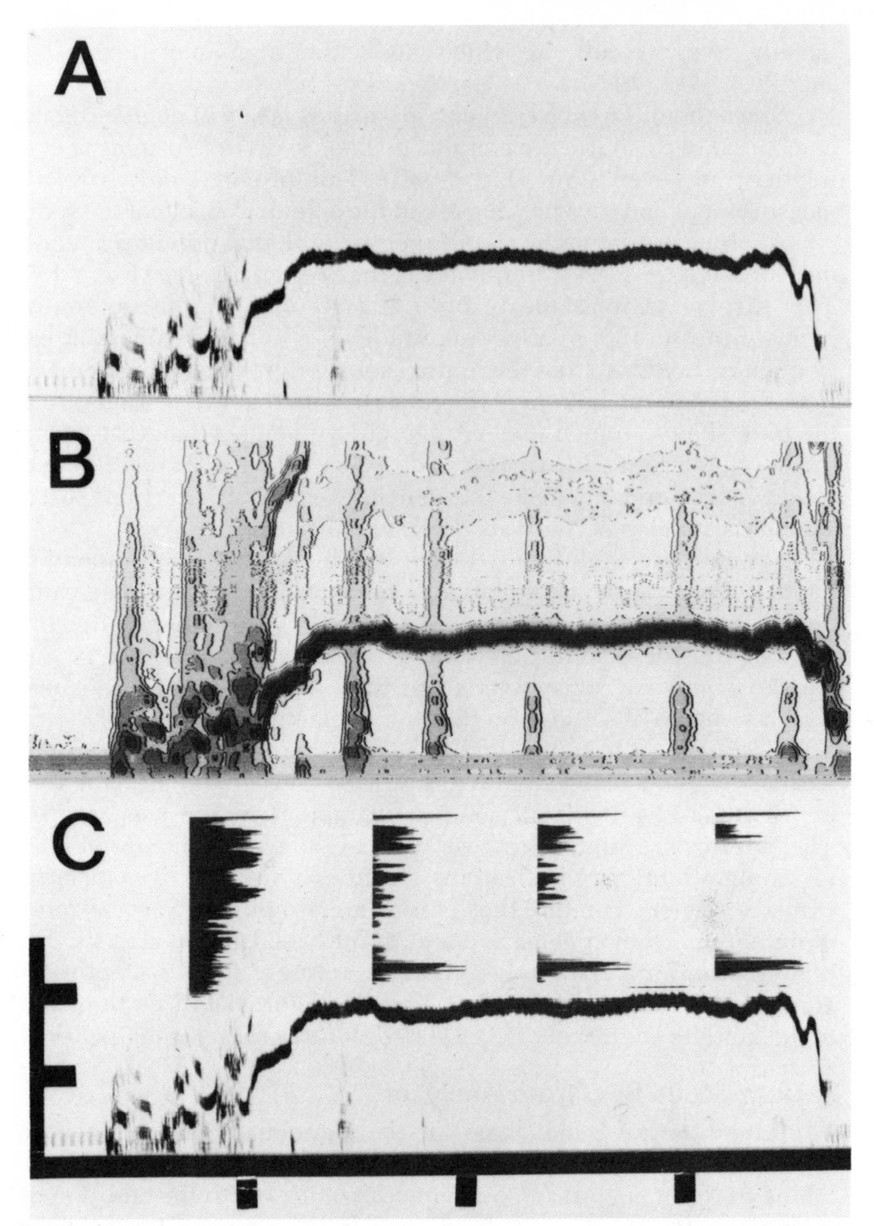

Figure 8-13 *A*, normal display sound spectrogram (sonagram) of a coyote howl; *B*, contour display of same howl; *C*, section display above the normal display. Time is marked on the horizontal axis in one-second intervals. Frequency is marked on the vertical axis in one kH_Z intervals. (See text for an explanation of the types of sound spectrograms.)

imately two seconds, at which point the frequency dropped off sharply.

Since the difference between intensities is one of degree (relative blackness), we can use the *contour display* to delineate areas of equal intensity in seven steps (Fig. 8-13*B*). This provides only a relative measurement and says nothing about the actual intensity of the sound.

Another feature of the sound spectrograph that provides a relative measure of intensity vs. frequency is the *section display* (Fig. 8-13*C*). This display samples the recording at six or fewer predetermined points and presents relative amplitude as a horizontal mark at each frequency, inverted from the normal sonagraph (frequency increasing from the top to the bottom of the paper). Note that the section through the bark shows a much wider range of frequencies than that through the howl. Marler and Isaac (1960) describe a device for modifying the sound spectrograph to make frequency-vs.-amplitude sections serially through a syllable at intervals down to 2.5 msec.

Sections are useful for determining the relative amplitude of frequencies in a particular syllable or sonagram. However, they cannot be used to make absolute measurements (*e.g.*, number of decibels) without considerable difficulty, and they should not be used for comparisons between sonagrams since they are affected by the investigator's choice of settings on the sound spectrograph.

Vocalization terminology has been rather inconsistently applied, with few authors using similar terms. Kroodsma (1977) used the terms in Figure 8-14*A* to detail song development in the song sparrow. These terms are similar to those used by Rice and Thompson (1968) for indigo-bunting vocalizations. Although not totally satisfactory (Kroodsma, pers. comm.), these terms are applicable to vocalizations of numerous other species and are useful in sonagram analysis. Temporal patterns are extremely important in insect sounds. Bentley and Hoy (1972) developed the terminology in Figure 8-14*B* for their study of the genetic control of cricket (*Teleogryllus gryllus*) song patterns.

2. Ubiquitous Spectrum Analyzer

Ubiquitous is the trade name for the Federal Scientific Spectrum Analyzer. It is a real-time, time-compression scanning analyzer which can be used for analysis of animal vocalizations with the addition of a display system (Hopkins et al. 1974). A digital system is used to speed up the signal, and an analog system sweeps the time-compressed signal through a filter. Spectrograms can be displayed on a storage oscilloscope or photographed for permanent copy. Narrow and wide bandwidth analyses are possible, and section displays can be made at intervals as short as 3.125 msec.

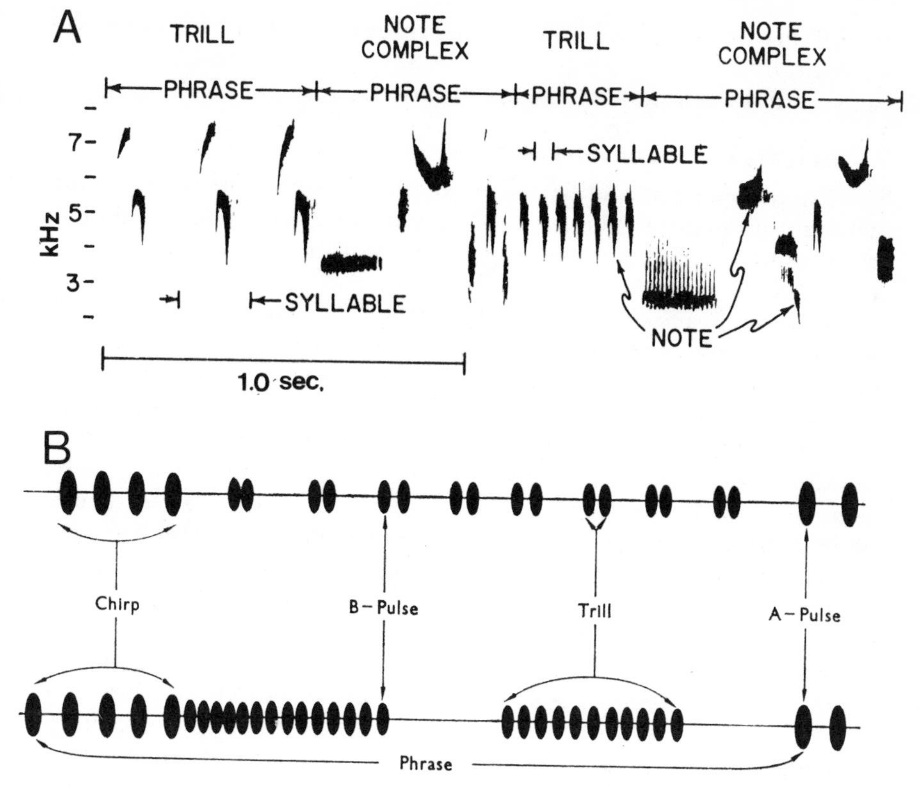

Figure 8-14 *A*, terminology used in the study of song development in the song sparrow; *B*, diagram of structural components and terminology of *Telegryllus* songs: upper line = *T. oceanicus*; lower line = *T. commodus*. Interchirp interval = interval between onset of A-pulses. Intratrill interval = interval between onset of B-pulses. Intertrill = interval between onset of last B-pulse in one trill and the first pulse of the next trill. (*A* from Kroodsma, D.E., 1977, A re-evaluation of song development in the song sparrow, *Anim. Behav.* 25(2):391, Fig. 1. *B* from Bentley, D.R. and R.R. Hoy, 1972, Genetic control of the neuronal network generating cricket *(Teleogryllus gryllus)* song patterns, *Anim. Behav.* 20(3):479, Fig. 1.)

One advantage of this system is the speed at which spectrograms can be produced. Hopkins et al. (1974) report that a 2.4-second-long spectrogram takes approximately 1.3 minutes to analyze on the Kay Sonagraph and only 9.6 seconds on the Ubiquitous. Another advantage is the relative ease with which section displays can be produced. Spectrograms produced by the Ubiquitous are grainier than those produced by the Sonagraph; however, this apparently does not affect interpretation (Hopkins et al. 1974).

Narins and Capranica (1977) described a method for electronically preprocessing tape-recorded sounds to enable direct computer analysis of the temporal characteristics of the sounds.

3. Analysis of Sound Spectrograms

Spectrograms are generally used to measure 1) frequencies (Hz), both dominant frequencies and harmonics; 2) durations of sounds and time intervals between them; and 3) relative intensities of portions of the sound.

Frequencies are measured from either a *narrow band filter* display on a *normal* spectrogram or from a *section* display. Transparent overlay grids are useful in the making of these measurements. Frequency measurement is more accurate when lower-frequency spreads are used for display (*e.g.*, 20–2,000 vs. 160–16,000 Hz). Contour displays are often useful to determine more accurately the dominant frequencies when large areas are burned. Horii (1974) described a method for producing digital sound spectrograms with simultaneous plotting of intensity and fundamental frequency. A digitizer system (*X*, *Y* cursor, Teletype, and computer) was used by Field (1976) to analyze sound spectrograms of wolf vocalizations. The cursor is moved along a selected frequency band (*e.g.*, dominant frequency). The *X*, *Y* axes of the cursor's plane of movement represent time and frequency, respectively. The operator depresses a button at predetermined points along the trace, and the *X*, *Y* coordinates are transmitted to the computer for storage and printed out on the teletype. Field (1976) used an overlaying grid to locate coordinate sample points every 0.05 second. The greater the variation in frequency of the sound, the more often the coordinates should be sampled. The coordinate sets can then be used to redraw the trace and compare for its accuracy in representing the original spectrogram.

Duration measurements are generally made from *wide band pass* spectrograms. However, the mark intensity can affect the measurements if they are either under- or over-burned.

H. PATTERN PLAYBACK

The Pattern Playback was devised to synthesize human speech for research on the recognition of consonants. It is essentially the opposite of a sound spectrograph. A sound spectrogram pattern is drawn on a piece of paper that is then run through the Pattern Playback, which converts the images drawn on the paper into sound played through a loudspeaker.

This device is used to segregate the relative importance of components of sound in transmitting information. This is generally measured by the effectiveness of the sound in stimulating a predictable behavior.

I. PHOTOGRAPHY

Ethologists should make a photographic record of their research. Pictures should at least be clear, well-composed still photos. Prints are necessary for publication, while color transparencies (slides) are very useful for oral presentations.

Your photos should depict 1) study site, 2) animals studied, 3) equipment and methodology, 4) results of data analysis (tables and figures), and 5) your interpretation of the results (*e.g.*, models). As photos are taken a log should be kept of photo number, date, time, location, subject matter, and why the photo was taken—what you were trying to depict and what in particular you should note when you see the transparency or print. In addition, in order to improve future photos, you may record environmental conditions, film type, shutter speed, lens opening, and filters. This log may be kept as part of your field notebook.

The techniques of good photography are beyond the scope of this book; complete and useful discussions can be found in Blaker (1976) and Anonymous (1970).

1. Still Photography

The most useful *still camera* for the ethologist is the 35-mm single lens reflex (SLR) camera (Fig. 8-15A). It is compact, light, and versatile. It accepts a large variety of film types (see next sections), although the most commonly used is color-slide film. Ideally, an ethologist entering the field in an unfamiliar area should be prepared with two types of cameras (normal lens and telephoto) and two types of film (color reversal and black-and-white negative). It can be argued which type of film should be used in each camera; however, I would use the color-slide film in the normal lens camera.

There are a large number of makes and models of 35-mm SLR on the market today, most of which have their own group of ardent supporters. If you can afford a Nikon or Leica, buy one; if not, look carefully at a Minolta, Pentax, Miranda, or Canon. The following features should be considered necessities in any camera you use or purchase:

1. Maximum shutter speed of at least 1/1000 second
2. Maximum lens opening of at least f1.9; that is, the f stops to go as low as f1.9
3. Through-the-lens light meter
4. Black camera body to reduce glare and reflections directed to the animal

Figure 8-15 *A*, above, Minolta single-lens reflex (SLR) camera with normal 55-mm lens. *B*, opposite, Bolex H-16 16-mm camera with 75-mm telephoto lens.

Another feature which is often handy, but not necessary, is a mount for an electronic flash on the camera. They all should have a socket for a synchronized electronic flash. The new automatic cameras may find widespread use by ethologists although their reliability under continuous field use remains to be demonstrated. Some cameras such as the Pentax ME allow you to select the lens opening you prefer, and the camera automatically selects the proper shutter speed for the available light. In contrast, with automatics like the Canon AE1 you select the shutter speed and it automatically adjusts the lens opening. The Minolta XD-11 allows you to select either the lens opening or shutter speed and it automatically sets the other; both can also be set manually. The Konica 35AF features an automatic focusing mechanism; but its maximum lens opening is only f2.8 and its maximum shutter speed is 1/250 second. When selecting one of these cameras for field use weigh the need for automatic features against the possibility of their failure.

An accessory which many ethologists find useful is a motor-drive unit which advances the film after each exposure. It allows you to take pictures as rapidly as five frames per second while maintaining continual observation of the animal(s); you can then concentrate on the

animal's behavior and photograph carefully selected behavior units, especially sequences, for later analysis or demonstration.

You should choose a camera that is ruggedly built and handles well. Cock the shutter, focus on an object, set the lens opening by adjusting the pointers in the through-the-lens light meter, and release the shutter. This whole sequence should come easily. On some cameras you have to turn the light meter on after each picture; if you are absent-minded, this may be a good feature for you, if not, you may

Table 8-5. Selected Kodak color-reversal film for use in 35-mm still cameras.

Film	Daylight Speed ASA	Definition			Type of Pictures and Degree of Enlargement Allowed	Suggested Uses
		Graininess	Resolving Power	Sharpness		
KODACHROME 25	25*	Extremely fine	High	High	Slides*** High	Has high color quality and a wide exposure latitude. It should be used under most daylight conditions when sufficient light is available and fast motion does not need to be stopped.
KODACHROME 64	64	Extremely fine	High	High	Slides*** High	Combines good definition with relatively high speed. It does not have the color quality of Kodachrome 25, so it should be used only when extra speed is necessary.
EKTACHROME 64	64	Extremely fine	Medium	Medium	Slides Moderate	Should be used as a substitute for Kodachrome-64 when you expect to do your own processing.
EKTACHROME 200	200**	Extremely fine	Medium	Medium	Slides Moderate	For use in dim light, shade, or to stop rapid movement; also with telephotos lacking large lens openings, in order to increase depth of field.

*This film may produce more desirable results at ASA 32 or 40.

**This ASA can be pushed to 400 with special processing.

***High quality prints can also be obtained through an additional process.

find it an inconvenience. Talk to several ethologists whose pictures you have admired and find out what cameras they recommend.

The most common photo format in ethology is the color slide. It can be projected for oral presentations or for use in analyzing postures (*e.g.*, facial expressions). With special processing color or black-and-white prints can be produced from color slides. The second most common photograph is the black-and-white print. If the purpose is publication, use film which is designed for black-and-white prints (Table 8-5), rather than having such prints made from color slides.

There are several characteristics which you should consider when selecting the proper film for your intended use (Table 8-6).

Film speed is represented by an ASA (American Standards Association) number. DIN is the German standard of sensitivity to light; it corresponds to, but is not the same as, the ASA number. These numbers are relative and comparable from one film to another. For example, a film with ASA 80 is twice as fast as one of ASA 40. The speed indicates the length of exposure time necessary for the film. Increased film speed 1) allows use of faster shutter speeds to stop motion; 2) allows for smaller lens opening, which gives greater depth of field (range of distances over which objects will be in focus); and 3) generally produces a grainier photograph (see next section).

You may want to modify the film speed rating recommended by the manufacturer for the results obtained with your camera. If your photos are consistently overexposed or underexposed you may want to adjust the film speed as follows:

$$\text{Overexposed: recommended film speed} \times 2 = \text{1 f stop less exposure}$$

$$\text{Underexposed: } \frac{\text{recommended film speed}}{2} = \text{1 f stop more exposure}$$

The ASA film speed rating for Tri-X (Table 8-6) can vary from 160 to 800 depending on the particular camera used. Many people shoot Kodachrome 25 at ASA 32 or 40 rather than the recommended 25. If you expect to use a camera much (and you should), experiment with different ASA ratings, photographing the same scene under the same conditions with the same camera.

Definition is the clarity of detail in the picture, *i.e.*, how closely it represents the actual scene as viewed by the human eye (aside from color rendition). Definition is the result of several factors, the three most important for our purposes being graininess, resolving power, and sharpness.

Table 8-6. Selected Kodak black-and-white film for use in 35-mm still cameras.

Film	Speed ASA	Definition			Degree of Enlargement Allowed	Suggested Uses
		Graininess	Resolving Power	Sharpness		
PANATOMIC-X	32	Extremely fine	Extremely high	Very high	Very high	For prints. With a special reversing process it will produce slides. It should be used when the emphasis is on very-high-quality prints for publication or enlargement
PLUS-X Pan	125	Very fine	High	Very high	High	For prints. A good all-around film that combines reasonable speed with high definition qualities.
TRI-X Pan	400	Fine	Medium	Very high	Moderate	For prints. Its major quality is high speed which can be pushed to ASA 800 in some cameras. It can be used in very low light (*e.g.*, forest) or to stop motion (*e.g.*, running antelope).
2475 Recording (ESTAR-AH Base)	1000	Coarse	Low	Very high	Low	For prints. This is a poor quality film which has only its speed to recommend it. It should be used only when very low light or high speed call for it.
High Contrast Copy Film 5069	64	Fine	High	Very high	Very high	For copying printed materials (*e.g.*, photos, charts, tables, drawings, etc.). Useful in preparing visual aids for presentations and field trips.

Graininess refers to the clumped appearance of photographs when viewed closely. A photo that is very coarse-grained will appear like a series of dots rather than a continual gradation of colors and shades. Graininess is particularly important when you consider photo enlargement, since it magnifies the grainy appearance. Generally, graininess increases with increased film speed and with overexposure.

Resolving power is the ability of a film to reproduce fine detail, such as two lines very close together. This is generally not of great importance in most ethological work, but it should be considered in photos of equipment, charts, etc.

Sharpness refers to the definition of edge in a photograph. For example, the definition of the side of an elk's antler against a background. If sharpness is poor the antler will appear "fuzzy." Sharpness is also affected by lens quality.

Exposure latitude is the amount of overexposure or underexposure which the film will take and still produce acceptable photographs. These exposure errors arise from mistakes in setting f stops (lens openings) or shutter speeds. Continuous-tone negative films have greater exposure latitude than slide, or reversal, films. However, Kodachrome 25 has a relatively wide exposure latitude for a slide film.

Color sensitivity is the range of wavelengths of light to which the film is sensitive. All the films listed in Tables 8-5 and 8-6 are *panchromatic*; they are sensitive to all visible colors as well as ultraviolet radiation.

In addition, *infrared films* are available for special uses. Kodak High Speed Infrared film is available in 20-exposure rolls for 35-mm cameras. It is moderately coarse-grained and has medium resolving power and low sharpness. It can be used to photograph through haze or to record behavior of nocturnal animals lighted by infrared bulbs. The speed of the film is highly variable, depending on the ratio of visible to infrared light available.

Storage is an important consideration for all types of film. All films are damaged by high temperature and high humidity. Films can be obtained in vapor-tight packaging if you anticipate working in areas of high humidity. It is best to keep film refrigerated until it is used. Kodak provides the following storage recommendations for black and white film:

For storage periods of up to:	2	6	12	months
Keep film below:	75°F	60°F	50°F	

Keep film away from industrial gases, motor exhaust, and vapors of mothballs, formaldehyde, solvents, cleaners, and mildew or fungus preventatives. Static electricity caused by rewinding film too rapidly in the camera will cause streaks, dots, or fogging. Also, X-ray inspection units in some airports may still ruin film; check it through by hand or protect it in special containers available at camera stores.

Prints and slides should be stored where they will be safe from damage, but where they can be easily retrieved. A cataloging system may be based on 1) separate research projects, 2) behavior types, 3) species, or 4) field season. Each ethologist must develop the system which he finds most useful. In addition, attempt to reduce possible losses in the mails by sending photos in separate packages; if possible, send duplicates.

2. Motion-Picture Photography

The obvious advantage of both motion pictures and videotape is that they allow you to record a two-dimensional visual representation of entire behavior patterns. The two-dimensional restriction can be overcome, in part, by the use of two or more strategically located cameras. In addition, it provides the capacity to record sound (produced by the animals, environs, or dictated by the observer) synchronously.

Hutt and Hutt (1974) list five situations in which motion pictures and videotape are especially useful: 1) swift action, 2) complex action, 3) subtle behavioral changes, 4) complex behavioral sequences, and 5) need for precise measurements of parameters.

The first choice you have to make is the film-size format you want to work in. The two basic choices are 16 mm and Super 8 mm; standard 8-mm film is rapidly disappearing. The relative advantages of each are listed in Table 8-7; however, it is basically a choice between cost (Super 8 mm) and quality (16 mm). If you don't intend to

Table 8-7. Relative advantages of Super 8-mm and 16-mm filming.

Super 8 mm	16 mm
1. Cheaper cameras, film and processing	1. Pictures with greater sharpness, resolution, and definition
2. Lighter equipment	2. Pictures brighter when projected to same size
3. Convenience of cartridge film	3. Cameras often more durable
	4. Film often easier to handle for editing and analysis
	5. Larger film capacity
	6. Better for sync sound

do much filming, borrow or rent a Super 8-mm camera. If you intend to make filming an integral part of your studies and can afford it, use 16 mm (Fig. 8-15).

In selecting a camera you will be confronted with a trade-off between cost and certain features (*e.g.*, lenses, built-in exposure meter, filming speeds, durability, etc.). Some of these features are discussed below. Remember to purchase what you need, but not more than you need. Also, if possible, try before you buy.

Lenses should be selected with an eye toward the uses you intend for your equipment. If the camera has a lens turret, then you might select three lenses, such as 10 mm (wide-angle), 26 mm (standard), and 75 mm (telephoto). It may be necessary to use a telephoto lens as large as 600 mm (Dane and Van der Kloot 1964) or 1000 mm. If the camera will handle only one lens, then a zoom (26 mm to 75 or 100 mm) is very useful. If you are working with insects, a close-up lens and extension tubes are often desirable. Select high-quality lenses with large apertures approaching f/1.1.

The diversity of *films* available for movie-making is so great as to confuse the neophyte. Selection of the proper film is generally a trade off between film speed (amount of light necessary for proper exposure) and picture quality. Black-and-white film is cheaper to purchase but more expensive to process, while color film provides an additional dimension which is not only esthetically pleasing but often necessary in some ethological studies. Table 8-8 provides a list of Kodak films that are most useful for filming animal behavior. Additional, more specialized films can be found in Eastman Kodak's publication R-31, Kodak Photographic Materials Guide, as well as from other manufacturers. For example, infrared film can be used in conjunction with infrared lighting (*e.g.*, photofloods and infrared filters) to obtain motion pictures under nocturnal conditions (Delgado and Delgado 1964).

The *filming speed* you choose will depend on the purpose of the filming. Normal projection speeds for Super 8 mm and 16 mm are 18 and 24 frames/second, respectively. The effect of accelerated motion is produced by filming at slower speeds (*e.g.*, 2–10 frames/second), and slow motion is produced with greater filming speeds (*e.g.*, 32–64 frames/second). If you are interested in frame-by-frame analysis (see below) then the faster you film, the smaller the change in the animal's position from frame to frame. Faster filming speeds also allow for unsteadiness by the cameraman; but it means changing film more often and increased costs in purchasing and processing the larger amount of film.

Various filming speeds and the authors' rationale for their use can be found in the literature. For example, in terms of frames per second,

Table 8-8. Selected Kodak Reversal Motion-Picture Films.

Film	Color (C) or Black and white (BW)	Daylight Speed ASA	16 mm/ Super 8	Characteristics	Suggested Uses
PLUS-X	BW	50	16/8	High degree of sharpness, good contrast, and excellent tonal gradation	General outdoor photography
TRI-X	BW	200	16/8	Excellent tonal gradation	Under adverse lighting conditions
4-X	BW	400	16		Adverse lighting and high-speed photography
KODACHROME 25	C	25	16/8	Excellent color rendition	General outdoor photography
KODACHROME 40	C	25	8	Good color rendition	With tungsten lighting
EKTACHROME 160	C	160	8		Adverse lighting
EKTACHROME MS	C	64	16	Excellent color rendition and sharp images	General outdoor photography
EKTACHROME EF (tungsten)	C	80	16/8		Adverse lighting and high-speed photography; with tungsten lighting
EKTACHROME EF (daylight)	C	160	16		Adverse lighting and high-speed photography

2–7 or 48 (Eibl-Eibesfeldt 1972), 16 (Clayton 1976), 22 (Diakow 1975), 24 (Kruijt 1964; Dane and Van der Kloot 1964), 32 (Duncan and Wood-Gush 1972), 64 (Bekoff 1977*a*), and 128 (Hildebrand 1965) have all been used.

Both Super 8-mm and 16-mm films and cameras are available for simultaneous recording of a *sound track*. The sound reproduction is generally not of high quality, but can be useful for recording the observer's commentary during filming. Good-quality sound recordings are best made with 16-mm cameras (*e.g.*, Bolex H-16, Fig. 8-15) that will synchronize with a high-quality tape recorder, such as the Nagra IV-L.

3. Film Analysis

Ethologists take motion pictures for basically two purposes: 1) to have a visual record of the behavior for illustrative purposes (presentations and publications, *e.g.*, J.M. Davis 1975); and/or 2) for analysis of a) individual movements (Hailman 1967), including locomotion (Hildebrand 1965; see also p. 204) and social displays (Barlow 1977; Bekoff 1977*a,b*); b) intra-individual sequences (Tinbergen 1960);

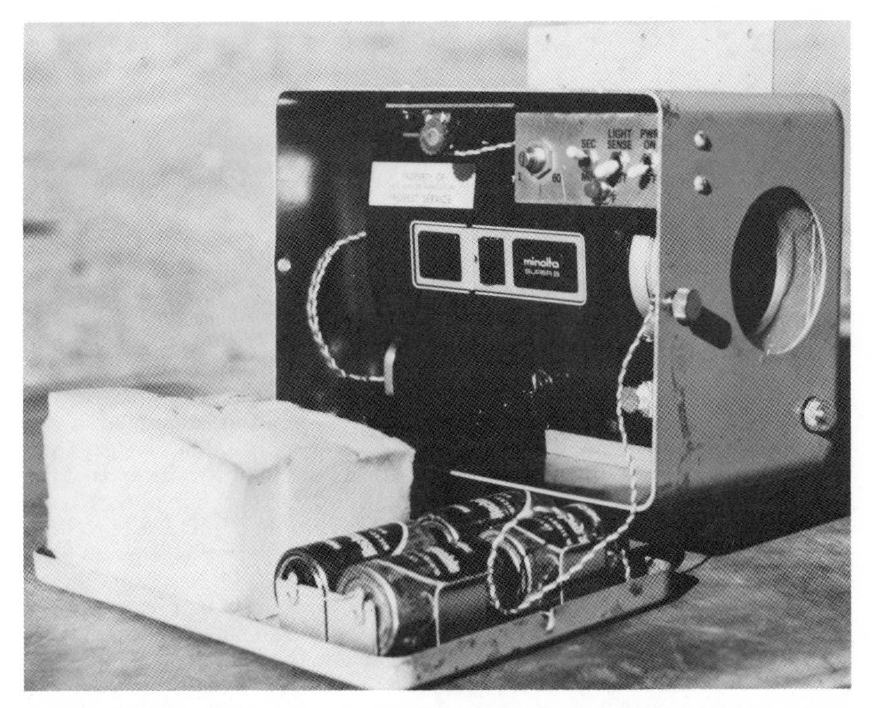

Figure 8-16 An 8-mm sequence camera and intervalometer inside a weatherproof housing.

Balgooyen 1976); c) interindividual sequences (Diakow 1975); d) spatial relationships (Dane and Van der Kloot 1964).

Analysis of film is conducted either frame by frame or by sampling frames at regular intervals, *e.g.*, every 24th frame (Golani 1973). If frames are to be selected at intervals for analysis, an intervalometer can be coupled with the camera to expose frames at set intervals (Fig. 8-16). This provides a more efficient use of film. Analyses are

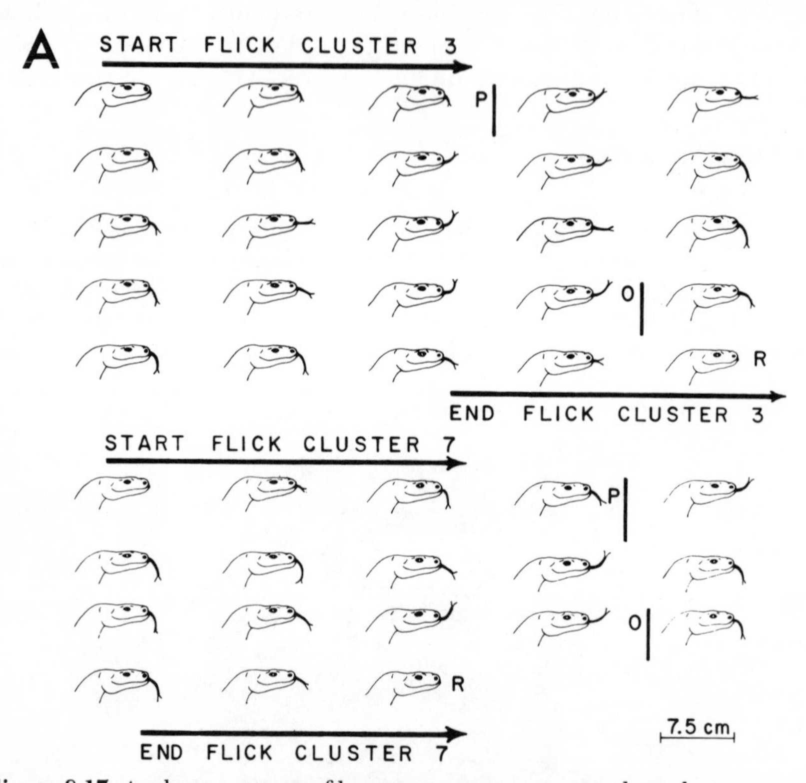

Figure 8-17 *A*, above, pattern of boa tongue movements in lateral view. Tracings of each frame in motion pictures of two complete flick clusters are illustrated. Successive pictures are about 42 ms apart in time. The ends of the protrusion phase (*P*), the oscillation phase (*O*), and the retraction phase (*R*) are indicated by vertical lines. The figures should be studied from left to right in each line. (From Ulinski, P.S., 1972, Tongue movements in the common boa *(Constrictor constrictor)*, *Anim. Behav.* 20(2):375, Fig. 1.) *B*, opposite, the movements of the foot of the bivalve mollusc, *Cardium echinatum,* during a single leap, shown with reference to the shell as a fixed object. The positions shown are taken from a cinefilm of the movement, the figures indicating the number of the frame corresponding to each position (16 frames per second). Active (frames 12 to 23) and recovery movements (frames 23 to 50) are shown separately. (From Ansell, A.D., 1967, Leaping and other movements in some cardiid bivalves, *Anim. Behav.* 15(4):422, Fig. 1.)

generally conducted with either film editors that have a built-in projection screen (Hutt and Hutt 1974) or an analyzer-projector (*e.g.*, Lafayette Analyzer; Lafayette Instrument Co., Lafayette, Indiana). The latter projects the film onto a large screen (Dane and Van der Kloot 1964). Whichever system is used, it should have a reverse and a frame counter. The frame counter, coupled with the filming speed provides a time base for measuring the latencies, durations, and inter-act periods of behaviors.

Frame-by-frame analysis has been used to measure the movements of the tongue of boas *(Constrictor constrictor)* and the foot of a mollusc *(Cardium echinatum)* by Ulinski (1972) and Ansell (1967), respectively. Illustrations of the results of their analyses are shown in Figure 8-17. Head movements relative to particular be-

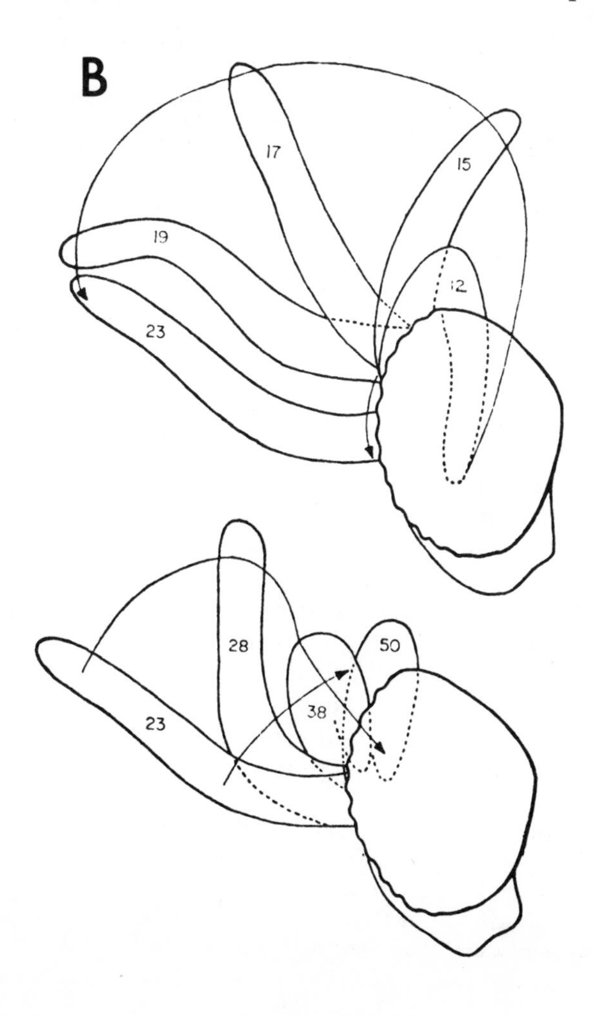

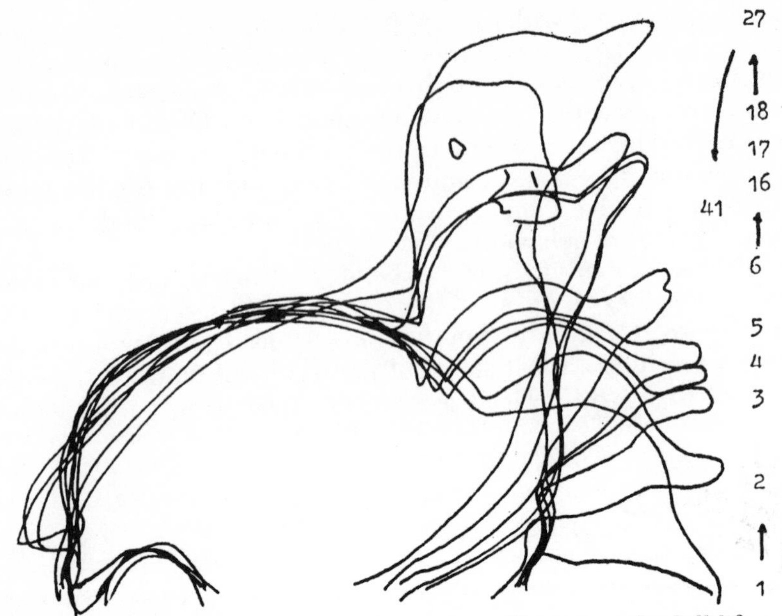

Figure 8-18 The duckling's drinking response illustrating the bill-lift element. This composite line drawing is based on stills taken from a 16 foot-per-second film. The sequence of numbers corresponds to the frame numbers beginning as the bill leaves the water. (From Clayton, 1976, The effects of pre-test conditions on social facilitation of drinking in ducks, *Anim. Behav.*, 24(1):127, Fig. 2.)

haviors have been analyzed frame by frame for the Burmese red jungle fowl *(Gallus gallus spadiceous)* (Kruijt 1964), laughing gull chick *(Larus atricilla)* (Hailman 1967), and domestic duck *(Anas platyrhynchos)* (Clayton 1976) (See Figure 8-18).

Spatial relationships between courting goldeneyes were measured by Dane and Van der Kloot (1964) by projecting film frame by frame onto a screen that they had divided with 16 equally spaced vertical lines. Distances perpendicular to the camera's line of sight are relatively easy to measure; but the perspective of depth is lost in measurements parallel to the line of sight. Dane and Van der Kloot list other complications and restrictions which are common to similar types of film analysis:

(1) Birds are often passing in and out of the field of view of the camera. When the final analysis is undertaken, there is always the chance that an action given by a bird outside of the field is affecting those recorded on film. This problem was minimized by analyzing discrete groups. (2) Computing the distance between birds, and thus the relative position of each individual in the

flock, is difficult when using a telephoto lens. (3) When testing for a relationship between the actions of two birds, there is always the possibility that one is not distinguishing the pair which is actively interacting. If this were the case, correlations which actually exist might be overlooked. (4) Finally, though unlikely, a movement which was too subtle to be detected on the film, might be a stimulus for another individual. [From DANE and VAN DER KLOOT 1964:285]

A computer system for frame-by-frame analysis of film, FIDAC (Ledley 1965), has been described by Watt (1966). The system consists of a cathode-ray-tube generator which projects an ordered array of rows and columns of spots of light through the film frame, where the intensity of the light transmitted is measured by a photocell as one of seven different levels of gray. This information is then transmitted to a digital computer. The computer can be programmed to control the location of the array of spots of light, their density in the array, and the area covered. The system has both high speed and high resolution. This system, or a similar one, may find useful application in ethological studies of movement where the animal is filmed against a light background.

Table 8-9. Relative advantages and disadvantages of videotape and movie film for ethological studies.

Videotape	Movie Film
Advantages	
1. Immediate playback	1. Better quality
2. Reusable	2. Easily analyzable frame-by-frame, providing an accurate time base for studies of movements
3. Tape relatively inexpensive	3. With wind-up cameras, time in the field limited only by the amount of film
4. Easily duplicated	4. Equipment generally light
Disadvantages	
1. Poorer-quality picture with less expensive video recorders	1. Time delay for developing
2. Most not analyzable frame by frame,* but stop action possible	2. Film usable only once
3. Equipment run off batteries with limited chargeable life (20 min.—3 hours)	3. Film relatively expensive
4. Equipment generally heavy	4. Duplicating more expensive

*Panasonic has developed a time-lapse videotape recorder (NV-8030) that will record single fields (pictures) at rates of from 1.8 to 60 per second. Single fields can then be played back individually for analysis in the same way that movie film can be analyzed frame by frame.

In summary, I have not touched upon the vast array of additional equipment (*e.g.*, light meters, filters, tripods, gunstocks, etc.) that may be necessary for proper filming. These items should be discussed with your local camera dealer. Likewise, the various techniques which will improve your motion pictures and their analysis can best be gathered through discussions, experience and literature (Dewsbury 1975; Matzkin 1975; Wildi 1973).

J. VIDEOTAPE RECORDING

Videotape has several advantages and disadvantages relative to movie film (Table 8-9).

Figure 8-19 A Panasonic NV-3085 portable video recorder (*a*), camera (*b*), battery charger (*c*), and carrying case (*d*).

If permanent documentation is desired or if an accurate time base is necessary, movie film is suggested. However, if you want to record behavior to be reviewed again and again soon after it occurs, then videotape using a system such as shown in Figure 8-19 is recommended.

Behavior is usually thought of as an animal doing something. Only movie film or videotape will accurately capture that activity. High-speed motor-driven slide cameras are often an acceptable alternative.

Videotape (or film) can be used to simply gain experience with an animal. By viewing the same footage several times you learn to anticipate behaviors; you see subtleties in behavior which you often miss in a single observation.

Video cameras can be made sensitive to *infrared* by replacing the normal vidicon tube with an infrared tube. Wells and Lehner (1978) flooded a large room with infrared light and used an infrared-adapted video camera to study the predatory behavior of coyotes in the dark. Conner and Masters (1978) described a video system for viewing in the near infrared (700 to 1000 mm), which was used to observe the nocturnal courtship of an arctiid moth and nocturnal predatory behavior of the Florida mouse *(Peromyscus floridanus)*.

K. STOPWATCHES

Stopwatches are a time-honored piece of equipment in ethological studies. They are used primarily to measure durations and latencies. The new *electronic digital stopwatches* (Fig. 8-20) are rapidly displacing the old *mechanical stopwatches* (Carpenter and Grubitz 1961) in many ethologists' pockets.

All the electronic digital stopwatches are comparable to good mechanical stopwatches in price, but they are larger. Some researchers are prone to question the dependability of electronic devices (even though their accuracy is greater) and do not like to tie themselves to batteries.

Electronic digital stopwatches are easier to read, and many have several functions that are useful to the ethologist. For example, the Heathkit stopwatch (Fig. 8-20*C*) provides the five functions listed in Table 8-10 plus two programmable functions which might prove useful for laboratory work.

Kits for digital stopwatches can be purchased from several manufacturers (*e.g.*, James Electronics, San Carlos, California). Wolach et al. (1975) described an economical method for converting an elec-

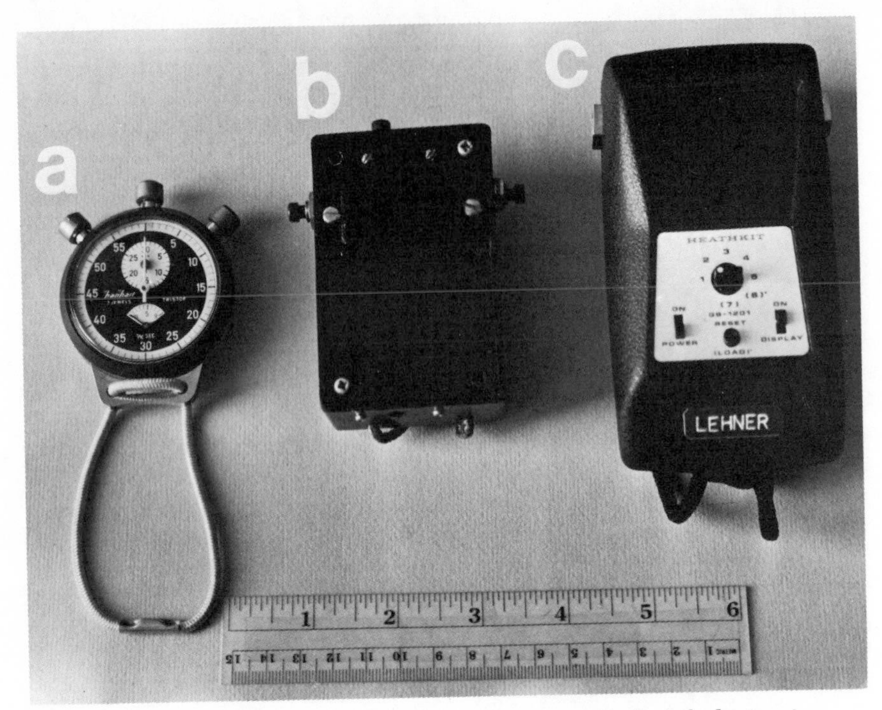

Figure 8-20 *a*, mechanical stopwatch; *b*, home-made digital electronic stopwatch; *c*, Heathkit programmable digital electronic stopwatch.

tronic handheld calculator into a digital stopwatch with increments in 0.10 second intervals.

L. METRONOMES

Metronomes provide a time base for field observations. They allow the observer to enter a time point (*e.g.*, every ten seconds put a slash) in his notes, time instantaneous samples (p. 122), and provide an electronic beep at intervals (*e.g.*, 1 second) which can then be counted in order to determine durations of behaviors.

An electronic metronome constructed by Jim Starkey and used in our studies can be set to beep at 1- or 10-second intervals through a small earphone. It is both small and light, which makes it easily adaptable for field work. Lockard (1976) described a metronome which has a pulse rate continuously adjustable from 0.5 to 20 seconds and includes a light-emitting diode providing a visual signal at the set intervals (Fig. 8-21). Wiens et al. (1970) designed an electronic metronome which emits tone pulses through a small earphone at intervals which

Table 8-10. Functions provided by Heathkit Model GB-1201E digital stopwatch.

Function	Description	Illustration
1	Duration of separate behaviors plus total duration of session	A B C; 0; total duration + 0 through C
2	Time from one event to another; latencies; plus total time	→ A → B → C; 0; + total time from 0 to C
3	Accumulated time for several occurrences of a particular behavior; plus total time of session	A_1 A_2 A_3; 0; A_1+A_2; $A_1+A_2+A_3$; total time + 0 through A_3
4	Latencies for events from a single starting point; plus total latencies	A B C; 0; total latencies + $A + B + C$
5	Duration of separate occurrences of a behavior; plus total duration of all occurrences	A_1 A_2 A_3; 0; + total duration $A_1 + A_2 + A_3$

can be varied from 1 to 20 seconds. Their metronome was used by Dwyer (1975) in his study of time budgets in gadwall ducks *(Anas strepera).* Reynierse and Toeus (1973) describe a metronome which produces pulse rates of from 5 per second to 1 per 30 seconds and can be built using the circuit diagram they provide. Figure 8-22 includes two schematics for constructing your own electronic metronome.

M. BIOTELEMETRY

Radio transmitters have been used for marking and locating a wide variety of species over the last two decades. Many of these species have been listed by Brander and Cochran (1969). Much has happened in biotelemetry since Slater (1965) reported on studies that employ biotelemetry. For the latest compendiums of biotelemetry studies see Long (1977) and Fryer et al. (1976).

Biotelemetry has been used for one or more of the following three purposes:

1. Location of an animal for plotting its movements and generally for calculating its home range (p. 297).

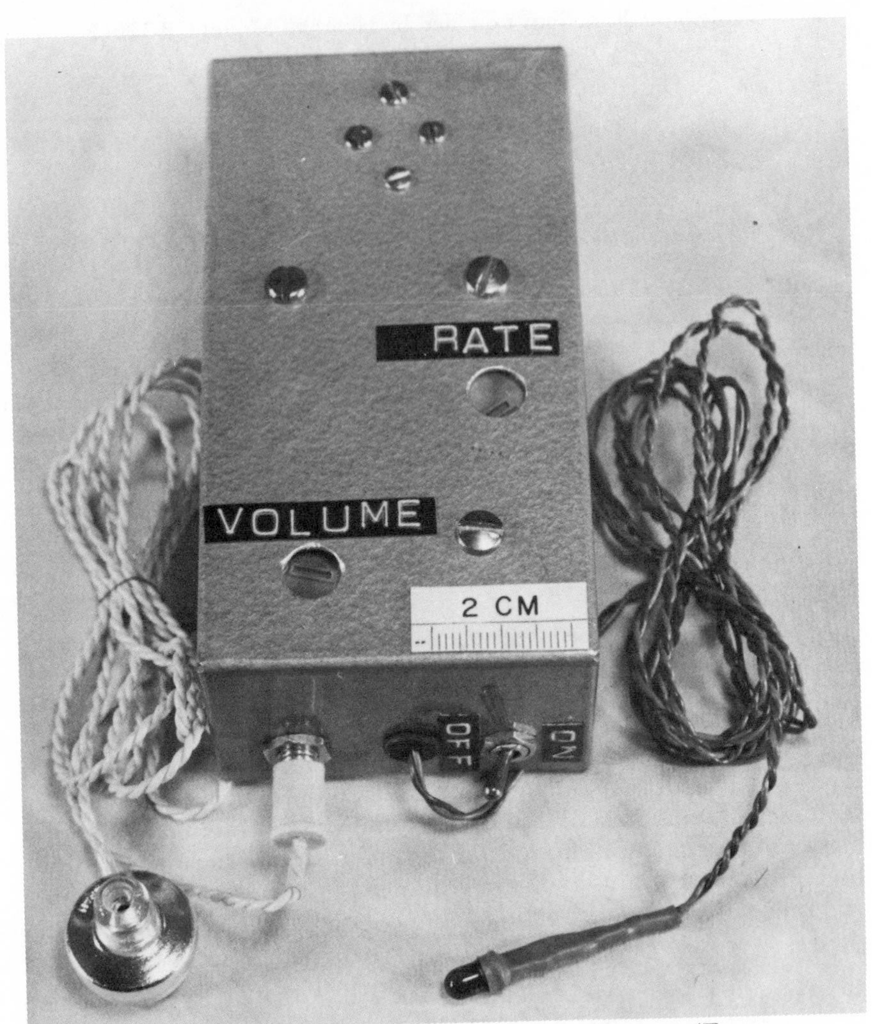

Figure 8-21 Photograph of Lockard's electronic metronome. (From Lockard, J.S., 1976, Small interval timer for observational studies, *Behav. Res. Meth. and Instrum.* 8(5):478, Fig. 1.)

Figure 8-22 Two schematics for constructing an electronic metronome. (*A* from J. Starkey, personal communication; *B* from Lockard, J.S., 1976, *op. cit.*, 478, Fig. 2.)

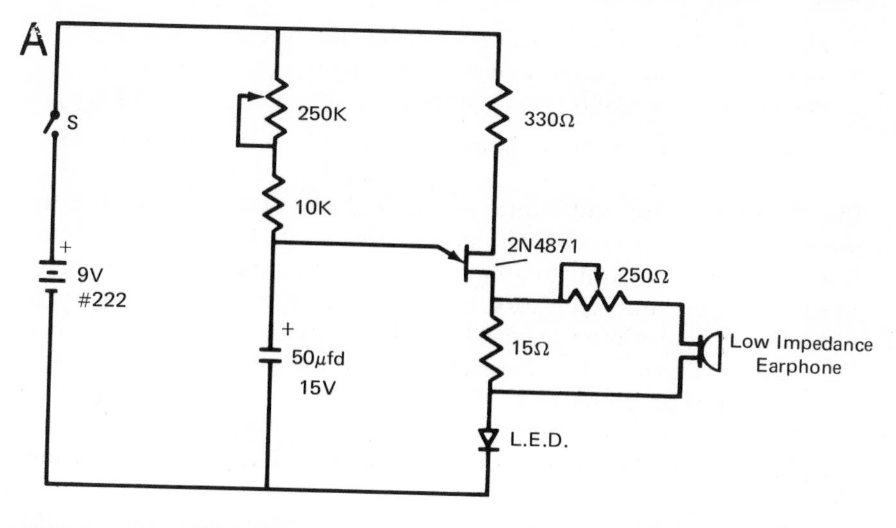

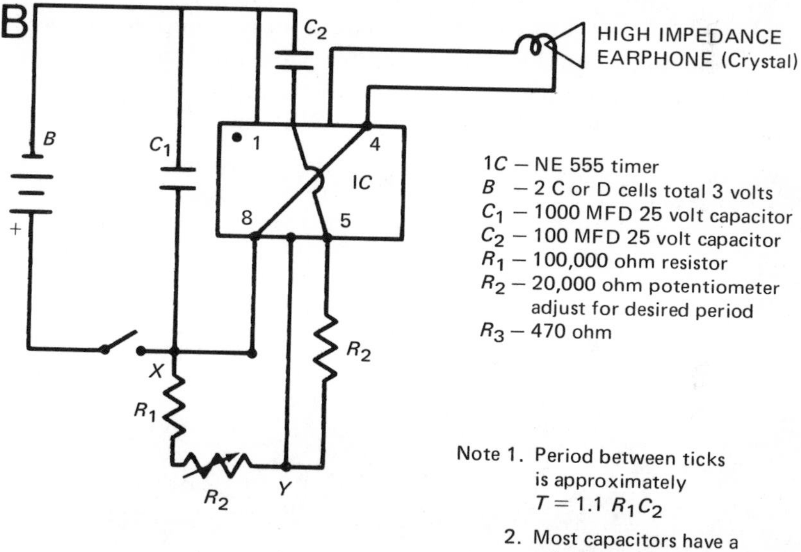

$1C$ – NE 555 timer
B – 2 C or D cells total 3 volts
C_1 – 1000 MFD 25 volt capacitor
C_2 – 100 MFD 25 volt capacitor
R_1 – 100,000 ohm resistor
R_2 – 20,000 ohm potentiometer
adjust for desired period
R_3 – 470 ohm

Note 1. Period between ticks is approximately $T = 1.1\ R_1C_2$

2. Most capacitors have a tolerance of +20 to –50% usually requiring higher values of $R_1 + R_2$ than calculated

Alternate switchable range selection substitute for $R_1 + R_2$

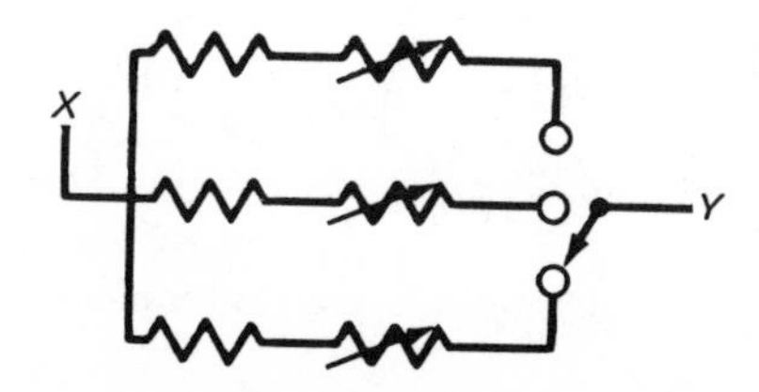

Select resistor and potentiometer for desired range, that is, 8000 ohm resistor and 5000 ohm pot for a period of 1 second

2. To assist in location of an animal for direct observation.
3. To record physiological data such as EEG, EKG, respiratory rate, and internal or surface body temperature.

Typical biotelemetry field equipment is shown in Figure 8-23.

Location of an animal is accomplished by setting the receiver to the channel for the individual animal's signal and then rotating the antennae until the maximum signal is received. Simultaneous directional information from two or more receiving stations can then be used to triangulate the location of the animal (Fig. 8-24) within a margin of error determined primarily by 1) the distance between receiving stations, 2) strength of the signal, 3) topography, and 4) vegetative characteristics of the habitat.

A change in characteristics of the signal received can often be correlated with the behavior of the animal (*e.g.*, walking, wing-flapping).

Radio tracking systems have been made completely automatic by transferring time and directional information into a computer where it can be stored and plotted by an X-Y plotter (Cochran et al. 1965).

Useful bibliographies on wildlife biotelemetry have been prepared by Schladweiler and Ball (1968) and Will and Patric (1972).

Figure 8-23 Radio-collared coyote, receiver, and handheld yagi antennae. (Photo by A. Olsen.)

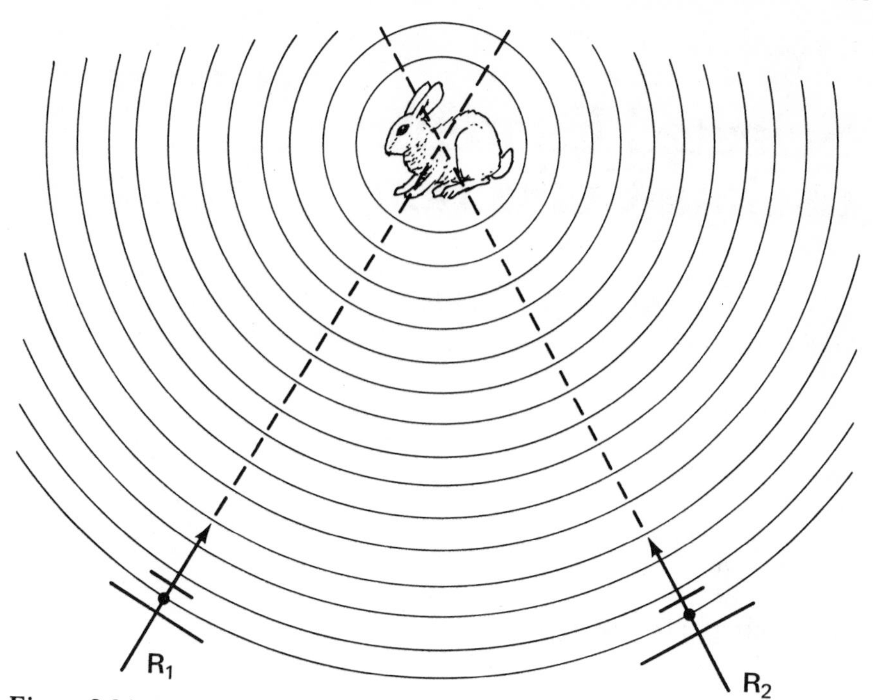

Figure 8-24 Diagram of location of animal at *A*, drawn by using directional information received from receiving stations R_1 and R_2.

9
Selected Examples of Data Collection

A. INDIVIDUAL BEHAVIOR

It is generally preferable to study individual behavior before examining interactions between two or more individuals. Knowledge of individual behavior, experience in data collection, and discipline in concentrating on particular units of behavior can then be applied to the study of social behavior.

1. Terrestrial Tetrapod Locomotion

Studies of terrestrial tetrapod locomotion begin with descriptions of the spatial and temporal relationships of limbs. The position of the limbs is sampled at high rates of speed, generally with the use of motion pictures. Videotape generally does not provide the frame-by-frame analysis possible with movie film.

Bullock (1974) studied locomotion in the pronghorn antelope. He took movies with a Pathe Professional Reflex 16-mm movie camera mounted on a modified rifle stock, using black-and-white Kodak Plus-X Reversal film exposed at 80 frames per second. He studied the film frame by frame with a Zeiss-Ikon Moviscop Viewer fitted with a 2× magnifying lens. To facilitate more in-depth study, Bullock also projected the film onto a solid screen with a 35-mm film-strip projector; he then traced the sequences in silhouette form on the screen. These analyses generated footfall formulas and gait diagrams for the pronghorn's various gaits. These descriptions derive from determining when each foot is on and off the ground. Bullock was able to do this at

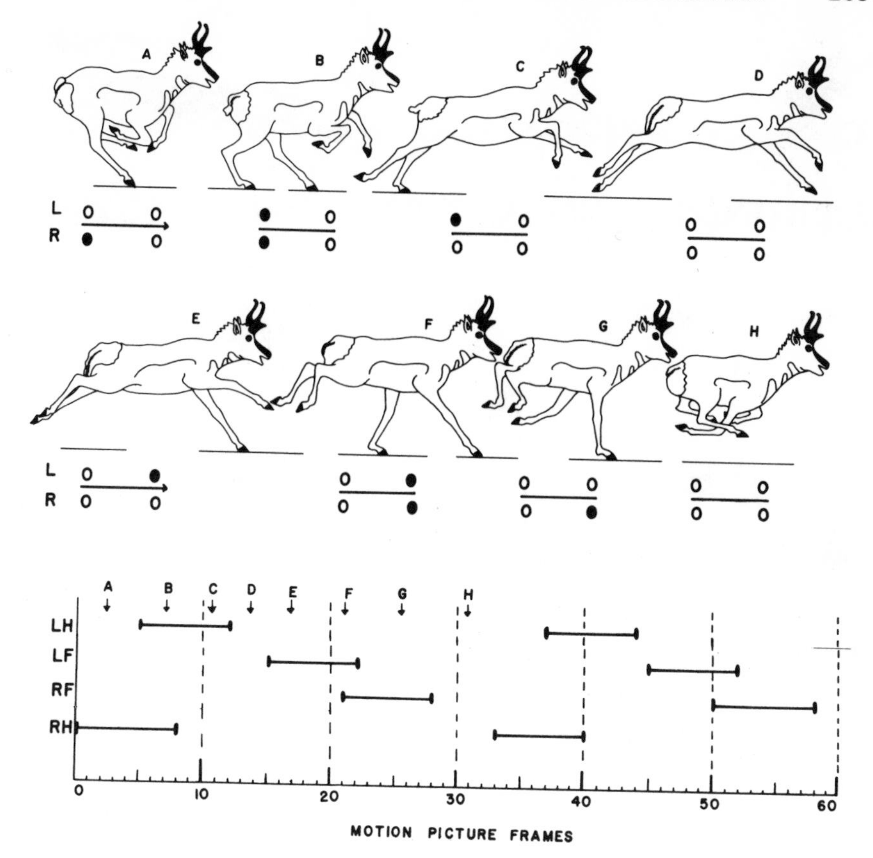

Figure 9-1 Footfall formula (above) and gait diagram of a pronghorn employing a lateral (rotary) gallop. (From Bullock, R.E., 1974, Functional analysis of locomotion in pronghorn antelope, p. 285, Fig. 6 *in* Geist, V., and F. Walther, (eds.), The behaviour of ringulates and its relation to management, IUCN Publ. 24, Vol. 1.)

1/80th-second intervals. For example, Figure 9-1 illustrates the footfall formula and gait diagram for the lateral gallop.

Since the movie was taken at 80 frames per second, only 0.75 sec. is depicted in the gait diagram. From the information in Figure 9-1 it can be shown that this gait is both rapid and asymmetrical (see Hildebrand 1977; Müller-Schwarze 1968). Bullock's (1794) analysis also included 1) support intervals of fast gaits, 2) leads 3) turning, 4) change of gait, 5) synchronization of gait and lead, and 6) speed.

Figure 9-2 Choking: *a*, herring gull; *b*, common gull; *c*, and *d*, black-headed gull; *e*, kittiwake (*b* after Weidmann 1955). (From Tinbergen 1959.)

B. SOCIAL BEHAVIOR

1. Displays

A display is "a behavior pattern that has been modified in the course of evolution to convey information" (Wilson 1975:528; Beer 1977). Displays are often dramatic, eye-catching behaviors and have attracted the attention of ethologists since the inception of the discipline. The classical studies of the comparative behavior of the Anatidae (O. Heinroth 1911; Lorenz 1941) were based primarily on displays.

a. Description of Displays

The first step in the study of a behavior pattern is description. The components of the display must be described clearly and completely, without bias as to interpretation (Chap. 3, Sec. B). Many hours of observation are generally necessary before you will feel comfortable with your description.

> Only by watching, writing down, drawing, realizing how much you are not certain about, watching again, and thus completing your description step by step, can you attain a reasonable accuracy and completeness. [TINBERGEN 1953:131]

Tinbergen's study of the comparative behavior of gulls *(Laridae)* has long served as a model of careful description of displays. As an example the following is his description of the choking display.

CHOKING [Fig. 9-2]. In this posture, the bird squats and bends forward. The tongue bone is usually lowered, the neck is held in an S-bend, and the bill is pointing down. In this position the head makes rapid downward movements, usually however without touching the ground. The carpal joints are often

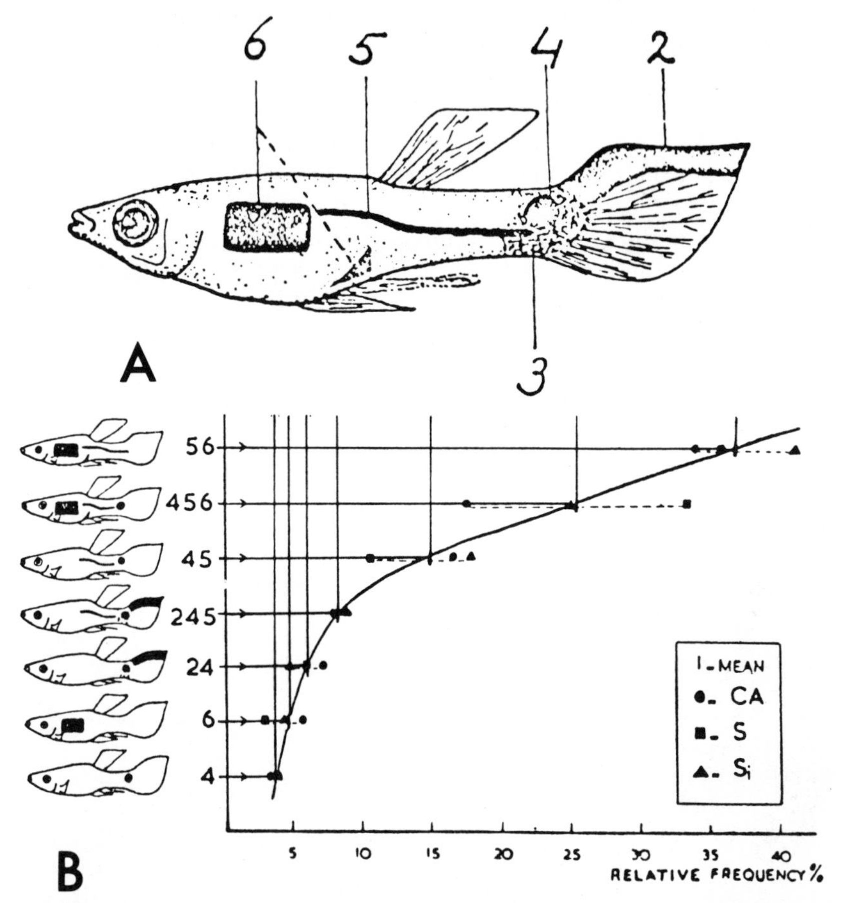

Figure 9-3 *A*, the black markings that may develop during courtship in male guppies. Number 1 (not shown) referred to the overall darkness of the entire body. *B*, relative frequency of occurrence of marking patterns associated with copulatory attempts (CA-○) sigmoid postures (*S*-■), and sigmoid intention movements (S_i-▲). (From Baerends, G.P., et al., 1955, Ethological studies on *Lebistes reticulatus* (Peters): An analysis of the male courtship pattern, *Behaviour* 8:275, 307, Figs. 10, 24.)

raised, and the wings may even be raised and spread, and kept stationary for seconds. A muffled, rhythmical sound is given which may or may not be in time with the pecking movements. The breast is "heaving" strongly, particularly in large gulls. Often the lateral ventral feathers are raised. The bird may be facing another bird, or face away from it, or take up an intermediate orientation. [TINBERGEN 1959:16-17]

b. Quantifying Displays

Baerends et al. (1955) studied the courtship displays of the guppy *(Lebistes reticulatus)* by observing them in large aquariums. Besides describing and quantifying various postures and movements, they also studied the occurrence of the black markings on the males. For ease in description and recording data, they assigned a number to each of the markings (Fig. 9-3A) and then measured the frequency of occurrence of the various patterns in different behavioral contexts (Fig. 9-3*B*).

Bekoff (1977*a*) used movement along a single coordinate to measure the variability in the duration and form of the play bow in three canid species and one hybrid. Movies (Super 8 and 16 mm) were made of individuals at 64 frames/second and then analyzed frame by frame. Duration was determined according to the number of frames during which the bow was maintained. Form was measured as a declination of the shoulders relative to standing height on a grid system (Fig. 9-4*A*). Similar techniques were used by Hausfater (1977) to study tail carriage in baboons (Fig. 9-4*B*).

In Bekoff's (1977*a*) study, each individual's shoulder height was divided into 10 equal segments to normalize individual differences. Each of the 10 segments was then divided into fourths to increase the resolution of measurement.

Tail position as a component of pelecaniform displays was measured by Van Tets (1965). He divided the vertical component into nine 30-degree sectors and measured the frequency of occurrence of the tail elevations during different displays (Fig. 9-5*A*, *B*).

Golani (1976) used several limb and body axes (Fig. 9-6) in applying the Eshkol-Wachmann (E-W) movement notation to a description of the motor sequences in the interactions of golden jackals *(Canis aureus)* and Tasmanian devils *(Sarcophilus harrisii)*. The E-W notation system uses a coordinate system (Fig. 9-7) from which to describe a movement of any part of the body. The coordinate system can be applied relative to 1) the individual's body, 2) the environment, or 3) a social partner. The illustration below shows the coordinate system centered on the owl's head in order to describe head movements.

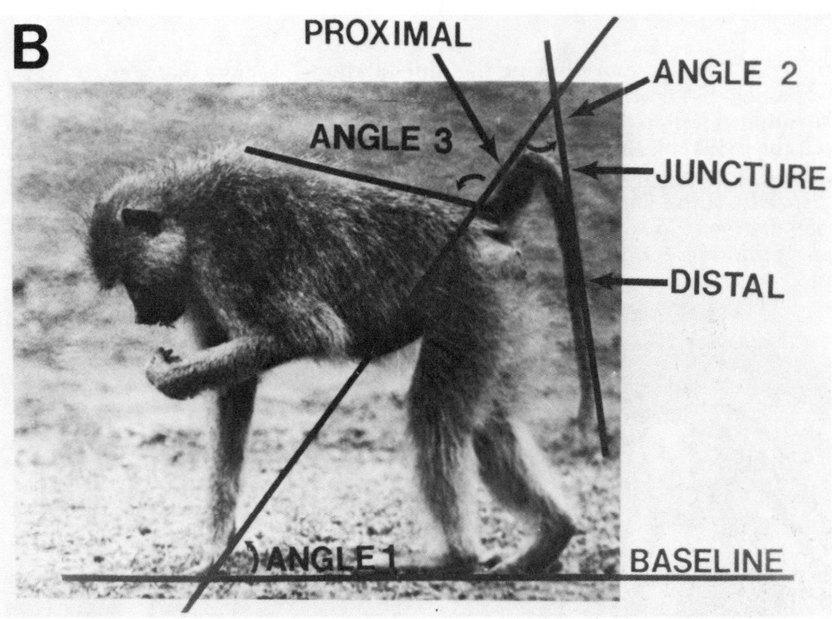

Figure 9-4 *A*, coordinates for measuring form of the canid play bow. *B*, coordinates for measuring tail carriage in baboons. (*A* from Bekoff, M., Social Communication in Canids: evidence for the evolution of a stereotyped mammalian display, *Science* 197:1098, Fig. 1. Copyright 1977 by the American Association for the Advancement of Science. *B* courtesy of G. Hausfater.

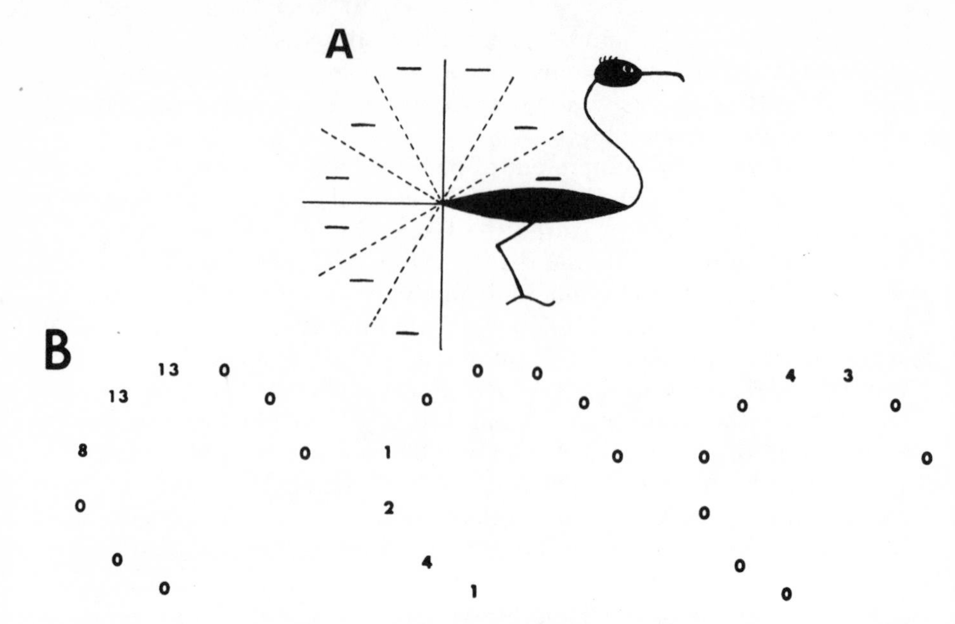

Figure 9-5 *A*, diagram showing the nine 30-degree sectors that were used for frequency distributions of tail elevations. *B*, frequency distribution of tail elevations for nine 30-degree sectors between downward and forward for (left to right): Sky-pointing of *Sula sula* at Half Moon Cay, B.H.; wing-waving of *Anhinga anhinga* at Avery Island, Louisiana, and throw-back of *Phalacrocorax aristotelis* at the Farne Islands, England. (From Van Tets, G.F., 1965, A comparative study of some social communication patterns in the pelecaniformes, Ornithological Monographs No.2:27, Fig. 16.)

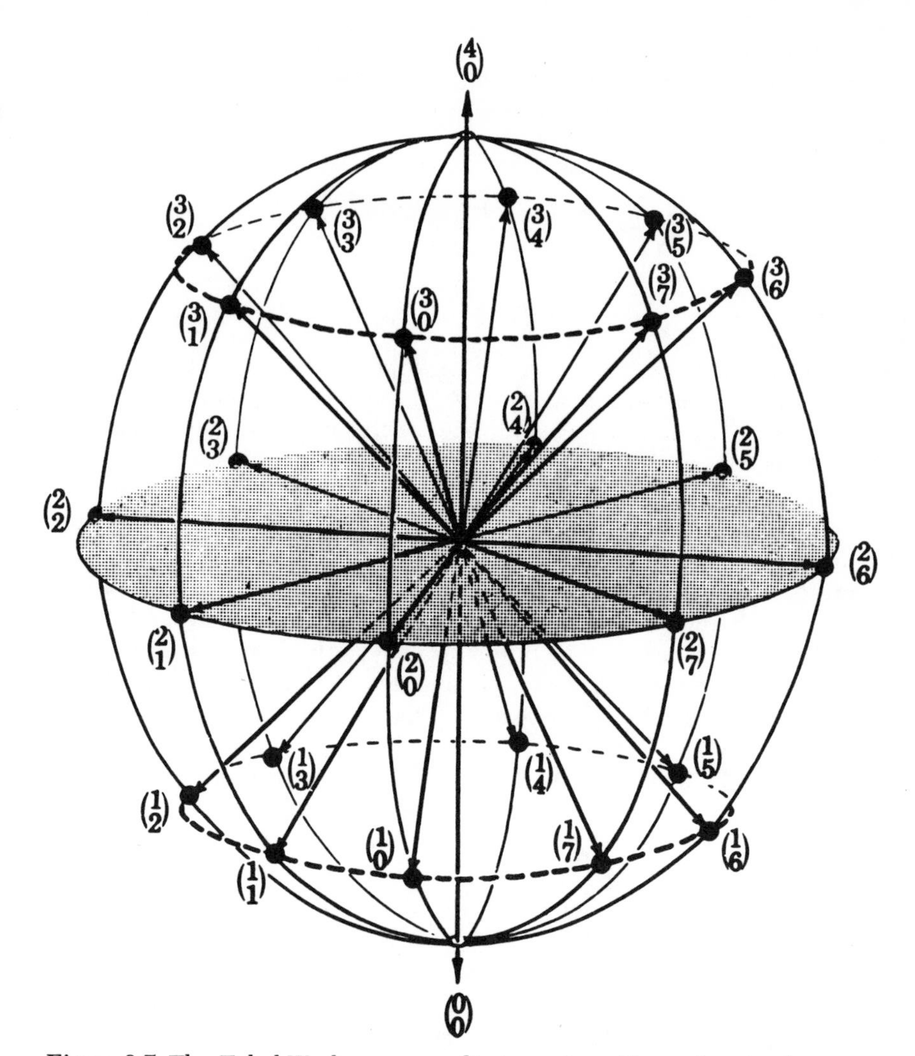

Figure 9-7 The Eskol-Wachmann coordinate system. For each pair of numerals, the lower indicates the horizontal, and the upper the vertical, coordinate. In Golani's study one unit of displacement equals 45°. (By permission of the Movement Notation Society, Israel.)

←

Figure 9-6 A pair of golden jackals during precopulatory behavior. The superimposed bars indicate the body parts that were considered as separate limb segments during the present study. (From Golani, I., 1976, Homeostatic motor processes in mammalian interactions: A choreography of display, page 73, Fig. 1 *in* Bateson, P.P.G., and P.H. Klopfer, (eds.), Perspectives in ethology, Vol. 2, Plenum Press, New York.)

Each movement is noted on a "score page" in terms of: 1) point of beginning, 2) point of ending, and 3) spatial and 4) temporal units of movement. For example, the owl's head movement illustrated in Figure 9-8 would be noted as follows:

2 0	2 →				2 6

In this case we are using the beak position as an index of head position. The beak started at point $\frac{2}{0}$ and stopped at point $\frac{2}{6}$. This represents two ($\xrightarrow{2}$) units of measurement since the coordinates are set at 45° angles. Each block represents 0.2 second; the movement thus took 1.2 seconds.

What has been presented is a simplified version of Golani's movement notation method, and his paper should be consulted for more detailed descriptions of additional notations and more sophisticated uses. However, be cautioned that many studies will not require this intense an analysis in order to answer the research questions. Weigh the increased resolution obtained with the method against the fact that unfamiliarity will make it initially tedious and perhaps introduce errors of recording.

Trochim (1976) described a three-dimensional method for quantifying body position. Height, width, and depth of various body points are taken from videotape. These coordinates can then be graphed by a computer (Fig. 9-9) and analyzed for various parameters, including interpersonal distance and body activity. Trochim describes the collection and analysis of data, as well as the computer programs which he developed.

2. Dominant-Subordinate Relationships

Another example of data collection and organization is illustrated here for dominant-subordinate relationships. The concept of one (or more) individuals dominating one (or more) other individuals has been studied for many years by ethologists using several techniques. The behavioral units selected and the criteria of dominance have varied widely. When individuals are ranked by different criteria, the rankings are often not comparable (Bekoff 1977*b*; Bernstein 1970; Syme 1974). However, Richards (1974) used the ten factors listed below to assess dominance rank in six groups of captive rhesus monkeys and found that they produced comparable results.

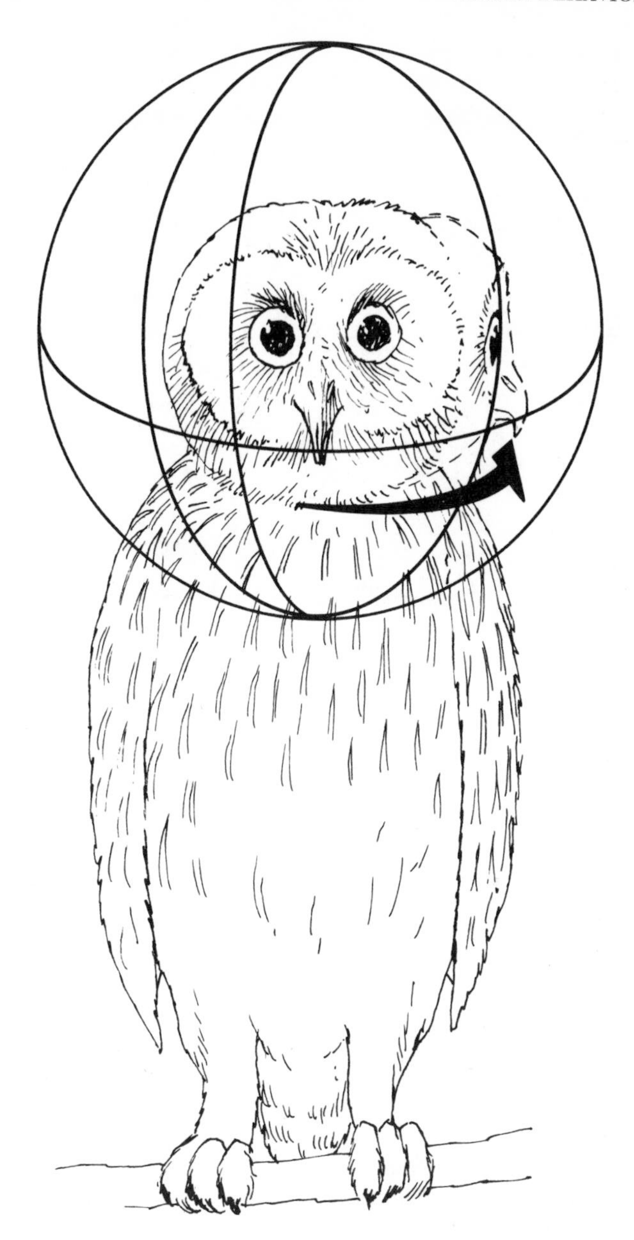

Figure 9-8 The Eskol-Wachmann coordinate system superimposed on an owl's head in order to note movement. In this example the head has moved 90° to the left from point $\frac{2}{0}$ to point $\frac{2}{6}$. (See text for explanation.)

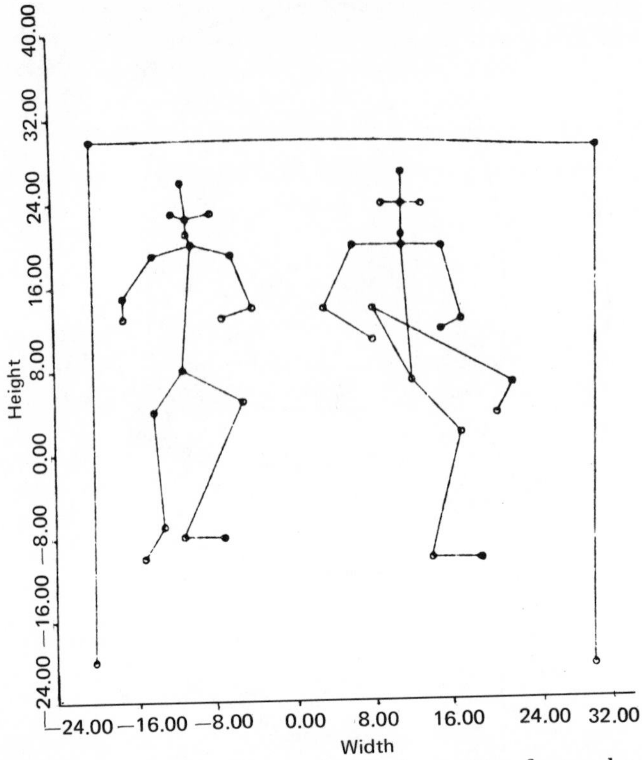

Figure 9-9 Front-view Cal-comp graph of body position for two human subjects. (From Trochim, W.M.K., 1976, The three-dimensional graphic method for quantifying body position, *Behav. Res. Meth. and Instrum.* 8(1):2, Fig. 1.)

1. Priority to food incentives
   a. Order to daily food ration
   b. Order to milk bottle
   c. Interactions at milk bottle
   d. Order to approach experimenter during food offers
2. Agonistic encounters
3. Displays
4. Gestures for fear-submission
   a. Yielding ground/avoidances
   b. Cautious approaches
   c. Nonsexual presentations/mountings
   d. Fear-grins

Dominant-subordinate relationships in groups generally can be divided into two types—linear and nonlinear. Murchison (1935) found that the ranking in a group of six domestic fowl roosters changed from a nonlinear to a linear hierarchy as they matured (Fig. 9-10).

a. Measuring Dominant-Subordinate Relationships

The most common criterion for the expression of a dominant relationship is priority of access to a limited resource (*e.g.*, food, shelter, space, estrus female, etc.). Access is established through fighting or threats and the supplanting of one individual by another (see Richards' 10 measures above).

Berger (1977) devised a dominance coefficient to measure the relative dominance of feral horses within their respective bands. Aggressive bouts were recorded during periods when individual bands were using a limited water supply in a small basin. Berger assumed that the horses with the highest frequency of successful bouts and the lowest number of bouts initiated by them were dominant within the band. He then calculated the intraband dominance as follows:

$$\text{dominance coefficient} = \frac{(a/b \times 100)}{(i + 1)}$$

where: a = no. successful bouts
b = total no. bouts
i = no. interactions initiated

$$\text{maximum dominance coefficient} = 100$$

Data can be gathered by observing naturally occurring conflicts in wild or captive groups or by manipulating the environment to encourage conflict (*e.g.*, introducing a limited amount of a preferred food). Among captive animals, the researcher can artificially stage an equal number of encounters between all possible pairs of a group. The observer records the winner and the loser (supplanter and supplantee) of each encounter and organizes his data into a matrix.

Brown (1975) provided the following list of steps to follow in the construction of a dominance matrix:

1. *Observations:* $B > D$, $C > A$, $B > A$, $C > B$, $B > D$, etc.: $B > D$ means B won an encounter with D. In most cases, these encounters take the form of supplanting rather than fighting.
2. *Starting Order:* Choose an arbitrary order, *e.g.*, *DEACB*.
3. *Starting Matrix:* Enter the number of wins and losses observed in the matrix.

		Loser D	E	A	C	B
Winner	D		24	3	0	0
	E	0		13	0	0
	A	21	11		0	0
	C	12	16	17		14
	B	37	31	41	0	

4. *Treatment of Reversals:* A win by one individual over another that has won the majority of encounters with the first is termed a *reversal*. Rearrange the order so that only reversals fall below the diagonal, so far as possible; that is, change the above order to *CBDAE* or *CBEDA* or *CBAED*.
5. *Treatment of Nonlinearity:* An order in which an individual dominates another (wins the majority of encounters) that dominates the first is termed *nonlinear* or circular. Rearrange to minimize the inevitable ambiguity. From the circular relationship diagrammed below there are three main alternatives, as shown. In the three alternatives not shown, the departure from linearity involves two individuals rather than one.

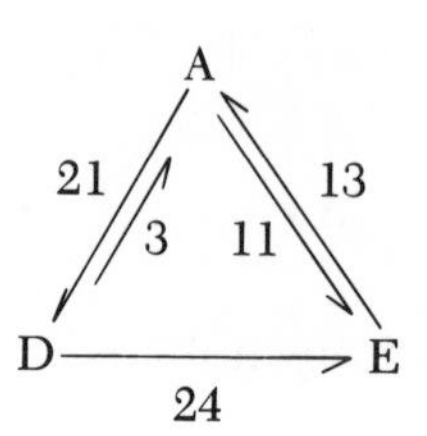

Best

	A	D	E
A		21	11
D	3		24
E	13	0	

	E	A	D
E		13	0
A	11		21
D	24	3	

	D	E	A
D		24	3
E	0		13
A	21	11	

Place the individuals that are in the least ambiguous relationships (lowest proportion of reversals) in linear order. This procedure tends to minimize the total number of encounters entered below the diagonal.

6. *Final Matrix:* The one order that best reflects the order of dominance within the group is then *CBADE*. A matrix may then be constructed.

	C	B	A	D	E	Wins	Losses
C		14	17	12	16	59	0
B	0		41	37	31	109	14
A	0	0		21	11	32	74
D	0	0	3		24	27	70
E	0	0	13	0		13	82

b. Analysis of Linearity

Perfectly linear hierarchies are relatively rare, making most hierarchies technically nonlinear. Perfectly linear hierarchies are unidirectional. They can contain *reversals* (*i.e.*, a subordinate wins an occasional encounter with a dominant individual), but they cannot contain any individuals of *equal status* or have any *circularity* such as: $\overleftarrow{A \rightarrow B \rightarrow C}$. The nonlinearity is, however, of varying degrees, and some so closely approximate perfectly linear hierarchies that they should be considered linear.

Landau's index of linearity has been discussed by Bekoff (1977*b*) and Chase (1974). The index (h) is calculated accordingly:

$$h = \left(\frac{12}{n^3 - n}\right) \sum_{a=1}^{n} = 1 \left\{V_a - \frac{(n-1)}{2}\right\}^2$$

n = number of animals in the group

V_a = number of animals that an individual dominates

The term $\left(\frac{12}{n^3 - n}\right)$ normalizes the index so that it ranges from (nonlinear) to 1 (perfectly linear). Bekoff (1977*b*) agreed with Chase (1974) that $h \geqslant 0.9$ would be a reasonable (although arbitrary) cutoff criterion for "strong," nearly linear hierarchies.

We can calculate the Landau index of linearity for the rooster hierarchies in Figure 9-10A.

EXAMPLE 1: Landau index of linearity analysis for the dominance hierarchy of 16-week-old domestic-fowl roosters (Fig. 9-10A).

$n = 6$

YY	dominated	Blue, *G*, *R*, *W*, *Y*	;	$V_{YY} = 5$
Blue	dominated	*W*, *R*, *Y*	;	$V_{\text{Blue}} = 3$
G	dominated	Blue, *R*, *Y*	;	$V_G = 3$
R	dominated	*W*, *Y*	;	$V_R = 2$
W	dominated	*R*, *G*	;	$V_R = 2$
Y	dominated	none	;	$V_Y = 0$

$$h = \left(\frac{12}{n^3 - n}\right) \sum_{a=1}^{n} = 1 \left\{V_a - \frac{(n-1)}{2}\right\}^2$$

$$h = \left(\frac{12}{n^3 - n}\right) \{6.25 + 0.25 + 0.25 + 0.25 + 0.25 + 6.25\}$$

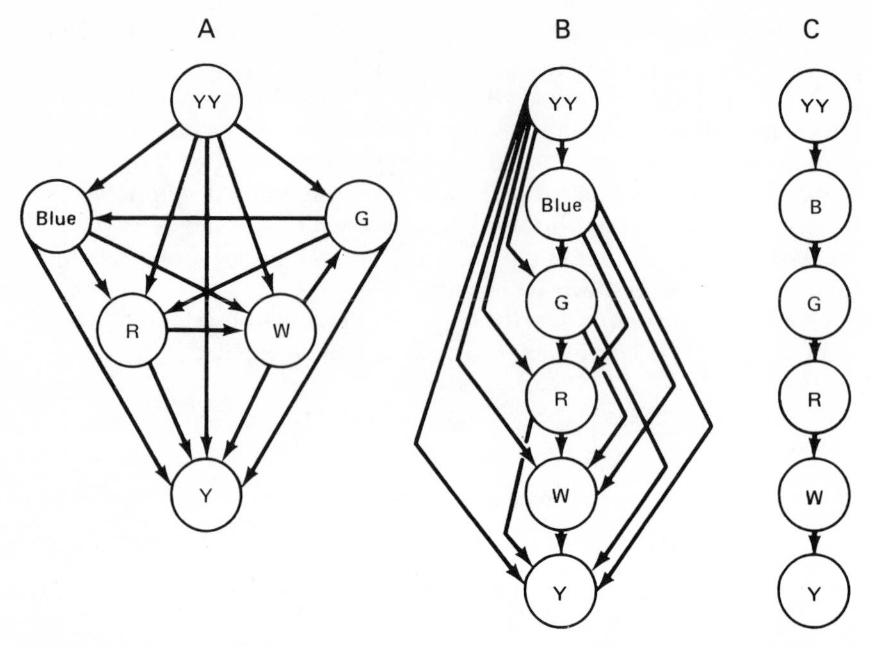

Figure 9-10 *A*, nonlinear dominance hierarchy in a group of 16-week-old domestic-fowl roosters. *B*, linear hierarchy in the same group of roosters at 32 weeks of age (*A* and *B* from Murchison 1935). *C*, shorthand diagram of linear hierarchy in *B*; all individuals above are assumed to dominate all those below.

$$h = \left(\frac{12}{n^3 - n}\right)(13.5) = \left(\frac{12}{210}\right)(13.5) = 0.057\,(13.5) = 0.77$$

The low h value of 0.77 reflects the high degree of nonlinearity.

EXAMPLE 2: Landau index of linearity analysis for dominance hierarchy of 32-week-old domestic-fowl roosters (Fig. 9-10*B*).

$n = 6$
$V_{YY} = 5$
$V_{\text{Blue}} = 4$
$V_G = 3$
$V_R = 2$
$V_W = 1$
$V_Y = 0$

$$h = (0.057)\,(6.25 + 2.25 + 0.25 + 0.25 + 2.25 + 6.25)$$

$$h = (0.057)\,(17.5) = 1.0$$

The h value of 1 is just as we would expect for a perfectly linear hierarchy. Bekoff (1978*b*) demonstrated an h value of 1 in litters of coyote pups (*Canis latrans*) at various ages. Calculating the index of linearity for the hierarchy in Brown's example (above) will generate an h of < 1.

When individuals are close in rank, that is they supplant each other approximately equally, then assigning clearcut dominance status to one may indicate more linearity to the hierarchy than truly exists. However, Landau's index can still be used when individuals are of *equal rank* by applying the following rule:

For individuals of equal rank:

$V_a = 1$ for each individual dominated $+0.5$ for each individual of equal rank

For example:

Individuals D and E of equal rank

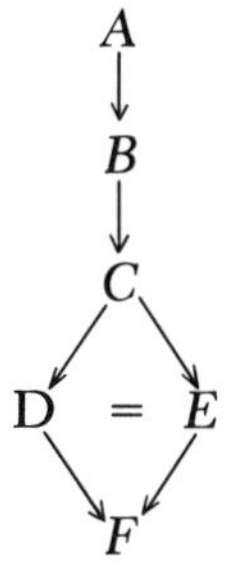

A → B → C; C → D, C → E; D = E; D → F, E → F

$V_A = 5$

$V_B = 4$

$V_C = 3$

$V_D = 1.5$

$V_E = 1.5$

$V_F = 0$

$$h = (0.057)\,(6.25 + 2.25 + 0.25 + 1 + 1 + 6.25)$$
$$= (0.057)\,(17) = 0.97$$

Individuals B, C, D and E of equal rank.

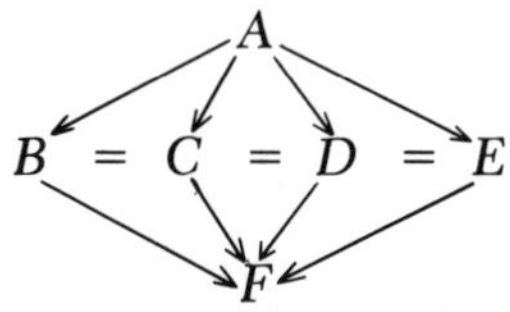

A → B, C, D, E; B = C = D = E; B, C, D, E → F

$V_A = 5$

$V_B = 2.5$

$V_C = 2.5$

$V_D = 2.5$

$V_E = 2.5$

$V_F = 0$

$$h = (0.057)\,(6.25 + 6.25)$$
$$= (0.057)\,(12.50) = 0.71$$

3. Social Organization

Society: a group of individuals belonging to the same species and organized in a cooperative manner. [WILSON 1975:7]

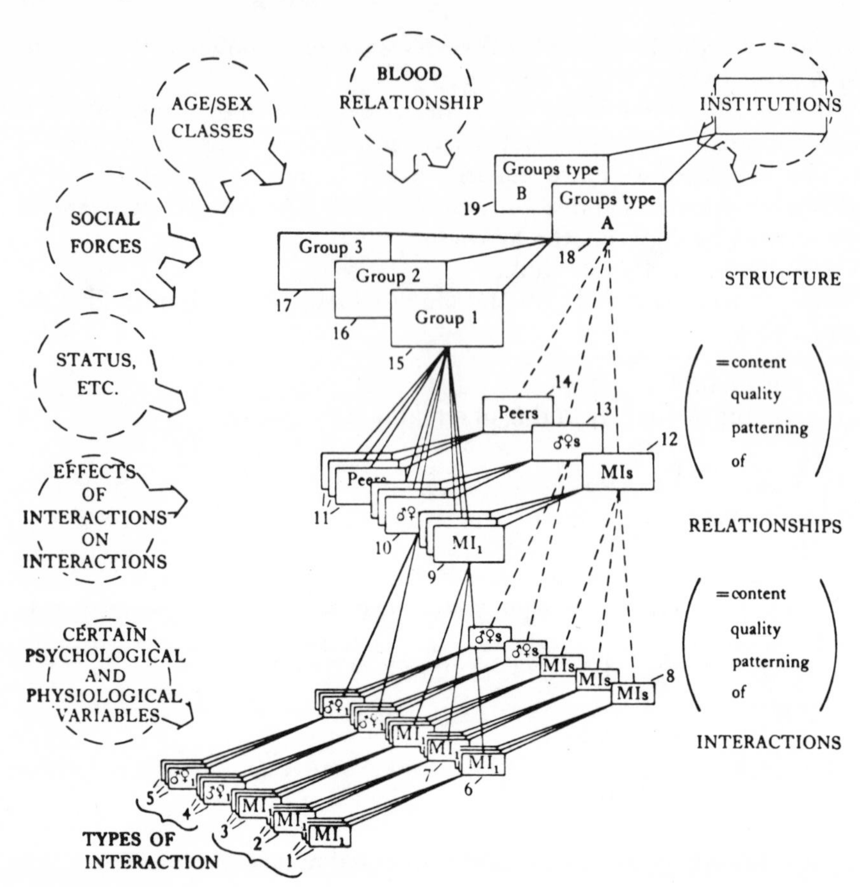

Figure 9-11 Diagrammatic representation of the relations between interactions, relationships, and social structure, shown as rectangles on three levels, with successive stages of abstraction from left to right. The discontinuous circles represent independent or intervening variables operating at each level. Institutions, having a dual role, are shown in both a rectangle and a circle.

In the specific instance of a nonhuman primate, the rectangles might represent (1) Instances of grooming interactions between a mother *A* and her infant *B*. (2) Instances of nursing interactions between *A* and *B*. (3) Instances of play between *A* and *B*. (4) Instances of grooming between female *A* and male *C*. (5) Instances of copulation between *A* and *C*. (6) First stage abstraction-schematic grooming interactions between *A* and *B*. Abstractions of grooming interactions between other mother-infant pairs are shown behind, but the specific instances from which they were abstracted are not shown. (7) First stage abstraction-schematic nursing interactions between *A* and *B*. Abstractions of nursing interactions of other mother-infant pairs are shown behind. (8) Second stage abstraction-schematic grooming interactions between all mother-infant pairs in troop. (9) Mother-infant relationship between *A* and *B*. Mother-infant relationships of other mother-infant pairs are

Social organization is the behavioral organization (type, temporal, and spatial) of the society's members. Analysis of social organization is the most complex endeavor an ethologist can undertake, because it necessitates the integrative analysis of social behavior both within and between group members. This includes the development of social behavior in the individual (socialization) and the interaction of group members over time (social phases).

Interactions between individuals serve as the basis for social relationships which are then integrated into the society's social organization. Hinde and Stevenson-Hinde (1976) have presented these relationships in diagrammatic, but rather complex, form (Fig. 9-11).

In order to understand fully the social-behavior matrix which is the structural basis for social organization, the researcher should begin the analysis of social organization at the level of interactions. However, a superficial knowledge of social organization can be gained by looking at relationships and perhaps even structure.

S.A. Altmann (1968) determined the relationships among and within sex/age classes of monkeys by observing interactions among individuals. This allowed him to describe the social organization of rhesus monkeys at the relationship level in the diagrammatic form of a sociogram (Fig. 9-12).

Interactions, relationships, and structure are specific to individual societies and should not be generalized to other species or even other populations of the same species without confirming evidence. Differ-

---

Figure 9-11 (continued)

shown behind (but connections to grooming, nursing, and other interactions are not shown). (10) Consort relationship between *A* and *C*. Other consort relationships are shown behind. (11) Specific relationship of another type (for example, peer-peer). (12) (13) (14) Abstraction of mother-infant, consort, and peer-peer relationships. These may depend on abstractions of the contributing interactions. (15) Surface structure of troop containing *A*, *B*, *C*, and so on. (16) (17) Surface structures of other troops (contributing relationships not shown). (18) Abstraction of structure of troops including that containing *A*, *B*, *C*, and so on. This may depend on abstractions of mother-infant, and other relationships. (19) Abstraction of structure of a different set of troops (from another environment, species, and so on).

Rectangles labelled MI refer to behavior of dyad female *A* and her infant *B*. Rectangles labelled ♂ ♀$_1$ refer to consort pair female *A* and male *C*. Rectangles labelled MIs, ♂ ♀s refer to generalizations about behavior of mother-infant dyads and consort pairs respectively. (From Hinde, R.A., and J. Stevenson-Hinde, 1976, Towards understanding relationships: dynamics stability, p. 452, Fig. 1 *in* Bateson, P.P.G., and R.A. Hinde, (eds.), Growing points in ethology, Cambridge Univ. Press, Cambridge.)

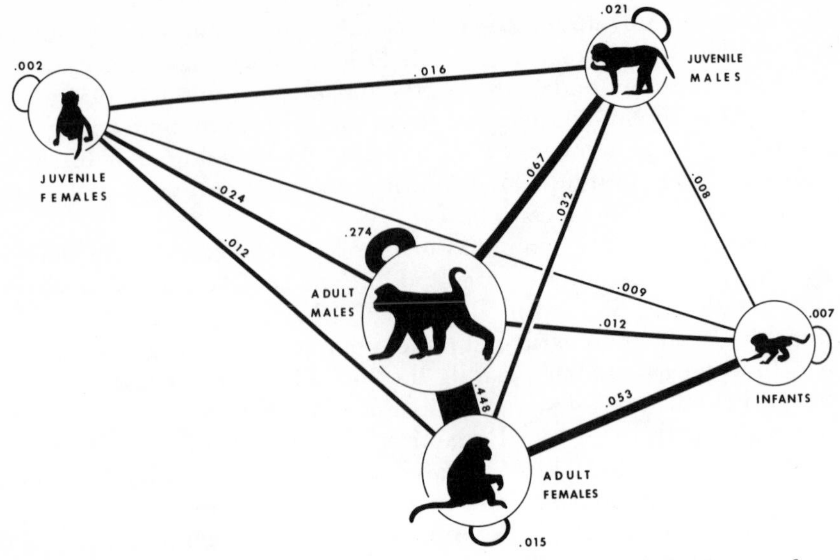

Figure 9-12 A model of social interactions between and within classes of rhesus monkeys. Probabilities of various interactions, shown by the figures and the thickness of the lines, are calculated from a hypothetical population with equal representation of age-sex classes. (From Altmann, S.A., 1968, Sociobiology of rhesus monkeys III: The basic communication network, *Behaviour* 32(1–3):26, Fig. 3.)

ences in group composition, group size, and habitat can all affect social organization. For example, at the level of interactions, Farr and Herrnkind (1974) measured the frequency of occurrence of courtship displays in the guppy *(Poecilia reticulata)* at different densities of pairs. The correlation between pair density and courtship interactions is shown in Figure 9-13.

The types of structures found in animal societies have been classified in diverse ways by several authors. One of the most useful classifications is that proposed by Brown (1975) in Table 9-1.

Within the conceptual framework discussed above, how do you go about studying social organization? Using the methods and equipment discussed previously, you should begin at the level of *interactions* (such as those discussed in the previous section on dominant-subordinate relationships) and build through *relationships* to the level of *structure*. This can be accomplished only through intensive and extensive observation spread over several seasons. One should focus on certain aspects of social interactions at each level in Figure 9-11. This can be accomplished by following the questionnaire compiled by McBride (1976) following a committee's discussion of the

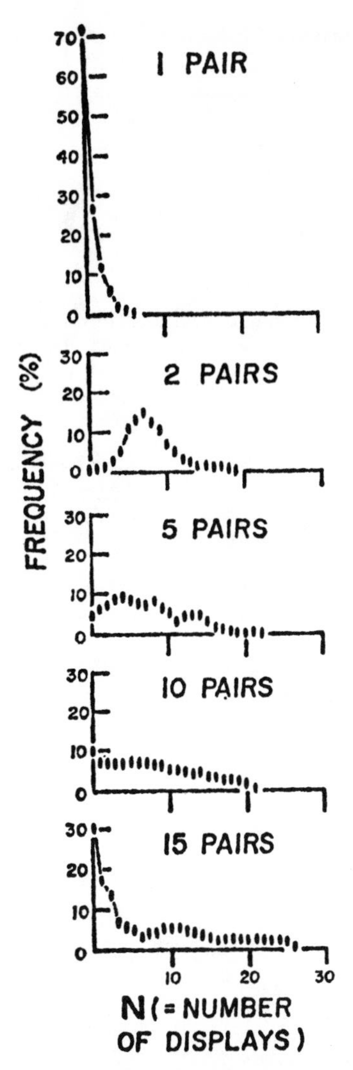

Figure 9-13 A running average of frequency of occurrence of N male courtship displays in an observation period. (From Farr, J.A., and W.F. Herrnkind, 1974, A quantitative study of social interaction of the guppy, *Poecilia Reticulata* (Pisces:Poeciliidae) as a function of population density, *Anim. Behav.*, 22(3):586, Figs. 4, 5.)

information necessary in order to adequately describe the social organization of a species. Numerous established ethologists contributed to the questionnaire prepared by McBride. They are listed at the beginning of the paper which is reprinted in Appendix D.

Table 9-1. Types of intraspecific animal groups (adapted from Brown 1975).

Types of Groups	Examples
1. Kin Groups	
Clones—groups formed by asexual reproduction of sessile colonial invertebrates, typically in permanent physical contact	colonial coelenterates
Families—groups formed by one or two parents and their most recent offspring	goose and swan families coyote (*canis latrans*)
Extended families—groups formed from families by failure of many offspring to leave parents	prairiedogs (*Cynomys*) some primate groups Mexican jay (*Aphelocoma ultramarina*) gray wolf (*canis lupus*)
2. Mating Groups	
Pairs—monogamous groups of two	scrub jay (*Aphelocoma coerulescens*) lar gibbon (*Hylobates lar*)
Harems—groups in which a male attempts to keep females together and away from other males, with or without cooperation of females	red deer (*Cervus elephas*)
Leks—groups formed by attraction of males (and subsequently females) to a communal mating ground; eggs or young produced elsewhere	lek birds and mammals hilltopping butterflies Hawaiian *Drosophila*
Spawning groups—groups of both sexes formed at localized spawning grounds; no provisioning of young	many fishes and amphibians
3. Colonial Groups	
Groups formed by colonial nesting of pairs or one-male harems; young provisioned at nest	tricolored blackbird (*Agelaius tricolor*) many sea birds some bats and seals
4. Survival Groups	
Groups formed by aggregation of randomly related, usually nonbreeding individuals who are mutually attracted by each other	foraging flocks night roosts of New World blackbirds ducks and geese herding mammals fish schools bachelor groups of wapiti
5. Aggregations (Coincidental Groups)	
Groups formed by physical factors acting on migrating or moving animals	hawks migrating along a mountain ridge land birds migrating through a mountain pass whelks on a sheltered ocean rock stream-surface insects on a calm eddy
Groups formed by attraction to a common resource, such as food or water	bears at a garbage dump

Owen-Smith's (1974) description of the social organization of the white rhinoceros *(Ceratotherium simum)* is a concise example of a "typical" social organization study. Although it is lacking in some respects, it can serve as a model for studies of this type. Also, consult Kummer's excellent study (1968) of the social organization of hamadryas baboons, Crook et al.'s excellent conceptual model of the structure and function of mammalian social systems (1976), and Eisenberg's discussion of the social organization in mammals (1966).

10
Introduction to Statistical Analyses

Analysis is the ordering, breaking down, and manipulation of data in order to obtain answers to research questions. What we will be concerned with in this book are 1) ways to initially look at data, 2) first approximation statistics, and 3) nonparametric statistical tests.

A. STATISTICAL PRINCIPLES

Statistics are measures computed from observations in a sample. *Statistical tests* are procedures whereby hypotheses are tested. Kerlinger has defined statistics in the way it is most commonly used:

> Statistics is the theory, discipline, and method of studying quantitative data gathered from samples of observations in order to study and compare sources of variance of phenomena, to help make decisions to accept or reject hypothesized relations between the phenomena so studied, and to aid in making reliable inferences from observations. [KERLINGER 1967:148]

In order to make statements concerning the results of their research, ethologists should support their conclusions with statistical tests. Like other biological scientists, the ethologist assumes that there is some order to animal behavior and therefore it is amenable to statistical testing.

> In biology most phenomena are affected by many causal factors, uncontrollable in their variation and often unidentifiable. Statistics is needed to measure such variable phenomena with a predictable error and to ascertain the reality

of minute but important differences. Whether biological phenomena are in fact fundamentally deterministic and only the variety of causal variables and our inability to control these make these phenomena appear probabilistic, or whether biological processes are truly probabilistic, as postulated in quantum mechanics for elementary particles, is a deep philosophical question. [SOKAL and ROHLF 1969:5]

Needless to say, and as intriguing as it is, our purpose here is not the philosophical question, but rather justification for demanding that ethologists use statistics.

B. STATISTICAL HYPOTHESES

When I discussed the scientific method in the Introduction, I stated that the scientific method is basically a matter of hypothesis testing. Also, when we examined the design of ethological research I stated that a *research hypothesis* is our best guess as to the answer to our *research question*. Research hypotheses refer to the phenomena of nature; that is, tentative predictions about the causation, function, ontogeny, or evolution of some aspect of behavior. Statistical hypotheses are statements about population parameters; they are amenable to evaluation by statistical tests.

A statistical hypothesis is either a *null hypothesis* (H_0) or an *alternative hypothesis* (H_1). Both need to be stated in order to conduct a statistical test. The statistical test is a procedure whereby a researcher chooses which one of the dichotomous set of mutually exclusive and exhaustive hypotheses (H_0 and H_1) is to be rejected and which one is to be accepted. This is done at some predetermined risk of making an incorrect decision (Type I and Type II errors, to be discussed later).

Statistical hypotheses are either *exact* (nondirectional) or *inexact* (directional). The hypothesis that the mean number (μ) of vocalizations (j) given by a population of quail each day is 100 is expressed by

$$H_0 : \mu j = 100$$

and is an exact null hypothesis. The hypothesis that the number is equal to, or greater than, 100

$$H_0 : \mu j \geqslant 100$$

is an inexact null hypothesis. The alternative hypothesis for the first H_0 is

$$H_1 : \mu j \neq 100$$

and is an inexact alternative hypothesis. Likewise the alternative hypothesis for the second H_0 is

$$H_1 : \mu j < 100$$

and is also an inexact alternative hypothesis. Hypothesis testing in ethological research usually involves either an exact or inexact H_0 and an inexact H_1.

C. HYPOTHESIS TESTING

The main purpose of inferential statistics is to test research hypotheses by testing statistical hypotheses. [KERLINGER 1967:173]

Statistical tests are designed to determine whether you can reject the null hypothesis and thus accept the alternative hypothesis. Therefore, the alternative hypothesis should closely approximate the research hypothesis (*i.e.*, what you believe the true situation to be). That is, if you are attempting to demonstrate statistically what you believe to be the case from observation, your H_1 should state what you have observed. For example, if you believe that a population of quail rarely if ever call more than 100 times a day, your statistical hypotheses should read as follows:

$$H_0 : \mu \geqslant 100$$
$$H_1 : \mu < 100$$

If our statistical test is significant and we reject the H_0, we then infer that the H_1 is correct. In this particular example we are interested in testing whether the population mean for the number of quail calls per day is *equal to* or *greater than* 100. That is, if it is significantly *less* than 100, we will reject the H_0 and accept the H_1. Since we are interested in only one side of the distribution, in this case the lefthand side of the distribution (below), our statistical test will be *one-tailed*. One-tailed tests are associated with inexact (directional) null hypotheses. At the 0.05 level of significance we would have that probability of committing a Type I error (see next section).

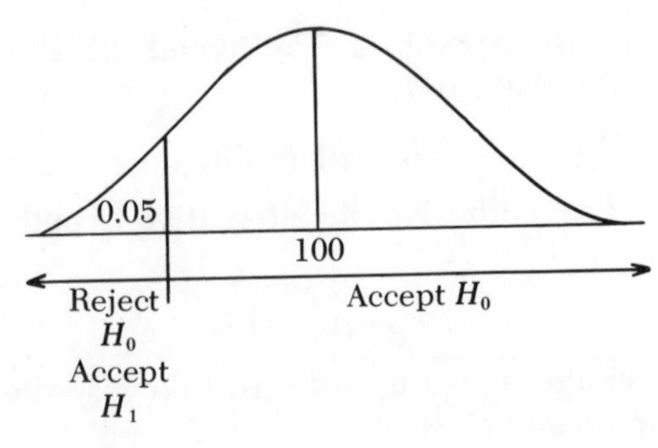

If we were simply testing whether the population mean for number of quail calls per day was significantly *different* from 100 ($H_0 : \mu = 100$; $H_1 : \mu \neq 100$), then half of the 0.05 probability of a Type I error would be associated with each tail of the curve (below). Therefore we would be conducting a *two-tailed* test which is associated with exact (nondirectional) null hypotheses.

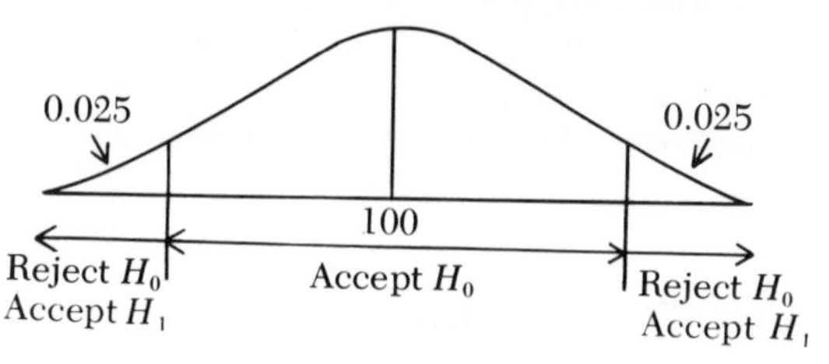

It is important to know which type of statistical test (one- or two-tailed) you are conducting in order to obtain the correct values from the statistical tables in Appendix A. Values in a statistical table for two-tailed tests are the same as for one-tailed tests with twice the level of significance (*i.e.*, one-half the alpha value); that is, if the level of significance for the two-tailed test is 0.05 then the tabular values are the same as for a one-tailed test with a 0.025 level of significance, and vice versa.

Hypothesis-testing procedures are important to the ethologist in designing, analyzing, and interpreting research. However, they should not be allowed to blind the careful observer or overshadow common sense.

> Hypothesis-testing procedures should be viewed as tools that aid an experimenter in interpreting the outcome of research. Such procedures should not be permitted to replace the judicial use of logic by an alert analytic experimenter. [KIRK 1968:33]

A statistical test is used to compare the null and alternative hypotheses and make a choice between them. The null hypothesis (H_0) is essentially a prediction of a sampling distribution of anticipated values for a *sample statistic* (*e.g.*, mean). If the sample statistic, generated by the data collected, falls within the sampling distribution of anticipated values, then we decide to accept the null hypothesis.

Sample statistic—statistic generated from data that are used to estimate population parameters (*e.g.*, mean, standard deviation)

Test statistic—computed from data; used to test a statistical hypothesis (*e.g.*, chi-square, t-test)

The following is the stepwise procedure used in hypothesis testing:

1. State a null hypothesis (H_0) and an alternative hypothesis (H_1).
2. Select an appropriate sample statistic and test statistic.
3. Select a level of significance (alpha level; see below) and a sample size (N).
4. Collect the data (Chapters 5–9).
5. Compute the sample statistic and the test statistic. If the test statistic's value falls in the region of rejection, the H_0 is rejected and the H_1 is accepted. Failure to reject the H_0 is not the same as accepting it, although this is often done.

D. TYPE I AND TYPE II ERRORS

Researchers take a chance in hypothesis testing. They can commit basically two types of errors in making a decision to accept or reject a null hypothesis.

Type I error—Reject H_0 when it is true. The probability α is the risk of making a Type I error.
Type II error—Fail to reject H_0 when it is false. The risk of marking a Type II error is designated as β.

The probability of making Type I and Type II errors is shown in Figure 10-1.

E. SELECTING THE ALPHA LEVEL

Selecting the probability level for a Type I error (α level) is generally done by convention. This is also called the *level of significance* for a particular statistical test. This level is generally set at 0.05 (significant) or 0.01 (highly significant) for no other reason than that it is generally accepted that those levels represent a reasonable risk. Generally, values that are greater than 0.05 are not considered to be *statistically significant* (Sokal and Rohlf 1969:161).

F. POWER OF A TEST

The *power* of a test is the probability of rejecting the null hypothesis when the alternative hypothesis is true; that is, the probability that you will make a correct decision in your favor. See Dixon (1954) and Siegel (1956) for a tabulation of the power and efficiency of several nonparametric tests. Remember that you should state your hypotheses

$H_0 : \mu = 100$

$H_1 : \mu = 150$

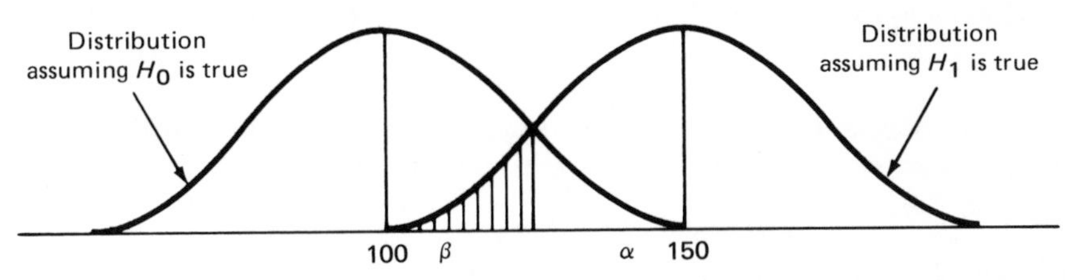

Figure 10-1 Example of regions at sampling distribution represented by Type I (α) and Type II (β) errors.

in such a way that in order tö support your research hypothesis you must reject the null hypothesis. Increased power of a test increases the probability of your rejecting the null hypothesis if the research hypothesis is correct. For example, if you believe that a particular male goldfish spends more time in shaded areas than in sunlit areas, you make several sample observations and record the length of time that the goldfish is in the shade and in the sunlit area. From this you can compute a mean time (sample statistic) for each area ($\hat{\mu}_{\text{shade}}$, $\hat{\mu}_{\text{sun}}$). You then state your hypotheses in such a way that you expect to reject the null hypothesis.

$$H_0 : \hat{\mu}_{\text{shade}} \leqslant \hat{\mu}_{\text{sun}}$$

$$H_1 : \hat{\mu}_{\text{shade}} > \hat{\mu}_{\text{sun}}$$

The power of a statistical test, for any given level of significance, can be increased in basically two ways:

1. Increase the sample size.
2. Select an experimental design that more precisely measures treatment effects and has a smaller error effect.

G. PARAMETRIC VS. NONPARAMETRIC METHODS

There are generally three assumptions inherent in parametric statistics that cannot always be met in ethological research, hence necessitating the use of nonparametric statistics.

1. Samples are drawn from populations that are *normally distributed* (bell-shaped curve, *e.g.*, Fig. 10-1).
2. In the analysis of the variance it is assumed that there is *homogeneity of variance*; that is, the variances between groups are homogeneous within the bounds of random variation. The F-max Test (p. 247) can be used to determine homogeneity of variance.
3. The measures to be analyzed are *continuous measures* with *equal intervals* (*i.e.*, interval or ratio scales of measurement).

Nonparametric tests do not demand that these assumptions be met; they are distribution-free tests which are relatively quick and easy to perform.

A nonparametric statistical test is a test whose model does not specify conditions about the parameters of the population from which the sample was drawn. [SIEGEL 1956:31]

Since there are fewer constraints on nonparametric tests, they are also less powerful. Therefore, many researchers use parametric tests without having necessarily satisfied the three criteria listed above.

. . . it is common practice in the behavioral sciences to use the more powerful parametric tests even though the assumptions are only approximately fulfilled. [KIRK 1968:493]

Also, some parametric tests are *robust*; that is, they can be used with reasonable validity even when some of the parametric test assumptions (listed above) are violated. For example, the t-test can be used even when there is considerable deviation from normality and/or homogeneity of variance, except in an independent-samples design with unequal numbers of scores. Researchers can also transform (*e.g.*, square-root, logarithmic) data to meet the homogeneity of variance assumption.

H. SAMPLE STATISTICS

Sample statistics are used to define the nature and distribution of the data. They should be calculated immediately to give the researcher a first approximation look at his results. Sample statistics will often provide knowledge as to whether the results are significant (statistically), or not. Many pocket calculators have provisions for calculating sample statistics and, in some cases, are pre-programmed to conduct selected statistical tests.

1. Sample Distributions

Sample data from a population show characteristics that reflect both the population's properties and the sampling methods used. Proper sampling methods (p. 111) must be selected so that a valid measure of the population can be made.

The choice of appropriate statistical tests will be based to a large extent on the distribution of the sample data.

EXAMPLE: We measure the duration of fighting a mirror image in 20 male siamese fighting fish *(Betta splendens)*.

Duration in seconds:

3.7	4.2	3.8	7.7	5.6
9.5	3.7	8.6	3.3	2.5
10.8	4.5	5.9	4.7	4.1
2.4	6.9	4.4	4.4	4.5
5.8	1.6	11.7	6.6	7.6

2. Sample Mean and Medians

We compute the *sample mean* ($\bar{X}$) by summing (Σ) the sample measurements (x_i) and dividing by the sample size (N).

$$\bar{X} = \frac{\Sigma x_i}{N} = \frac{142.5}{25} = 5.7$$

The *sample median* is the measurement with an equal number of measurements (scores) on either side of it. It can be determined by arranging the measurements in order. For example:

1. 1.6
2. 2.4
3. 2.5
4. 3.3
5. 3.7
6. 3.7
7. 3.8
8. 4.1
9. 4.2
10. 4.4
11. 4.4
12. 4.5

*13. 4.5 Median

14. 4.7
15. 5.6
16. 5.8
17. 5.9
18. 6.6
19. 6.9
20. 7.6
21. 7.7
22. 8.6
23. 9.5
24. 10.8
25. 11.7

With 25 measurements, the median value will be the 13th measurement (4.5).

In order to plot the frequency distribution, the measurements are placed into intervals—in this case one-second intervals.

Interval (sec.)	No. Occurrences
0–1	0
1–2	1
2–3	2
3–4	4
4–5	7
5–6	3
6–7	2
7–8	1
8–9	1
9–10	2
10–11	1
11–12	1

These measurements are then plotted in a histogram.

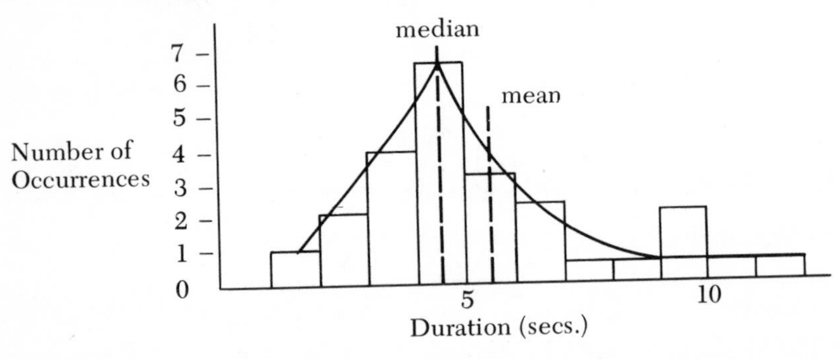

The difference between the median and mean in the above frequency distribution demonstrates that the sample data are not normally distributed. That is, sample data are normally distributed when their frequency distribution is the same on either side of the mean (see below).

3. Skewness

When sample data are not normally distributed, they are skewed, either positively (the curve tailing off to the right toward higher values) or negatively (the curve tailing off toward lower values).

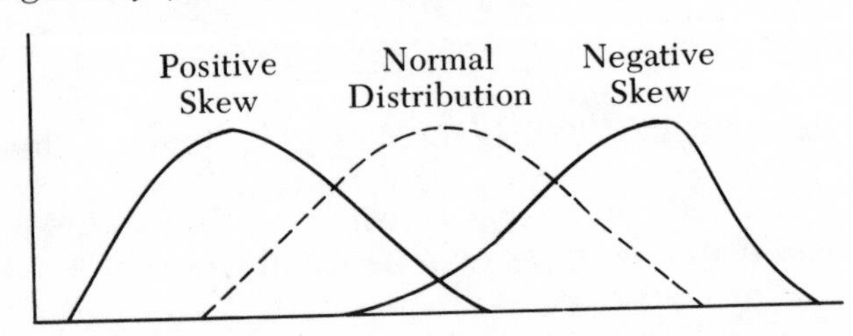

4. Location

Data sample distributions may be alike in form but may differ in location. The two curves below are both skewed positively and have the same variability but differ in their location on the scale of measurements.

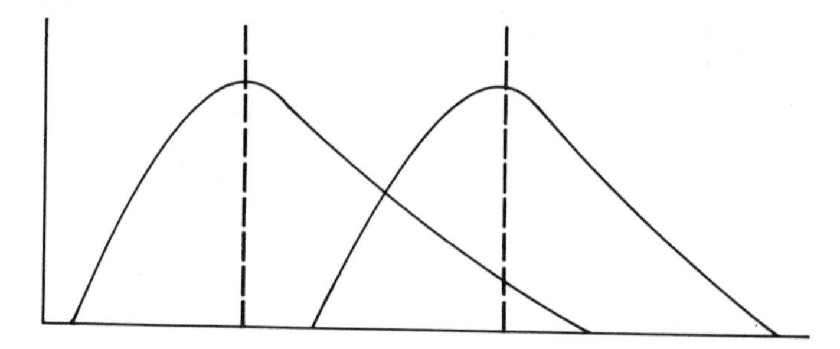

The location of a sample distribution is specified by quantities such as the mean and the median. These are referred to as *location parameters*.

Sample distributions are often worth plotting in order to obtain a visual image of their skewness + variability (= form) and locations.

5. Variability

Data samples from two populations may both be normally distributed and have the same location, but may differ in variability. That is, the frequency distribution of the data on either side of the mean may be the same within each population, but be different between the populations. Curves *A* and *B* (below) are two data samples which are both normally distributed. However, curve *A* represents much more variable data spread over a larger range (10–50).

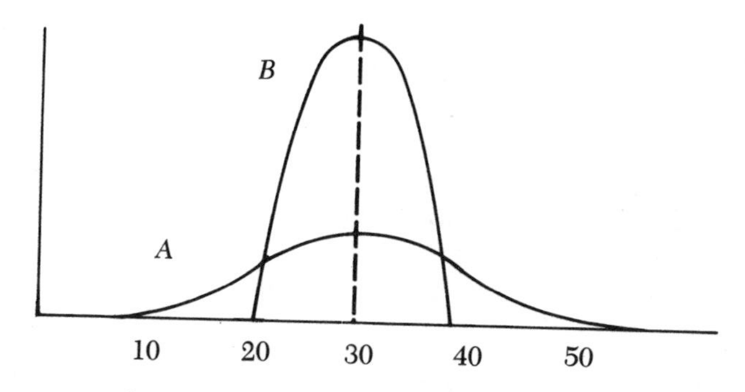

Skewness and variability are characteristics which combine to determine the *form* of a sample distribution.

6. Standard Deviation

When means are compared, it is also important to know how much variability there is in the original measurements (x_i) from which those means were derived. The standard deviation (s) is a measure of that variability about the mean and is represented by the formula:

$$s = \sqrt{\frac{\sum(x_i - \bar{X})^2}{N - 1}} \quad \text{or} \quad \sqrt{\frac{\sum x_i^2 - \frac{(\sum x_i)^2}{N}}{N - 1}}$$

and is computed as follows:

1. Compute the sample means, ($\bar{X}$)
2. Calculate the deviation from the mean for each measurement, ($x_i - \bar{X}$)
3. Square each of the deviations, $(x_i - \bar{X})^2$
4. Sum all the squared deviations, $\sum(x_i - \bar{X})^2$
5. Divide the sum of the squared deviations by the sample size minus one,

$$\frac{\sum(x_i - \bar{X})^2}{N - 1}$$

6. Take the square root of the number computed in Step 5.

The standard deviation can be used to reflect the distribution of the data. If the sample data are normally distributed (Fig. 10-2), the range included in the mean ±1 standard deviation includes approximately 68% of the data, the mean ±2 standard deviations includes about 96% of the data, and the mean ±3 standard deviations includes 99.7% of the data (Fig. 10-2).

7. Sample Mean Confidence Interval

The sample mean only approximates the true population mean, since it is based only on a sample from the entire population. We can, however, calculate a range around the sample mean in which we feel confident the population mean lies.

The confidence interval (C) is computed by dividing the standard deviation of the sample mean (s) by the square root of the number of measurements (N) and multiplying by a factor (t) based on the confidence level (probability level) desired and the number of measurements.

$$C = \pm t \left(\frac{s}{\sqrt{N}}\right)$$

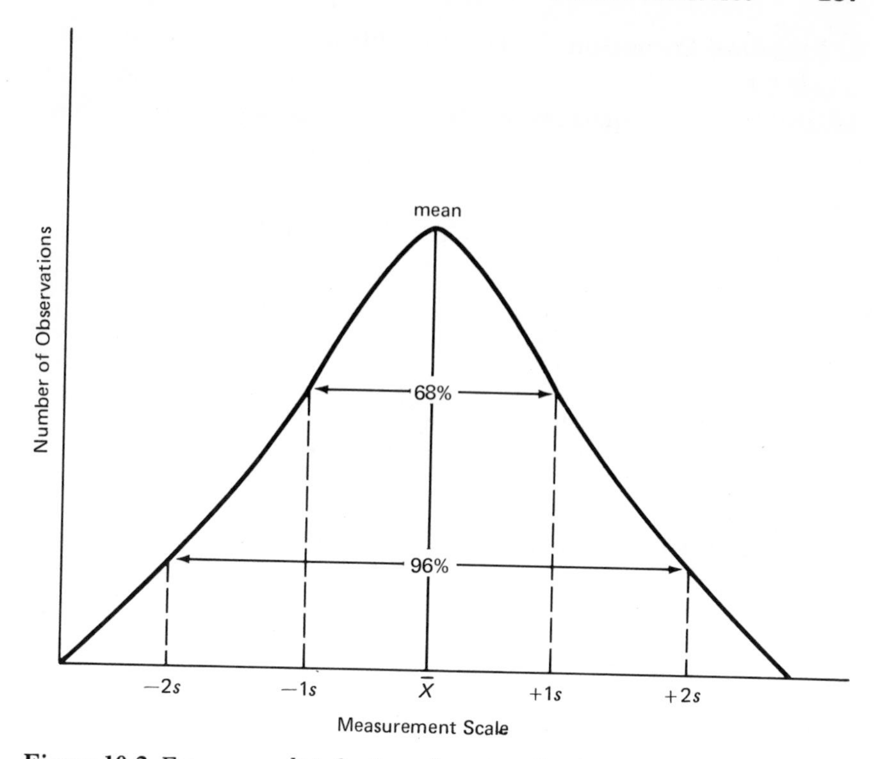

Figure 10-2 Frequency distribution of a *normally distributed* sample of measurements (observations) and the percentage of those measurements encompassed by the mean ± 1*s* and 2*s*.

The value $s/\sqrt{N}$ is also referred to as the *standard error of the mean* ($S.E._{\bar{X}}$):

$$S.E._{\bar{X}} = \frac{s}{\sqrt{N}}$$

Therefore the confidence interval (C) can also be calculated by multiplying the standard error of the mean by t:

$$C = t(S.E._{\bar{X}})$$

The value for t is obtained from Table A1. The confidence level is determined first (*e.g.*, 90% confidence level = 10 in Table A1) and the degrees of freedom = $N - 1$.

EXAMPLE: We measure the duration of 10 singing bouts in a male bird as follows:

Bout No.	Duration (sec.)
1	4.6
2	5.3
3	4.4
4	3.1
5	6.4
6	5.3
7	4.7
8	4.8
9	5.0
10	4.4
Total	= 48.0
$\bar{X}$	= 4.8

We calculate the standard deviation as follows:

Bout No.	$(x_i - \bar{X})$	$(x_i - \bar{X})^2$
1	−0.2	0.04
2	0.5	0.25
3	−0.4	0.16
4	−1.7	2.89
5	1.6	2.56
6	0.5	0.25
7	−0.1	0.01
8	0.0	0.00
9	0.2	0.04
10	−0.4	0.16
	Total	6.36 $= \sum(x_i - \bar{X})^2$

$$\frac{\sum(x_i - \bar{X})^2}{N - 1} = \frac{6.36}{9} = 0.71$$

The square root of this number provides the standard deviation.

$$s = \sqrt{\frac{\sum(x_i - \bar{X})^2}{N - 1}}$$

$$= \sqrt{0.71}$$

$$= 0.84$$

We can look for normality in the data by ranking the observations:

			Range
6.4			
5.3	4 observations above $\bar{X}$		1.6
5.3			
5.0			↑
$4.8 = \bar{X}$	———————	$\bar{X}$	———
4.7			↓
4.6			
4.4	5 observations below $\bar{X}$		1.7
4.4			
3.1			

The measurements appear normally distributed with 4 observations above the mean and 5 below, and the ranges above and below the mean are almost equal, being 1.6 and 1.7, respectively. However, we do not know if the data are really normally distributed without also knowing the actual frequency distribution on each side of the mean. Nevertheless, we can observe how our data are being distributed with regard to the standard deviation as follows:

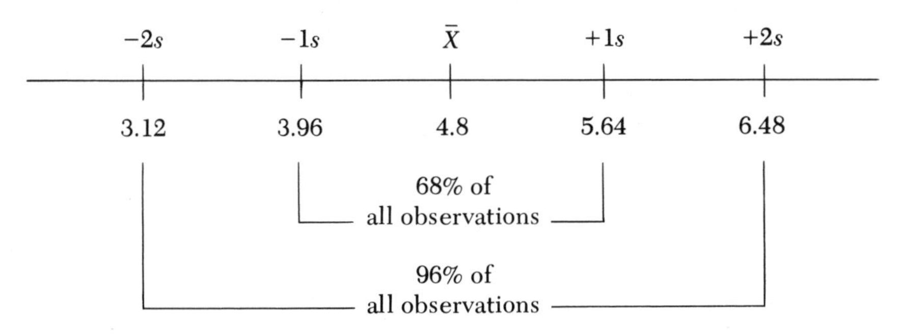

To define the confidence limits for the mean we begin by calculating the standard error of the mean:

$$S.E._{\bar{X}} = \frac{s}{\sqrt{N}} = \frac{0.84}{3.16} = 0.27$$

The confidence interval is then calculated by multiplying the $S.E._{\bar{X}}$ by t. We set our confidence level at 95% (0.05) and our degrees of freedom are $N - 1 = 9$. The value for t is 2.262 (Table A1).

$$C = \pm t\,(S.E._{\bar{X}}) = 2.262\,(0.27) = \pm 0.61$$

Then we are confident at the 95% level that the population mean lies between:

4.19 and 5.41 *i.e.* $4.19 \xleftarrow{-0.61} 4.8 \xrightarrow{+0.61} 5.41$

8. Coefficient of Variation

We may wish to compare the amount of variation about the mean for two or more samples of data. For example, is the variation in song-bout duration for male *B* different from that of male A, measured previously?

Male *A*: $\bar{X} = 4.8$ $s = 0.84$

Male *B*:

Bout No.	Duration (sec.)
1	5.7
2	3.2
3	7.5
4	6.9
5	3.4
6	4.9
7	7.7
8	6.9
9	3.8
10	4.5
Total	54.5
$\bar{X}$ =	5.4 $s = 1.72$

By comparing the 10 bouts for each individual we can see that the durations for male *B* are more variable, and our sample statistic ($s = 1.72$ vs. 0.84) bears this out. However, the mean for male *B* (5.4 seconds) is also larger than for male A and may contribute to the larger variation. That is, it is possible to have greater variation around larger means than smaller means. We therefore generate the sample statistic *Coefficient of variation* (*CV*) that expresses the standard deviation as a percentage of the mean. The greater the *CV*, the greater the variability in the data.

$$CV = \frac{s}{\bar{X}} \times 100$$

$$\text{Male A:}\quad CV = \frac{0.84}{4.80} \times 100 = 0.175 \times 100 = 17.5\%$$

$$\text{Male } B\text{:}\quad CV = \frac{1.72}{5.40} \times 100 = 0.318 \times 100 = 31.8\%$$

Even after adjusting for the differences in means, the *CV*'s demonstrate that male *B*'s song duration is much more variable than male *A*'s song duration.

Significant differences between CV's can be determined using the test statistic C (Dawkins and Dawkins 1973):

$$C = \frac{(CV_1 - CV_2)}{\sqrt{Scv_1^2 = Scv_2^2}}$$

where $Scv = \frac{CV}{\sqrt{2N}}$

The probability associated with C is obtained from the table for the distribution for t (Table A1). Using this method we can test for a significant difference between the CV's for song duration between Male A and Male B.

$$C = \frac{(0.175 + 0.318)}{\sqrt{Scv_1^2 + Scv_2^2}} \qquad N = 10$$

$$Scv_1^2 = \left(\frac{0.175}{\sqrt{2N}}\right)^2 = \left(\frac{0.175}{\sqrt{20}}\right)^2 = (0.039)^2 = 0.001$$

$$Scv_2^2 = \left(\frac{0.318}{\sqrt{2N}}\right)^2 = \left(\frac{0.318}{\sqrt{20}}\right)^2 = (0.071)^2 = 0.005$$

$$C = \frac{(0.175 + 0.318)}{\sqrt{0.006}} = \frac{0.493}{0.077} = 6.40$$

Since 6.40 exceeds the tabular value of 2.26 (9 d.f., 0.05 level), we conclude that there is a significant difference in the duration of songs between Males A and B. The coefficient of variation has been used to measure how "fixed" or "stereotyped" a behavior, particularly a display, is relative to other behavior patterns. Barlow (1977) has proposed a measure related to the coefficient of variation which he calls *Stereotypy* (ST).

$$ST = \frac{\bar{X}}{s} + 0.01\bar{x}$$

The maximum values of ST that are allowable in order to refer to a behavior pattern as "stereotyped" are undecided and relatively arbitrary. Since the communicative value of many displays varies with context, guidelines for the use of ST measures are difficult to formulate (Bekoff 1977*b*).

11 Statistical Tests, Analytical Methods and Their Selection

A. TEST STATISTICS

1. Selection of a Statistical Test

Test statistics are used to test hypotheses about one or more samples of data. The statistical test you choose for analysis of your data is dictated to a great extent by your experimental design, and the type of analysis should be considered when designing the data collection format (*i.e.*, experimental design). Neither the statistical test nor the experimental design should entirely dictate the other, but they should be coordinated.

Table 11-1 will assist in selecting appropriate nonparametric statistical tests for completely randomized or randomized block designs (p. 79).

The flow diagram (Fig. 11-1) is also provided to aid in the selection of appropriate parametric and nonparametric statistical tests.

2. Standard Error of the Difference Between Means

We can compare means from two samples and determine if they are significantly different, that is, whether they came from separate populations or whether there was a significant treatment effect.

The standard error of the difference of the means is computed according to the following formula:

$$S.E._{\bar{x}_1 - \bar{x}_2} = \sqrt{\frac{s_1}{N_1} + \frac{s_2}{N_2}}$$

The symbols s_1, s_2 and N_1, N_2 equal the standard deviations and sample sizes of groups 1 and 2, respectively. If the difference between the two means is larger than two times the standard error of the difference, $S.E._{\bar{X}_1 - \bar{X}_2}$, they are significantly different.

EXAMPLE: We want to test the hypothesis that the mean duration of the song bout in population *A* of a bird species is longer than it is in population *B*.

	Duration (sec.)	
Bout No.	Population A	Population B
1	5.2	4.7
2	4.8	5.1
3	6.4	3.2
4	5.3	4.2
5	3.1	3.7
6	5.0	4.1
7	4.4	4.5
8	5.2	3.6
9	4.9	2.9
10	4.7	3.0
Total =	49.0	39.0
$\bar{X}$ =	4.9	3.9

We calculate the standard deviation for population *A*:

Bout No.	$(x_i - \bar{X})$	$(x_i - \bar{X})^2$	
1	0.3	0.09	
2	−0.1	0.01	
3	1.5	2.25	
4	0.4	0.16	
5	−1.8	3.24	
6	0.1	0.01	
7	−0.5	0.25	
8	0.3	0.09	
9	0.0	0.00	
10	−0.2	0.04	
	Total =	6.14	$= \sum(x_i - \bar{X})^2$

$$\frac{\sum(x_i \times \bar{X})^2}{N - 1} = \frac{6.14}{9} = 0.68$$

$$s_A = \sqrt{\frac{\sum(x_i - \bar{X})^2}{N - 1}} = \sqrt{0.68} = 0.82$$

We then calculate the standard deviation for population B:

Bout No.	$(x_i - \bar{X})$	$(x_i - \bar{X})^2$	
1	0.8	0.64	
2	1.2	1.44	
3	−0.7	0.49	
4	0.3	0.09	
5	−0.2	0.04	
6	0.2	0.04	
7	0.6	0.36	
8	−0.3	0.09	
9	−1.0	1.00	
10	−0.9	0.81	
	Total	5.00	$= \sum(x_i - \bar{X})^2$

$$\frac{\sum(x_i - \bar{X})^2}{N - 1} = \frac{5.00}{9} = 0.55$$

$$s_B = \sqrt{\frac{\sum(x_i - \bar{X})^2}{N - 1}} = \sqrt{0.55} = 0.74$$

We can then calculate the standard error of the difference of the means:

$$S.E._{\bar{X}_A - \bar{X}_B} = \sqrt{\frac{s_A}{N_A} + \frac{s_B}{N_B}} = \sqrt{\frac{0.83}{10} + \frac{0.74}{10}} = \sqrt{0.16} = 0.4$$

The difference between the means, 4.9 − 3.9 = 1.0, is larger than twice the $S.E._{\bar{X}_A - \bar{X}_B} = 0.4 \times 2 = 0.8$; therefore, the difference between the means is statistically significant.

Table 11-1. Relationship between nonparametric statistical tests and experimental designs.

Experimental Design	No. of Samples	Statistical Test	Tests for Differences in
Completely randomized	2	Mann-Whitney U test	Location
	2	Kolmogorov-Smirnov test	Form or location
	2	Wald-Wolfowitz runs test	Form or location
	>2	Kruskal-Wallis one-way analysis of variance (ANOVA)	Location
Randomized block	2	Wilcoxen signed rank test	Location
	2	Sign test	Form or location
	>2	Friedman two-way analysis of variance (ANOVA)	Location

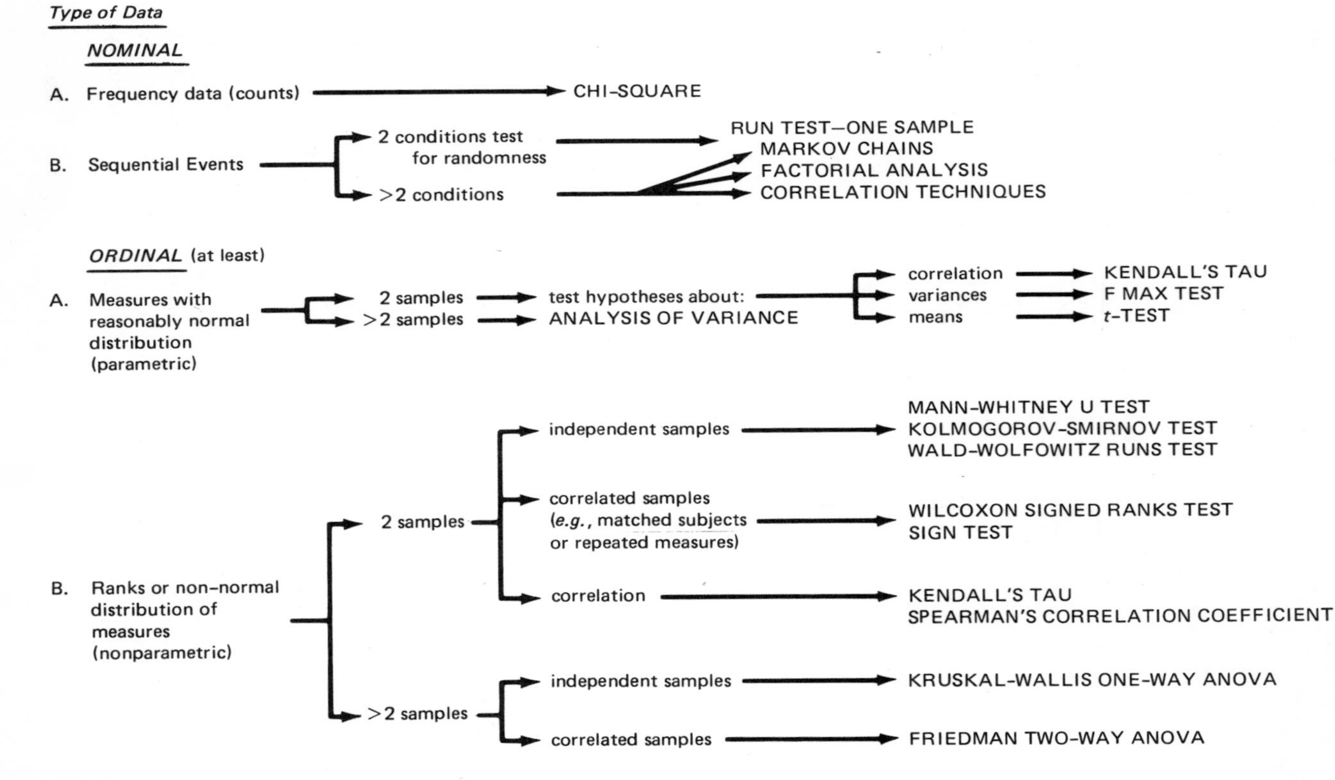

Figure 11-1 Selection of a statistical test (modified from Robson 1973).

3. *t*-Test

The *t*-test is also used to test for significant differences between two sets of data and is based on a comparison of means. We will use the same data on song duration from the two populations. Also, all the factors in the formula have already been calculated.

$$t = \frac{(\bar{X}_A - \bar{X}_B)\sqrt{\dfrac{N_A N_B}{N_A + N_B}}}{\sqrt{\dfrac{(N_A - 1)(s_A^2) + (N_B - 1)(s_B^2)}{N_A + N_B - 2}}}$$

$$= \frac{4.8 - 3.9\sqrt{\dfrac{10 \times 10}{10 + 10}}}{\sqrt{\dfrac{(10 - 1)(0.71) + (10 - 1)(0.56)}{10 + 10 - 2}}}$$

$$= \frac{0.9\sqrt{\dfrac{100}{20}}}{\sqrt{\dfrac{(9)(0.71) + (9)(0.55)}{18}}}$$

$$= \frac{(0.9)(2.24)}{\sqrt{\dfrac{6.39 + 4.95}{18}}} = \frac{2.02}{\sqrt{0.63}} = \frac{2.02}{0.79} = 2.56$$

$$t = 2.56$$

We then obtain the tabular value for *t* from Table A1 for 9 degrees of freedom (*d.f.*) and a significance level of $P < 0.05$ (95%).

$$\text{tabular } t = 2.262$$

Since our calculated *t* of 2.56 is larger than the tabular *t* value of 2.26, we conclude that the data are from two distinctly different populations. That is, song duration in Population *A* is statistically greater than it is in Population *B*.

Although there are several assumptions about the data (*e.g.*, normality, homogeneity of variance) which should be met before the *t*-test is used, the *t*-test is sufficiently *robust* so that these assumptions can be violated to a reasonable extent without affecting the validity of the test.

B. NONPARAMETRIC STATISTICAL TESTS

1. *F*-Max Test

This test is used to determine whether there is homogeneity of variance between two or more sample populations. For our previous example we can determine whether there is a significant difference in song-duration variability between Population *A* and Population *B*. Like the *t*-test, the *F*-max test assumes that the data are normally distributed; however, the *F*-ratio test is also *robust* and is valid even when this assumption is violated slightly.

Determine the variance for each set of scores (populations) by squaring the standard deviation:

$$\text{variance of Pop. } A = s_A^2 = (0.82^2) = 0.67$$

$$\text{variance of Pop. } B = s_B^2 = (0.74^2) = 0.55$$

$$F = \frac{\text{largest variance}}{\text{smallest variance}} = \frac{0.67}{0.55} = 1.21$$

Obtain the tabular value of *F* from Table *A*2. Two different figures are needed. Those across the top of the table refer to the sample which had the larger variance; those on the side are for the sample with the smaller variance. In our case both degrees of freedom are 9 ($d.f. = N - 1$).

The tabular *F* value for the 95% confidence level is 4.03. Since our calculated value of 1.21 is not larger than the tabular value (4.03), we conclude that there is no statistically significant difference in the variances of the samples; that is, the song durations do not vary significantly more in one population than the other.

2. Mann-Whitney *U* Test

This is a nonparametric test for use with independent subjects designs. It is the nonparametric counterpart of the *t*-test. Whereas the *t*-test determines significant differences between means, the Mann-Whitney *U* test uses the medians to test for a significant difference in the location of the sample data.

The Mann-Whitney *U* test should be used if the data are in the form of ranks, not normally distributed or the variances of the two groups being greatly different. If the samples are correlated (paired or matched), use the Wilcoxon test.

Let us use two other samples of song durations from populations in which the variances are obviously different.

Song Duration (sec.)		Ranks	
Population A	Population B	Population A	Population B
4.7	8.1	10	15
5.3	4.2	12	7
3.6	6.7	3	13
5.1	9.5	11	16
4.0	2.7	5	2
4.1	1.8	6	1
3.8	7.8	4	14
4.3		8	
4.4		9	
		Total	68

The first step is to rank the data using both groups. The lowest score gets rank no. 1.

T = sum of ranks in smaller sample (in this case Population B) = 68; if both samples are the same size, sum the ranks of the first sample.

Calculate

$$U_S = N_S N_L + \frac{N_S(N_S + 1)}{2} - T$$

where: N_S = number of measures in the smaller sample; N_L number of measures in the larger sample.

$$U_S = (7)(9) + \frac{7(7 + 1)}{2} - 68 = 63 + 28 - 68 = 23$$

$$U_L = N_S N_L - U_S = (7)(9) - 23 = 40$$

Look up the tabular value for $N_S = 7, N_L = 9$ in Table A3.

tabular value = 12
$U_S = 23$
$U_L = 40$

There is a significant difference if either of the observed values (U_S or U_L) is *equal to* or *less than* the tabular value. Hence, in this case the song durations are not significantly different in the two populations. This you would suspect, since there is such variability in Sample *B* that it overlaps the values in Sample *A*; we can see this better in graphic form

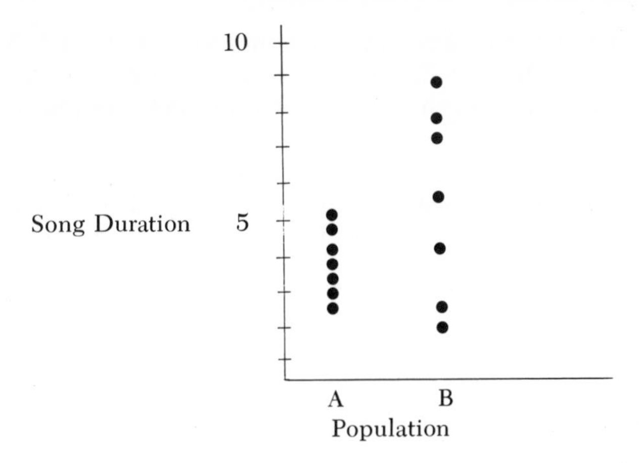

An F-max test would be expected to show a significant difference in the variation of bird-song duration from these two populations.

3. Kolmogorov-Smirnov Test

This test is used to determine whether two samples differ significantly in either form or location. That is, it tests the H_0 (null hypothesis) that the two samples are taken from the same population.

EXAMPLE: We wish to determine whether feeding bouts tend to be of different duration between two small herds of deer that differ in sex and age composition.

Feeding Bout Duration (min.)	
Herd A	Herd B
43	24
62	31
81	19
69	47
73	35
55	29
64	18
18	13
89	43
67	17
59	65
61	28

Since we are working with a continuous variable, the scale of observations must be divided into intervals. An interval should be selected such that no single interval contains more than 2–3 observations. Each sample is then arranged in order of magnitude, and the cumulative frequencies of observations for each sample up to each interval are determined.

Interval	Rearranged Observations		Cumulative Frequencies		
(No. Observations)	Herd A	Herd B	Herd A	Herd B	Difference
0–4			0	0	0
5–9			0	0	0
10–14		13	0	1	1
15–19	18	17, 18, 19	1	4	3
20–24		24	1	5	4
25–29		28, 29	1	7	6
30–34		31	1	8	7
35–39		35	1	9	8
40–44	43	43	2	10	8
45–49		47	2	11	9*
50–54			2	11	9*
55–59	55, 59		4	11	7
60–64	61, 62, 64		7	11	4
65–69	67, 69	65	9	12	3
70–74	73		10	12	2
75–79			10	12	2
80–84	81		11	12	1
85–89	89		12	12	0
90–94			12	12	0
95–100			12	12	0

Calculate the differences between the cumulative frequencies for the two samples (Herds *A* and *B*). The largest of these differences (*) is taken as the test statistic.

$$\text{calculated test statistic} = 9$$

Then obtain the tabular value from Table *A*4.

$$\text{tabular value } (N = 12) = 6 \text{ at } 0.05 \text{ level}$$

Since our calculated value (9) exceeds the tabular value (6), we conclude that the durations of feeding bouts differ significantly between the two herds.

4. Wald-Wolfowitz Runs Test

This test is used to test the H_0 that two independent samples have been drawn from the same population. It will reject H_0 if the two samples differ significantly in either form or location.

EXAMPLE: We wish to determine whether agonistic encounters between songbirds are more frequent at feeder Type *A* or Type *B*. We collect data during 5 sampling periods at each feeder type.

Feeder Type *A*	Feeder Type *B*
12	6
16 (N_1=5)	11 (N_2=5)
8	6
10	3
17	5

Rank all the scores in order of increasing size and cast them into a single order:

Identify their respective populations and determine the runs accordingly	3 5 6 6	8 10	11	12 16 17
	B B B B	A A	*B*	A A A
	1	2	3	4

r (no. runs) = 4

If the observed value (r = 4) is equal to or less than the tabular value (Table $A5_1$), then the H_0 is rejected at the 0.05 level.

$$\text{tabular value } (N_1 = 5, N_2 = 5) = 2$$

$r(4) > 2$ (tabular value); therefore the H_0 cannot be rejected. In this example the small sample size was reflected in our inability to demonstrate statistically what appears to be a real difference between the two types of feeders.

5. Kruskal-Wallis One-Way Analysis of Variance

The Kruskal-Wallis one-way analysis of variance is a nonparametric test for determining if several (>2) independent samples are from different populations. It tests for differences in location and requires at least ordinal measurement.

EXAMPLE: Once again we will use song durations. This time we will take samples from three different geographical areas. We want to determine if the song durations are significantly different in the different areas.

Song Duration (sec.)		
Area A	Area B	Area C
4.4	6.9	9.2
3.4	7.1	8.1
6.7	5.2	8.3
3.8	4.3	7.2
4.1	8.2	9.1
5.0		8.9

All of the scores are then ranked as a group beginning with rank 1 for the lowest score.

Ranks		
Area A	Area B	Area C
5	9	17
1	10	12
8	7	14
2	4	11
3	13	16
6		15
$R_A = 25$	$R_B = 43$	$R_C = 85$

Calculate the sum of the ranks in each column (R_J). Square each of the R_J's separately and divide by the number of scores in that group.

$$\frac{R_J^2}{N_J}$$

$$\frac{(25)^2}{6} = 104.17; \ \frac{(43)^2}{5} = 369.80; \ \frac{(85)^2}{6} = 1204.17$$

Sum the figures just calculated.

$$\sum \frac{R_J^2}{N_J} = 104.17 + 369.80 + 1204.17 = 1678.14$$

Now compute H from the following formula, where N = total number of observations.

$$H = \left(\frac{12}{N(N+1)} \times \sum \frac{R_J^2}{N_J}\right) - 3(N+1)$$

$$H = \left(\frac{12}{17(17+1)} \times (1678.14)\right) - 3(17+1) = 11.78$$

When there are more than 5 observations in each group, then the calculated H is compared to the chi-square value given in Table A6. Degrees of freedom equals the number of columns (or groups) minus one; $d.f. = 3 - 1 = 2$

$$\text{Tabular } x^2_{0.05,2} = 5.99$$
$$H = 11.78$$

Since our calculated H (11.78) is larger than the tabular chi-square value (5.99), we conclude that there is a statistically significant difference in song duration between the three groups.

6. Wilcoxon Test

This test should be used when the samples are correlated (matched or paired). This is also a nonparametric test and must be used if the parametric test assumptions (p. 232) are severely violated. It tests for significant differences in location of the sample data.

In this example we will use 16 adult male songbirds, eight of which have been raised in a sound-enriched environment and eight in a sound-impoverished environment. Then we look at their song durations when placed individually in an observation room. They have been matched according to age, four to each age group, and randomly assigned to one of the two treatments.

	Song Duration (sec.)			
Age (Yrs.)	Sound-Enriched	Sound-Impoverished	Differences	Rank
1	8.2	6.2	−2.0	4
	7.1	4.3	−2.8	6
2	7.5	5.4	−2.1	5
	6.8	1.3	−5.5	8
3	7.8	7.9	0.1	1
	8.1	7.1	−1.0	3
4	6.9	2.0	−4.9	7
	7.4	8.0	0.6	2

The first step is to obtain the differences between each pair of scores (right column score = left column score).

Next, rank the scores (ignoring the signs). The lowest score gets rank no. 1.

T = sum of ranks for differences with less frequent sign.
$T = 1 + 2 = 3$

Obtain a tabular value for T from Table $A7$. N = number of paired scores = 8

$$\text{Tabular } x^2_{0.05,2} = 4$$
$$\text{Calculated } T = 3$$

If the calculated T is equal to or less than the tabular value, then there is a significant difference between the two groups. In our example, since the calculated T (3) is smaller than the tabular T (4), we conclude that the type of sound environment had a significant effect on the birds' song duration.

7. Sign Test

The sign test is used with correlated samples to test for significant differences in form or location.

For our example we will use the same data that we compared with the standard error of the difference between the means (p. 242).

The first step is to give each pair of scores a plus (+) if the score in the left-hand column exceeds that in the right-hand column, a minus (−) if the reverse is true, or a zero (0) if there is no difference.

	Song Duration (sec.)		
Bout No.	Population A	Population B	Sign
1	4.6	4.7	−
2	5.3	5.1	+
3	4.4	3.2	+
4	3.1	4.2	−
5	6.4	3.7	+
6	5.3	4.1	+
7	4.7	4.5	+
8	4.8	3.6	+
9	5.0	2.9	+
10	4.4	3.0	+

$$L = \frac{\text{no. of times least}}{\text{frequent sign occurs}} = 2$$

$$T = \frac{\text{total no. of pluses}}{\text{and minuses}} = 10$$

Determine from Table A8 the probability of obtaining L or fewer of the less frequent sign out of a total of T signs.

$$p = 0.110$$

Since this probability is greater than the significance level decided upon (*e.g.*, 95% = 0.05), we conclude that the difference in song duration between the two populations is not significant.

This is really a first approximation test since it does not take into account the magnitude of the differences. Also, visual inspection of the data and the standard error of the difference of the means calculated earlier would suggest that perhaps song duration for population A is really greater. Another way to test for significant differences between song durations from the two populations is to utilize a test for significant difference between the means that also takes into account the variability in the data. One such measure is the standard error of the difference of the means (p. 242).

8. Friedman Two-Way Analysis of Variance

The Friedman Two-way Analysis of Variance is a nonparametric test used to determine if several (>2) correlated samples are from different populations. It requires at least ordinal measurements and tests for differences in location.

When we say the samples are correlated, we mean that they are grouped with regard to two variables, hence the two-way test.

EXAMPLE: Let us sample song durations again from four geographic areas; but this time we will block our samples into three time periods: sunrise, noon, and sunset.

	Song Duration (sec.)			
	Area A	Area B	Area C	Area D
Sunrise	6.7	8.6	10.2	6.5
Noon	3.4	4.3	6.5	2.5
Sunset	6.3	8.3	9.9	7.1

The scores for each *row* are then ranked separately. This minimizes the between-row differences and allows the between-column differences to be tested.

	Area A	Area B	Area C	Area D
Sunrise	2	3	4	1
Noon	2	3	4	1
Sunset	1	3	4	2
	$R_A = 5$	$R_B = 9$	$R_C = 12$	$R_D = 4$

Calculate each of the R_j's and then the separate R_j^2's.

$$R_A^2 = 25 \qquad R_B^2 = 81 \qquad R_C^2 = 144 \qquad R_D^2 = 16$$

Find the sum of R_j^2:

$$\Sigma R_j^2 = 25 + 81 + 144 + 16 = 266$$

Calculate x_r^2 according to the formula:

$$x_r^2 = \left(\frac{12}{kN(N+1)} \Sigma R_j^2\right) - 3k(N+1)$$

k = number of rows, or the number of times the rank order system was used

N = number of columns, or number of ranks

$$x_r^2 = \left(\frac{12}{12(5)} \times 266\right) - 9(5) = 53.20 - 45 = 8.20$$

Obtain the tabular chi-square value from Table A6 for d.f. $= k - 1 = 2$

$$\text{tabular } x^2_{0.05} = 5.99$$
$$\text{calculated } x_r^2 = 8.20$$

Since our calculated x_r^2 (8.20) is larger than the tabular value (5.99), we conclude that there is a significant difference in song duration between the areas.

Friedman two-way analysis of variance with replications:

	Song Duration (sec.)			
	Area A	Area B	Area C	Area D
Sunrise	6.7	8.6	10.2	6.5
	6.4	6.9	8.4	5.8
	5.5	7.0	9.8	5.9
Noon	3.4	4.8	4.6	2.5
	2.7	4.7	7.5	0.5
	4.1	3.9	7.4	2.9
Sunset	6.3	8.3	9.9	7.1
	5.8	7.0	11.1	6.1
	4.9	6.6	8.1	6.7

Above, with scores ranked:

	Area A	Area B	Area C	Area D
Sunrise	2	3	4	1
	2	3	4	1
	1	3	4	2
Noon	2	4	3	1
	1	3	4	2
	3	2	4	1
Sunset	1	3	4	2
	1	3	4	2
	1	2	4	3
	$R_A = 14$	$R_B = 26$	$R_C = 35$	$R_D = 15$
	$R_A^2 = 196$	$R_B^2 = 676$	$R_C^2 = 1225$	$R_D^2 = 225$

$$\sum R_j^2 = 196 + 676 + 1225 + 225 = 2322$$

$$x_r^2 = \left(\frac{12}{kN(N+1)} \sum R_j^2\right) - 3k(N+1) = \left(\frac{12}{36(5)} 2322\right) - 27(5)$$

$$= 154.88 - 135.00 = 19.88$$

$$\text{tabular } x^2_{0.05} = 15.51$$

$$\text{calculated } x_r^2 = 19.88$$

Since our calculated x_r^2 (19.88) is larger than the tabular value (15.51), we conclude that there is a statistically significant difference in song duration between the areas.

9. Correlation Analysis

These analyses are used to test for the relationship between two variables. In many instances we want to analyze not only the distribution of the dependent variable scores (as we have done previously), but also to examine the relationship between the dependent and independent variables.

The researcher should first prepare a *scattergram* of his data for visual inspection. Scattergrams will provide a good indication of (1) whether a correlation exists and (2) whether it is positive or negative.

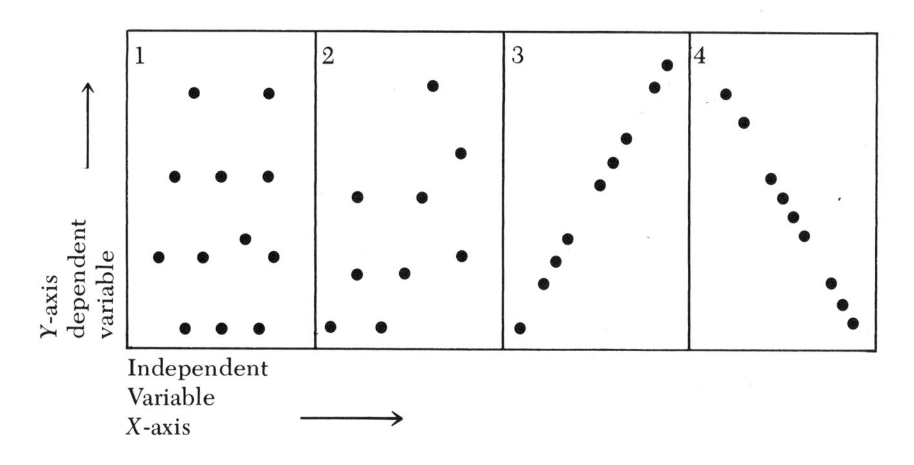

The graphs above show how the data are plotted with the dependent variable on the Y-axis and the independent variable on the X-axis. The examples illustrate (1) no correlation, (2) low positive correlation, (3) perfect positive correlation, and (4) perfect negative correlation.

a. Kendall's Tau

Kendall's tau (τ) is an example of a correlation coefficient. It is a nonparametric statistic which measures the tendency of two rank orders of data to be similar.

For an example we will use hypothetical data on activity level and frequency of song in the males of a species of territorial songbird. We believe that these are correlated so we begin by constructing a scattergram.

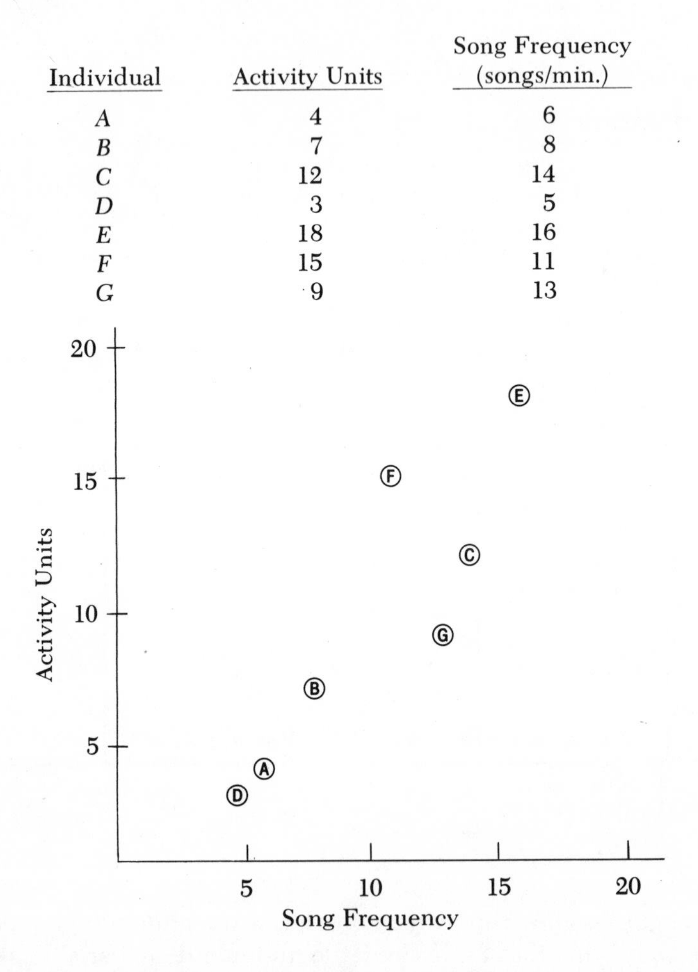

Individual	Activity Units	Song Frequency (songs/min.)
A	4	6
B	7	8
C	12	14
D	3	5
E	18	16
F	15	11
G	9	13

From the scattergram alone, there would appear to be a good positive correlation. However, we can proceed with the calculation of Kendall's tau and test for that correlation.

The first step is to arrange the measurements in order along one of the variables. In our case we will arrange them in order by the activity units, but either could be used.

Activity units:	3	4	7	9	12	15	18
Song frequency:	5	6	8	13	14	11	16
Individual:	*D*	*A*	*B*	*G*	*C*	*F*	*E*

For each individual (*A* through *G*) count the number of individuals to the right of it in the table with both larger activity unit and song frequency measurements: For example, to the right of individual *A* five individuals (*B*, *G*, *C*, *F*, and *E*) have both larger activity units and song frequencies.

Another method is to go to the scattergram and for each individual count the number of individuals which have larger dependent *and* independent scores (*i.e.*, those individuals to the right and above in the scattergram).

Individual	Number of individuals with larger dependent and independent scores
A	5
B	4
C	1
D	6
E	0
F	1
G	2
Calculate the total S^+ =	19

Calculate $S = 2S^+ - \dfrac{N(N-1)}{2}$

where N = no. of pairs of scores = 7

$$S = 2(19) - \frac{7(7-1)}{2} = 38 - \frac{42}{2} = 17$$

Next calculate tau:

$$\tau = \frac{S}{N(N-1)/2} = \frac{17}{7(7-1)/2} = \frac{17}{21} = 0.81$$

Obtain the tabular τ value from Table A9 at the appropriate level of significance.

$$\text{tabular } \tau = 0.63$$
$$\text{calculated } \tau = 0.81$$

Since our calculated τ (0.81) is larger than the tabular τ (0.63), we conclude that there is a significant positive correlation between activity and frequency of song. The size of the calculated τ is, of course, a measure of the strength of the correlation, with a τ of 1.00 indicating a perfect correlation. When τ is positive the relationship is direct (*i.e.*, the variables increase together). An inverse relationship is indicated by a negative τ.

b. Spearman's Rho

Spearman's rho (ρ_s) is another nonparametric correlation coefficient which measures the covariation between two rank-ordered variables. Only an ordinal scale of measurement is necessary, as it is with Kendall's tau.

As an example, we will again use the data from the previous example in order to compare Spearman's rho with Kendall's tau. Although they do not generate numerically equivalent correlation coefficients, they are equally powerful tests (Siegel 1956).

First, the measures for each variable (arbitrarily assigned X and Y) are ranked separately.

	X Variable		*Y* Variable	
Individual	Activity Units	Rank	Song Frequency	Rank
A	4	2	6	2
B	7	3	8	3
C	12	5	14	6
D	3	1	5	1
E	18	7	16	7
F	15	6	11	4
G	9	4	13	5

If two or more measures within a variable are equal, then assign to each of them the average of the ranks that would have been assigned had they not been equal.

Next, the difference between the ranks for each pair of measures (d) is calculated and squared.

Individual	*X* Variable Rank	*Y* Variable Rank	d	d^2
A	2	2	0	0
B	3	3	0	0
C	5	6	1	1
D	1	1	0	0
E	7	7	0	0
F	6	4	2	4
G	4	5	1	1
			Total ($\Sigma_{d}{}^{2}$) =	6

Spearman's rho is calculated according to the formula:

$$\rho_s = 1 - \frac{6\Sigma d^2}{N^3 - N}$$

where N = the number of measures per variable = 7

$$\rho_s = 1 - \frac{6(6)}{343 - 7} = 1 - \frac{36}{336} = 1 - 0.107 = 0.89$$

This calculated value of 0.89 is then compared to the values in Table A 10 for $N = 7$. Note that this is a one-tailed test. The calculated value is larger than the tabular value at the 0.05 level of significance, but not the 0.01 level. We accept the significance at the 0.05 level and conclude that there is a statistically significant correlation between activity and song frequency—just as we did using Kendall's tau.

When the sample size exceeds 30, the test statistic tau (τ) is calculated:

$$\tau = \rho_s \sqrt{\frac{N - 2}{1 - r_s^2}}$$

This calculated value is then compared to the tabular values in Table A1 where $d.f. = N - 2$. To obtain the level of significance for the one-tailed test, divide the probabilities shown (which are for two-tailed tests) by two.

c. *Chi-Square (2 × 2)*

Another test of association is the chi-square (χ^2) test. This is used with data in the form of frequencies (*i.e.*, number of occurrences). These data must be independent of each other; generally, this precludes taking more than one measurement from a single individual. It is possible to convert measurements into frequency data, but information is then lost. For example, we could group the data used in the example above into the following contingency table:

		Song Frequency	
		1–10	11–20
Activity Units	1–10	3	1
	11–20	0	3

It can also be seen that a small sample size does not lend itself well to a chi-square analysis (reflected in the scanty contingency table).

However, for an example let us assume that we have a much larger sample from 40 individual birds so that our contingency table is as follows:

		Song Frequency	
		1–10	11–20
Activity Units	1–10	17	4
	11–20	7	12

This may also be too small a sample size for chi-square analysis. This will be determined as we proceed with the analysis, since the chi-square test should not be used for a 2 × 2 when the sample size is 40 or less and one or more of the *expected frequencies* are smaller than five. For matrices larger than 2 × 2, do not proceed with the chi-square test if more than 20% of the cells have expected frequencies smaller than five or any cell has an expected frequency of less than one.

The number in each cell of the contingency table above is the *observed frequency* of each measurement.

Calculate the row and column totals:

	17	4	21	Row totals
	7	12	19	
Column totals	24	16	40	= Grand total

Calculate the *expected frequency* for each cell separately with the following formula:

$$\text{expected frequency } (E) = \frac{\text{row total} \times \text{column total}}{\text{grand total}}$$

$\frac{21 \times 24}{40} = 12.6$	$\frac{21 \times 16}{40} = 8.40$
$\frac{19 \times 24}{40} = 11.4$	$\frac{19 \times 16}{40} = 7.6$

Expected frequencies:

12.6	8.4
11.4	7.6

Since none of the expected frequencies are smaller than five, we conclude that the sample size is sufficiently large and continue with the analysis.

Calculate the difference between the observed and expected ($O - E$ or $E - O$) for each cell.

cell *A* 4.4	cell *B* 4.4
cell *C* 4.4	cell *D* 4.4

Subtract 0.5 from each cell:

Cell	$(O - E) - 0.5$	$\{(O - E) - 0.5\}^2$	$\frac{\{(O - E) - 0.5\}^2}{E}$
A	3.9	15.21	1.21
B	3.9	15.21	1.81
C	3.9	15.21	1.33
D	3.9	15.21	2.00

Square the results of the previous calculations: $\{(O - E) - 0.5\}^2$

Divide the results of the previous calculations by the E for each cell separately:

$$\frac{\{(O - E) - 0.5\}^2}{E}$$

$\frac{15.21}{12.6} = 1.21$	$\frac{15.21}{8.4} = 1.81$
$\frac{15.21}{11.4} = 1.33$	$\frac{15.21}{7.6} = 2.00$

Add together the four figures just calculated to obtain the χ^2.

$$\chi^2 = \sum \frac{\{(O - E) - 0.5\}^2}{E} = 1.21 + 1.81 + 1.33 + 2.00 = 6.35$$

Obtain the tabular χ^2 from Table A6 at the appropriate level of significance. Degrees of freedom = (no. of rows − 1) × (no. of columns − 1) = (2 − 1) × (2 − 1) = 1

tabular χ^2 = 3.84
calculated χ^2 = 6.35

Since our calculated χ^2 (6.35) is larger than the tabular χ^2 (3.84), we conclude that there is a significant correlation between activity and frequency of song.

Chi-Square analysis with contingency tables larger than 2 × 2 is essentially the same as for the 2 × 2. However, when the contingency table is larger, Yate's correction (which is the subtraction of 0.5 from $O - E$) is not needed. Therefore the formula becomes:

$$\chi^2 = \sum \frac{(O - E)^2}{E}$$

The expected frequencies are calculated in the same way:

$$E = \frac{\text{row total} \times \text{column total}}{\text{grand total}}$$

Degrees of freedom (*d.f.*) are calculated according to the formula:

$$d.f. = (\text{no. of rows} - 1) \times (\text{no. of columns} - 1)$$

Example:

4 rows × 4 columns

9 *d.f.*

$d.f. = (4 - 1) \times (4 - 1) = 9$

4 rows × 7 columns

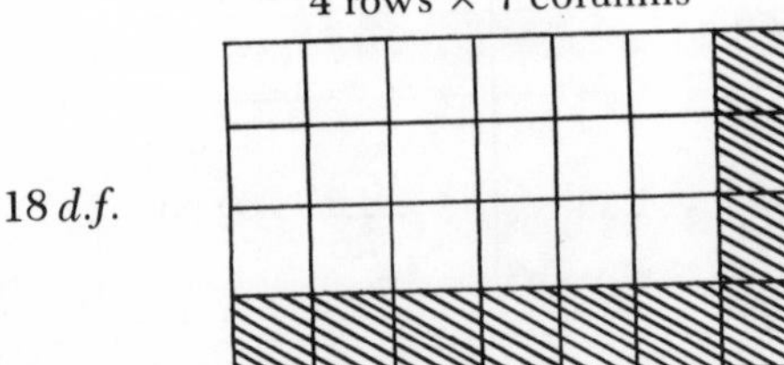

18 *d.f.*

$d.f. = (4 - 1) \times (7 - 1) = 18$

d. Chi-Square Goodness-of-Fit Test

The goodness-of-fit test is used to compare an observed frequency distribution with a hypothetical distribution. The same general rules apply as for the chi-square test discussed above; however, we supply the expected values in the goodness-of-fit test rather than generating them from the observed values.

For example, we observe male songbirds of species *A* singing from three different species of trees. We determine that the three species of trees are equally distributed throughout each male's territory, so that if there were no preference it would sing equally often from trees of each species. We gather the following data from 60 males and then assign equal numbers (20 = expected frequencies) to each tree species.

Tree Species	No. males observed singing (observed frequency *O*)	Expected frequency (*E*) based on equal distribution	$\frac{(O-E)^2}{E}$
A	38	20	16.2
B	10	20	5.0
C	12	20	3.2
	60	60	$\chi^2 = 24.4$

The number of degrees of freedom in a goodness-of-fit test equals the number of categories minus one. In our case we have three tree species (*A*, *B* and *C*). Therefore *d.f.* = 3 − 1 = 2. Looking at Table *A*6, with 2 *d.f.*, we see that our calculated χ^2 of 24.4 exceeds the tabular values even at alpha level 0.001. Therefore, we conclude that the male songbirds of species *A* do not sing equally from all three tree species.

Denenberg's (1976) and Meyer's (1976) textbooks should be consulted for further uses of the chi-square test.

e. Interpretation of an Association

Kendall's tau, Spearman's rho and chi-square measure degree of association. A significantly large tau (τ), rho (ρ_s) or chi-square (χ^2) indicate a high degree of association whether positive or negative. There are basically three interpretations for a high degree of association (*i.e.*, high correlation):

1. Change in variable *A* causes a change in variable *B*.
2. Change in variable *B* causes a change in variable *A*.
3. Neither interpretations 1 or 2 are correct, but rather a change in some other variable (*C*) causes a change in both *A* and *B*.

The statistical tests will not tell us which of these interpretations is correct. We can only judge (with varying degrees of validity) which is correct according to our prior knowledge of the variables and their relationships.

C. RATES OF BEHAVIOR

Problems associated with the analysis of *rates of behavior* were addressed by Altmann and Altmann (1977). They identified the following questions that commonly arise when dealing with rates of behavior:

1. Can the frequency distribution of the observations be accounted for by the population composition (age and sex class distribution)?
2. What are the expected values of these frequencies if the mean rate of behavior is independent of class?
3. How can reasonable estimates of class-specific behavior rates be obtained from a set of data and tested in a new sample?
4. What are the expected frequencies, for dyadic interactions, with pairwise independence?

EXAMPLE 1: Expected frequency of behavior per class when the rates are unknown but are assumed to be constant or independent of class, and the population composition is stable.

$$E_a = \frac{N\, m_x}{M}$$

where E_a = expected frequency of behavior a for members of class x
N = total number of occurrences of behavior a for all individuals of all classes in the sample
m_x = number of individuals in class x
M = number of individuals in the sampled population

In order to illustrate the use of the formula we will use the hypothetical data in Table 11-2. See Altmann and Altmann (1977) for an additional example.

Table 11-2. Hypothetical data on the number of threats in a herd of 50 mule deer.

	Adult Males	Adult Females	Immatures	Totals
Number of individuals	15	20	15	50
Observed number of threats	80	15	5	100

We can determine the expected frequency of threat behavior in adult males as follows:

$$N = 100$$
$$M = 50$$
$$m_x = 15$$
$$E_a = \frac{N\,m_x}{M} = \frac{100 \times 15}{50} = \frac{1500}{50} = 30$$

Therefore, the expected frequency of threats in adult males is 30 compared to the observed frequency of 80. This can be done for all age classes, and then a chi-square test can be conducted to determine whether the differences between the observed and expected frequencies are statistically significant.

EXAMPLE 2: Same as 1 above, but the population composition changes

$$E_a = N\,\frac{\Sigma_j t_j m_{xj}}{\Sigma_i \Sigma_j t_j m_{ij}}$$

E_a = expected frequency of behavior a for members of class x

N = total number of occurrences of behavior a for all individuals of all classes in the sample

$\Sigma_j t_j m_{xj}$ = total sample time for all individuals of class x in the entire study
= (time in sample period one × number of individuals in class y)+ (time in sample period two × number of individuals in class y) + . . . (time in final sample period × number of individuals in class y)

$\Sigma_i \Sigma_j t_j m_{ij}$ = total sample time for all individuals of all classes for the entire study
= (time in sample period one × number of individuals in all classes) + (time in sample period two × number of individuals in all classes) + . . . (time in final sample period × number of individuals in all classes)

As an illustration, we will use the hypothetical data in Table 11-3. See Altmann and Altmann (1977) for another example.

We can determine the expected frequency of threats in adult males in this population of changing composition as follows:

$$N = 260$$
$$\Sigma_j t_j m_{xj} = [(5 \times 10) + (11 \times 15) + (4 \times 12)]$$
$$= 50 + 165 + 48 = 263$$

$$\begin{aligned}\Sigma_i\Sigma_j t_j m_{ij} &= 263 + [(5 \times 18) + (11 \times 20) + (4 \times 16)] \\ &\quad + [(5 \times 13) - (11 \times 15) + (4 \times 12)] \\ &= 263 + 374 + 278 = 915\end{aligned}$$

$$E_a = N\,\frac{\Sigma_j t_j m_{xj}}{\Sigma_i\Sigma_j t_j m_{ij}} = 260\,(263/915) = 75$$

Therefore, the expected frequency of threats in adult males in this changing population is 75 compared to the observed number of 210. Once again we could make the same calculations for the adult females and immatures, and then use a chi-square test to determine if the differences are significant.

EXAMPLE 3: Hypothetical uniform class-specific rates of behavior expressed as a mean number of occurrences per individual per unit time.

$$E_a = \lambda_x \Sigma_j t_j m_{xj}$$

E_a = the hypothetical expected frequency of behavior a for members of class x

λ_x = hypothetical mean participation rate per member of class x

$\Sigma_j t_j m_{xj}$ = total sample time for all individuals of class x in the entire study

As an illustration we can obtain the hypothetical mean participation rate (λ) for adult males by referring to Table 11-3.

$$\lambda_x = \frac{80}{100} = 0.80$$

$$\Sigma_j t_j m_{xj} = (5 \times 10 + 11 \times 15 + 4 \times 12) = 263$$

$$E_a = 0.80 \times 263 = 210$$

Therefore, the hypothetical expected frequency of threats for adult males during the entire sample period (20 hours, Table 11-3) is 210. The total number of males observed was 37, so that the mean rate per individual was 210/37 = 5.7 threats for the 20-hour sample. This hypothetical mean rate can then be compared to the observed mean rate in this sample or from observations gathered later from the same or a different population.

Altmann and Altmann (1977) proceed to the consideration of the interactions between individuals. They provide procedures for calculating expected rates of behavior for symmetric and asymmetric interactions at constant rates and interactions with hypothetical class-specific rates of behavior. These procedures are more complex than those considered above and will not be described here.

Table 11-3. Hypothetical data on the number of threats in a herd of mule deer.

	Adult Males	Adult Females	Immatures	Totals
Sample Period 1 (5 hr.)				
Number of individuals	10	18	13	41
Observed number of threats	60	13	2	75
Sample Period 2 (11 hr.)				
Number of individuals	15	20	15	50
Observed number of threats	80	15	5	100
Sample Period 3 (4 hr.)				
Number of individuals	12	16	12	40
Observed number of threats	70	14	1	85
Total number of observed threats (Sample periods 1, 2 and 3)	210	42	8	260

D. ANALYSIS OF SEQUENCES

The temporal relationships between two behaviors (from the same or different individuals) are often complex and difficult to analyze. In some bird species the male and female of a pair sing separate portions of a duet, either simultaneously (polyphonically) or successively (antiphonally). Since the temporal patterning of each individual's contribution to the duet is quite constant, we assume that it has been selected for and that it is of adaptive value.

Golani (1976) developed a vocabulary to describe temporal patterns for two limb segments in his analysis of social interactions in golden jackals and Tasmanian devils. This same terminology can be applied to behavioral units performed by one or more individuals. This terminology has not received wide acceptance and use, but is included here as a descriptive format which is available for use or modification. Golani considered all of these relationships to be variations of "simultaneous" events, even when one immediately preceded (prevened) or succeeded (supervened) another.

We will use Golani's definitions; but let p and q stand for two behavioral units:

1. If p and q are temporally contiguous, *i.e.*, follow each other immediately, the relationships will be designated by the suffix "-vene."
2. If during every "instant" of the occurrence of p, q occurs, *i.e.*, p either starts together with or later than the start of q, and p ends together with or

Table 11-4. Vocabulary of terms used to describe the temporal patterning of two behavioral units (adapted from Golani 1976).

	P ends just before *q* starts (contiguous)	*P* ends after *q* started but before *q* ended	*P* ends together with *q*	*P* ends after *q* ended
P starts before *q* started	prevene	invade	convade	encase
P starts together with *q*		pridure	condure	concede
P starts after *q* started but before it ended		entdure	postdure	excede
P starts immediately after *q* ended (contiguous)				supervene

earlier than the end of *q*, the relationship will be designated by the suffix "-dure."

3. If only during the later part of the occurrence of *p*, *q* or part of it also occurs, *i.e.*, *p* starts before *q* and ends after the start of *q* but before or together with the end of *q*, the relationship will be designated by the suffix "-vade."
4. If *p* starts after or together with the start of *q*, but before the end of *q*, and ends after the end of *q*, the relationship will be designated by "-cede."
5. If *q* occurs only during the middle part of *p*, *i.e.*, *p* starts before the start of *q* and ends after the end of *q*, the relationship will be designated by the suffix "-case."

Variants within the five groups are distinguished by the prefixes pre- (before) and super- (after); in- (going in) and ex- (going out); pri- (prior), and post- (later); con- (together); end- (within); and en- (around). [GOLANI 1976]

These prefixes and suffixes can then be combined to describe the various temporal relationships shown in Table 11-4.

Most researchers have found it sufficient to examine sequences of behavior in a more general way than that proposed by Golani. For example, Bakeman (1978) has found it useful to make two basic distinctions in his data. First, are the behavior patterns *sequential* (mutually exclusive) or *concurrent*? Second, is the behavior recorded in an *event* or *time* base; that is, are they occurrences or occurrences plus durations, respectively? These two dichotomies combine to form four types of data:

1. *Type I Data* (event base, sequential). The observer records the order of mutually exclusive behavior units, ignoring their duration (Sackett 1974; Van Der Kloot and Morse 1975).
2. *Type II Data* (event base, concurrent). The observer records the order of behavior units, ignoring their duration; however, they can occur simultaneously (S.A. Altmann 1964).
3. *Type III Data* (time base, sequential). The observer records the order and duration of mutually exclusive behavior units (Dane and Van Der Kloot 1964).
4. *Type IV Data* (time base, concurrent). The observer records the order and duration of behavior units; however, they can occur simultaneously (Verberne and Leyhausen 1976).

The ethologist may suspect that behaviors occur in rather stereotyped sequences which may be the result of 1) a common causal factor, 2) one stimulating the other, 3) differing thresholds, or 4) one priming the other. Predictable sequences are to be expected since animals are certainly not random behavior generators. However, it is the *extent of predictability* that *A* will be quickly followed by *B* that is important to our analysis.

1. Intra-Individual Sequences

A simple form of analysis is to measure the *frequencies* or *conditional probabilities* associated with two or more behaviors. If the conditional probability that A is followed by B is 100%, then it is a *deterministic sequence*. This is rare in behavior. More often A is followed by B at some level of probability less than 100%. This is called a *probabilistic sequence* (or *stochastic sequence*).

Deterministic sequence (rare)	$A \xrightarrow{100\%} B$
Probabilistic sequence (Stochastic sequence) (common)	$A \xrightarrow{<100\%} B$

Kinematic graphs (p. 316) can be generated which show the conditional probabilities of several different behaviors. This is a useful procedure when sequential effects are strong.

There are several explanations for why one behavior is often found to follow another (see above). Another factor is the relative frequency with which particular behaviors occur in the animal's repertoire. The more frequent the two behaviors, the more likely they are to occur together. An extreme example of this is a hypothetical animal which is capable of only two behaviors A and B. Only four sequences are possible.

1. $A \rightarrow A$
2. $A \rightarrow B$
3. $B \rightarrow B$
4. $B \rightarrow A$

The frequency of occurrence of these sequences is determined not only by the size of the repertoire, but also by impossible combinations and the observer's criterion for determining the beginning and end of a sequence. It is also often difficult to determine whether a behavior was repeated ($A \rightarrow A$) or whether it was simply a single occurrence (A).

a. Markov Chains

The sequences discussed above considered only single transitions from one behavior to another. These are referred to as dyads. Higher level sequences are shown below:

Dyad	$A \rightarrow B$
Triad	$A \rightarrow B \rightarrow C$
Tetrad	$A \rightarrow B \rightarrow C \rightarrow D$

Markov chains are sequences of behaviors in which it can be shown that the transitions between two or more of the behaviors are dependent on one another at some level of probability greater than chance; also, the level of probability is assumed to be stationary (see below). Markov chain analysis should take into account the expected frequency of occurrence of sequences based on the frequency of occurrence in the repertoire.

Ashby (1963:165–166) has provided an illustrative example of the application of Markov chains.

Suppose an insect lives in and about a shallow pool—sometimes in the water (W), sometimes under pebbles (P), and sometimes on the bank (B). Suppose that, over each unit interval of time, there is a constant probability that, being under a pebble, it will go up on the bank; and similarly for the other possible transitions. (We can assume if we please, that its actual behaviour at any instant is determined by minor details and events in its environment.) Thus a protocol of its positions might read:

W B W B W P W B W B W B W P W B B W B W P W B W P W
B W B W B B W B W B W B W P P W P W B W B B B W

Suppose, for definiteness, that the transition probabilities are

↓	B	W	P
B	1/4	3/4	1/8
W	3/4	0	3/4
P	0	1/4	1/8

These probabilities would be found by observing its behaviour over long stretches of time, by finding the frequency of, say, B→W, and then finding the relative frequencies, which are the probabilities. Such a table would be, in essence, a *summary of actual past behaviour*, extracted from the protocol.

Markov chains are referred to in "orders" rather than dyads, triads, etc. (Table 11-5.)

Table 11-5. Orders of Markov chains.

Order of Markov Chain	Definition
"zeroth-order" (A,B,C)	The behavioral events are independent
first-order $(B \rightarrow C)$	The probability of occurence of a particular behavior is dependent on only the immediately preceding behavior
second-order $(A \rightarrow B \rightarrow C)$	The probability of occurrence of a particular behavior is dependent on the two immediately preceding behaviors
nth-order $(\ldots X \rightarrow Y \rightarrow Z)$	The probability of occurrence of a particular behavior is dependent on the "n" immediately preceding behaviors

b. Transition Matrices

Markov chains are generally analyzed through the use of a *transition matrix* and comparison to a *random model*. The random model is one that generates *expected* frequencies of occurrence based on number of occurrences. It should be noted that the transition matrix is not a true contingency table since the events included are not independent of each other.

Transition Matrix—Observed Occurrences

		Following Behavior			
		A	B	C	Row Totals
Preceding Behavior	A	10	20	5	35
	B	15	4	6	25
	C	10	2	8	20
Column Totals		35	26	19	80

The expected frequencies can then be calculated for each cell according to the formula:

$$\text{expected frequency} = \frac{\text{row total} \times \text{column total}}{\text{grand total}}$$

Transition Matrix—Expected Occurrences

		Following Behavior			
		A	B	C	Row Totals
Preceding Behavior	A	15.3	11.4	8.3	35
	B	10.9	8.1	6.0	25
	C	8.8	6.5	4.7	20
Column Totals		35	26	19	80

The individual cells in each table can then be searched for those where the observed greatly exceeds the expected. The *A*→*B* transition above is an example where the observed (20) is much larger than the expected (11.4).

This provides a first approximation look at the data. The researcher can then proceed with a chi-square test for the entire matrix (see analysis p. 264) or reduce it to the most important cells (*e.g.*, 2×2 matrix) and conduct a chi-square test (Stokes 1962).

Stationarity (the probability of one behavior following another not changing over time) is assumed in Markov chain analysis; however, *we know that stationarity rarely exists in the behavior of an animal.* For example, daily cycles (circadian rhythms) are likely to cause trends in the data, that is, the frequency of occurrence of most behaviors is not likely to be stable throughout the day. Staddon (1972) concluded that Markov chain analysis was not applicable in his study of behavioral sequences in *Columba livia* because of a lack of stationarity. Lemon and Chatfield (1971) tested their data for stationarity before proceeding with Markov chain analysis. They tested for significant differences in the probabilities of occurrence of different song types between the first and second halves of their sample periods.

Bekoff (1977*b*) lists the following five conditions in which social-behavior data are often collected and which generally do not satisfy the assumption of stationarity:

1. Two or more individuals interacting
2. Lumping data for different individuals (see Chatfield 1973)
3. Developmental studies; individuals forming social relations and perceptual motor skills being acquired
4. Motivational studies (see Slater 1973)
5. Studies of signals having a cumulative "tonic" effect (Schleidt 1973)

c. Lag Sequential Analysis

Sackett (1978) describes a technique called *lag sequential analysis* for measuring the frequency with which selected behaviors precede or follow each other at various lag steps. Lags are the number of event, or time-unit, steps between sequential behaviors. Bakeman (1978:71) describes Sackett's use of the analysis with sequential data from a crab-eating macaque mother and her infant:

> The analysis begins by designating one behavior the "criterion behavior" (this procedure can be repeated as many times as there are behaviors, so that each behavior can serve as the criterion). Then a set of "probability profiles" is constructed, one for each of the other behaviors. Each profile graphs the conditional probabilities for that particular behavior immediately following the criterion (lag 1), following an intervening behavior (lag 2), following two intervening behaviors (lag 3), and so forth . . . Peaks in the profile indicate

sequential positions following the criterion at which a given behavior is more likely to occur, whereas valleys indicate positions at which it is less likely to occur.

A manuscript describing the analysis and a listing of a computer program for performing lag sequential analyses are available from: Dr. Gene P. Sackett, Child Development and Mental Retardation Center, University of Washington, Seattle.

2. Interindividual Sequences

In the analysis of sequences discussed above we have been assuming that an individual's behavior is its primary source of stimulation for subsequent behavior. That is, the behavior of other individuals is disregarded as an important variable. That decision has traditionally been a subjective one left to the experience and discretion of the researcher.

a. Sociometric Matrices

The other side of the coin is where the researcher suspects that the most important stimuli are originating from other individuals—a situation we can loosely describe as communication. In this case our contingency table becomes a *sociometric matrix*, such as the one below with hypothetical data:

		Receiver's Behavior		
		A	B	C
Transmitter's Behavior	A	3	18	27
	B	24	7	35
	C	29	19	6

Once again we can use a chi-square test to analyze for significantly large correlations between the transmitter's behavior and the receiver's behavior. In interindividual matrices there is no problem involved with measurements found along the diagonal; however, stationarity is again a problem when the chi-square test is used.

As with the intraindividual analyses discussed before, correlations found between a transmitter's behavior and a receiver's behavior are only that—correlations. No causal relationship is explicitly demonstrated.

b. Information Analysis

Another way to analyze sequences, both intra- and inter-individual sequences, is through application of information theory. This approach consists of calculating the uncertainty of a series of behaviors given that another behavior or behaviors has occurred. Like Markov chain analysis, it assumes stationarity.

If all behavioral events (k) are not equiprobable, but are independent of each other, the uncertainty (U, often referred to as H) associated with any individual behavioral event (y) is given by the formula:

$$U\,(y) = -\Sigma_k\, p(k) \log_2 p(k)$$

where $p(k)$ is the probability associated with each of the k events. If we consider two events (dyads), the uncertainty of y when its antecedent x is known is given by:

$$U_x(y) = -\Sigma_{j,k}\; p(j,k) \log_2 p_j(k)$$

where $pj(k)$ is the probability that $y = k$ when $x = j$. Similarly, for triads:

$$U_{wx}(y) = -\Sigma_{i,j,k}\; p(i,j,k) \log_2 p_{i,j}(k)$$

and for tetrads:

$$U_{vwx}(y) = -\Sigma_{h,i,j,k}\, p(h,i,j,k) \log_2 p_{h,i,j}(k)$$

If the behavior of an animal depends in part upon the immediately preceding event, then $U_x(y)$ will be less than $U(y)$, and $U_{wx}(y)$ will be less than $U_x(y)$. The objective of information analysis is to determine the number of behavioral units which must be included in sequence (*e.g.*, dyads, triads, etc.) in order to reduce the uncertainty to an acceptably low level, perhaps <0.10. Thus the difference between the measures of uncertainty yielded by successive models will give a measure of the decrease in uncertainty (or conversely the extra amount of information) yielded by that model. However, soon the law of diminishing returns comes into play and we decrease the uncertainty only slightly for successive models. S.A. Altmann (1965) terminated his analysis of sequences of behavior in the rhesus monkey at quadrads (sequence of four). Altmann used Shannon's *measure of redundancy* to calculate an "index of stereotypy" for each order of approximation. This varied from 0 for the zero-order of approximation (all behaviors equiprobable) to over 0.9 for his fourth order of approximation (quadrad). Since a behavior of the rhesus monkey is almost completely determined by the three preceding behavioral events, Altmann chose to go no further.

Additional discussions of the use of information analysis in ethology can be found in Dingle (1972), Hailman (1977), Losey (1978), Steinberg (1977), and Wilson (1975). Note, however, that the utility and applicability of information analysis to ethological data has been questioned by some ethologists, including Bekoff (1976) and Fagen (1977).

E. MULTIVARIATE ANALYSES

Multivariate analyses treat several variables and compare two or more groups. They can be used for: 1) initial data exploration and hypothesis seeking, 2) classification (grouping according to similarities), and 3) hypothesis testing.

These techniques are useful in helping to clarify results through illustrative visual representations, such as dendrograms and three-dimensional diagrams. However, they should be used to express results, not to impress readers. In addition, the researcher should fit the method of analysis to the animal and the problem under investigation, not the animal to the method (Aspey and Blankenship 1977; Bekoff 1977*b*; Tinbergen 1951).

This section is a brief overview which explores selected multivariate analysis techniques by: 1) explaining what they will do for you in terms of how they treat your data, 2) illustrating how they have been used, and 3) describing or referencing the methodologies for this use (Table 11-6). For more extensive coverage see Aspey and Blankenship (1977; 1978), Morgan et al. (1976), Sneath and Sokal (1973), and Sparling and Williams (1978). Keep in mind that, overall, ". . . multivariate analyses are powerful diagnostic tools: (1) for uncovering homogeneous subgroups from naturally-selected heterogeneous samples; and (2) for identifying relationships among multiple variables when the underlying source, or biological basis, of individual variation is unknown" (Aspey and Blankenship 1977).

1. Matrices

Many ethological data are gathered or can be organized into matrices in which several individuals or behaviors are being studied (Fig. 11-2). These data may be widely variable, be scaled in arbitrary units, and frequently include interacting variables (Aspey and Blankenship 1977). Initial visual inspection of these matrices is often confusing; this is where multivariate analyses come into play.

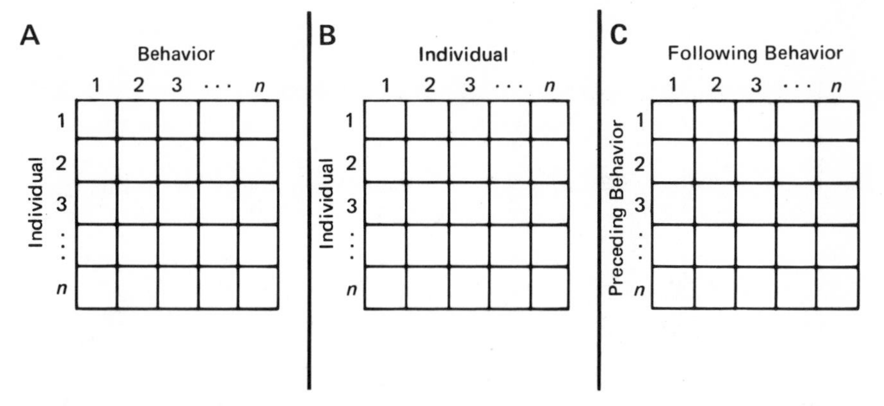

Figure 11-2 Three typical matrices in which ethological data can be organized: *A*, different behaviors given by different individuals; *B*, interactions between individuals; *C*, inter- or intra-individual behavior sequences.

Six multivariate techniques—*R*-factor analysis, *Q*-factor analysis, principal component analysis, cluster analysis, discriminant function analysis, and multidimensional scaling—will be discussed as they have been used in the analysis of data found in matrices like those in Figure 11-2. However, the application of these analyses is not restricted to those discussed, but may be used to analyze any correlation matrix. Nevertheless, the characteristics of the particular analysis should be matched to the objectives of the study (see Gottman 1978).

Since one of the advantages of these multivariate methods is that they allow visualization of the variables through graphical displays (Rohlf 1968), selected examples will be provided.

2. Grouping Behaviors

R-factor analysis organizes a large number of variables (*e.g.*, behaviors in Matrix *A*, Fig. 11-2) into a smaller number of "factors" based on their underlying similarities. When applied to Matrix *A*, this analysis extracts behavior-related factors which account for a large percentage of the total variance.

For example, Aspey (1977*c*) recorded the occurrence of 20 different behaviors in 40 individual spiders. He used R-factor analysis to extract four behavior-related factors (groups from the 20 different behaviors) which accounted for 74.3% of the total variance. Aspey then descriptively labeled the factors after examining the behaviors comprising them. Consequently Factor I was labeled "Approach/Signal," Factor II "Vigorous Pursuit," Factor III "Run/Retreat." Factor IV at

Table 11-6. Selected references on the use of multivariate analyses.

Matrix Type (see Fig. 11-2)	Multivariate Analysis	Example References	Procedure References
	Principal Components Analysis	Huntingford (1976) three-spined-stickleback reproductive behavior	Seal (1964)
		Halliday (1975) newt sexual behavior	Cooley and Lohnes (1971) FORTRAN program
		Aspey and Blankenship (1976) *Aplysia* burrowing behavior	
		Bekoff (1978*a*) ontogeny of Adelie penguin behavior	Overall and Klett (1972) computer program
		Sparling and Williams (1978) components of avian vocalizations	Frey and Pimentel (1978)
		Dudzinski and Norris (1970) rabbit behavior	
A	Discriminant Analysis	Falls and Brooks (1975) avian song recognition	Cooley and Lohnes (1971) FORTRAN program, BMD program
		Aspey (1977*c*) dominance in spiders	Nie et al. (1975) SPSS program
		Bekoff et al. (1975) behavioral taxonomy in canids	
		Bekoff (1978*a*) ontogeny of Adelie penguin behavior	Pimentel and Frey (1978)
		Sparling and Williams (1978) components of avian vocalizations	
	Cluster Analysis	Sparling and Williams (1978) avian vocalizations	DeGhett (1978)
	Factor Analysis	Svendsen and Armitage (1973) marmot response to mirror image	Fruchter (1954)
			Comrey (1973)
		Aspey and Blankenship (1976) *Aplysia* burrowing behavior	Schmitt (1977) BASIC program
			Overall and Klett (1972) computer program

Table 11-6. Multivariate analyses (continued).

Matrix Type (see Fig. 11-2)	Multivariate Analysis	Example References	Procedure References
	Multidimensional Scaling	Aspey and Blankenship (1976) *Aplysia* burrowing behavior	Overall and Free (1972)
B	Cluster Analysis	Morgan et al. (1976) chimpanzee social associations Ralston (1977) horse social associations	Morgan et al. (1976) SLCA Jardine and Sibson (1968) Sibson (1973) FORTRAN program
	Multidimensional Scaling	Morgan et al. (1976) chimpanzee social associations	Kruskal (1964) Shepard et al. (1972) computer program Nie et al. (1975) SPSS program
	Factor Analysis	Baerends and Van der Cingel (1962) heron snap display Baerends et al. (1970) herring-gull incubation behavior Wiepkema (1961) bitterling reproductive behavior VanHooff (1970) chimpanzee social behavior	Nie et al. (1975) SPSS program
C	Cluster Analysis	Dawkins and Dawkins (1976) grooming in flies Maurus and Pruscha (1973) squirrel monkey communication	DeGhett (1978)
	Multidimensional Scaling	Golani (1973) jackal precopulatory behavior Guttman et al. (1969) mouse behavioral sequences	Lingoes (1966) computer program Spence (1978)

first seemed biologically uninterpretable, but was then labeled "Non-linking" since none of the composite behaviors was significantly linked with any other behavior during interindividual interactions. Although it is commonly stated that use of factor analysis assumes the existence of a common "motivational state" underlying each extracted factor (Slater 1973), note that Aspey has provided fairly descriptive, rather than functional, labels. In contrast, Wiepkema (1961) labeled his three major extracted factors as "aggressive, flight and sexual tendencies."

3. Grouping Individuals

If it is suspected that groups of individuals are behaving in a similar way, then Q-factor analysis can be applied to the data in Matrix *A*. The analysis extracts individual-related factors based on their observed behaviors.

Principal-axis factor analysis utilizes an orthogonal rotation of the data. Cooley and Lohnes (1971) provide a FORTRAN program listing for Varimax rotation. Factors analyzed using rotation account for maximum possible variance among the observed behaviors.

Powered-vector factor analysis extracts factors without rotation. It places maximum emphasis on biological relevancy, whereas the principal-axis method emphasizes parsimony (*i.e.*, maximum variance accounted for by a few factors) (Aspey and Blankenship 1977).

Aspey and Blankenship (1976) studied burrowing behavior in the marine mollusc *Aplysia brasiliana*. They recorded the occurrence of 10 burrowing parameters in 32 individuals (Matrix Type *A*). Q-factor analysis extracted three factors (groups of individuals) which accounted for 80.2% of the burrowers. Burrowing characteristics, interpreted relative to efficiency, were examined and the three groups were labeled "inefficient burrowers, efficient burrowers, and intermediate burrowers." The three-dimensional representation of these 32 individuals relative to the three extracted factors illustrates their location into three distinct groups (Fig. 11-3).

Cluster analysis groups variables (*e.g.*, individuals or behaviors) on the basis of similarities (or differences) among common characteristics. Simple distance-function cluster analysis is sensitive to variability in the data and tends to "split" the variables into more groups (Aspey, pers. comm.). Cluster analysis often makes fewer assumptions than other methods and therefore is easier to understand (Morgan et al. 1976; Sparling and Williams, 1978).

Cluster analyses begin with a matrix of similarities (or dissimilarities), and variables are sequentially joined on the basis of their

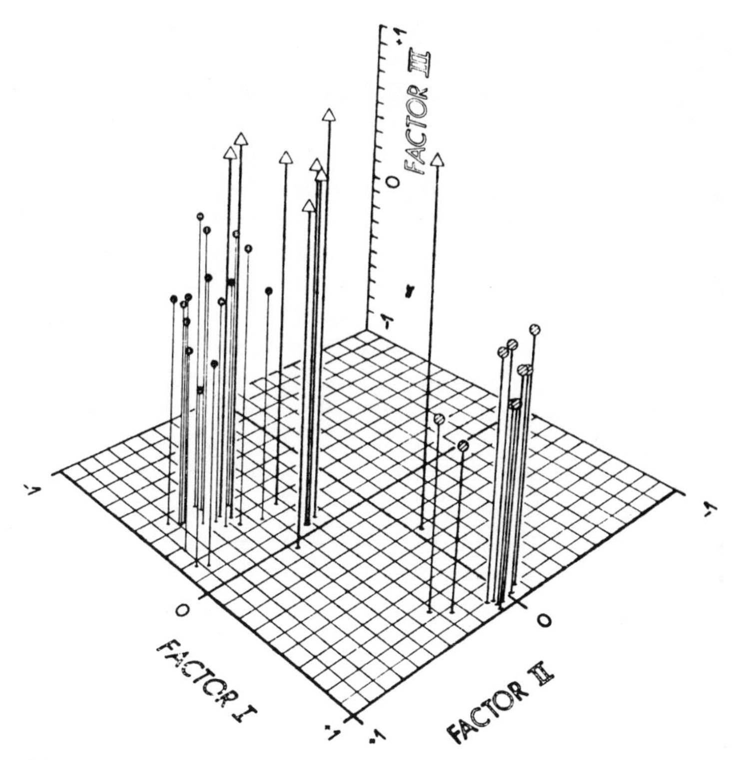

Figure 11-3 Factor loadings projected onto coordinate axes corresponding to the three factors extracted by Q-factor analysis. The origin falls in the center of the three-dimensional space. Factor *I* (large circles) represents "Inefficient Burrowers"; Factor *II* (small stippled circles), "Efficient Burrowers"; and Factor *III* (triangles), "Intermediate Burrowers." (From Aspey, W.P., and J.E. Blankenship, 1976, *Aplysia* behavioral biology: I. A multivariate analysis of burrowing in *A. brasiliana*, *Behav. Biology,* 17:292, Fig. 3.)

relative similarities into *dendograms* (Fig. 11-4), which are simple, visually interpretable representations of the results of the groupings.

Morgan et al. (1976) used cluster analysis on data from a matrix of Type *B* and provided a straightforward description of the mechanics of conducting a single-link cluster analysis (*SLCA*), as well as discussing the positive and negative attributes of the method. Their paper formed the basis for the discussion which follows.

In conducting a *single-level cluster analysis*, let us assume that we want to determine the relative association among individuals in a herd of eight adult doe deer in a wildlife preserve. We observe them for a total of 500 hours and record the amount of time that individuals are closer than four meters through scan sampling every minute. Each occurrence of an association is assumed to represent one minute of association (see section on sampling for a discussion of the hazards of this assumption). The procedure goes as follows:

1. The data are first organized into the association shown in Table 11-7. This table, a matrix of Type *B*, can be read both vertically and horizontally in order to determine associations. However, it is still difficult to see much pattern of associations from the data in the table. Cluster analysis will provide a better visual presentation of the associations.

2. The next step is to generate a triangular table of *similarities* (associations) among deer. Since each deer was seen for different total periods of time, we must first *normalize* the data to adjust for those differences.

Morgan et al. (1976) describe a method for normalizing data for time observed, which they derived from Dice's coefficient of association (1945). Note that this is the same as Cole's coefficient of association (1949) (p. 295).

$$\text{similarity} = \frac{XY}{X + Y}$$

where: XY = total time when individuals X and Y were observed together, X = total time X was observed, and Y = total time Y was observed.

Therefore the similarity for individuals A and B in Table 11-7 would be:

$$\text{similarity } A + B = \frac{21}{191 + 202} = 0.053$$

Table 11-7. Hypothetical association data for a herd of eight adult doe deer. Data are the number of hours observed in association.

	A	B	C	D	E	F	G	H
A								
B	21							
C	18	31						
D	30	21	48					
E	16	10	22	28				
F	38	22	11	17	11			
G	37	67	48	41	47	31		
H	31	30	25	23	34	27	77	
Total hours observed	191	202	203	208	168	157	348	247

Table 11-8. Similarities for the associations in the hypothetical herd of eight adult doe deer in Table 11-7.

	A	B	C	D	E	F	G	H
A								
B	53							
C	45	76						
D	75	51	117					
E	44	27	59	74				
F	109	61	30	46	34			
G	69	122	87	74	91	61		
H	71	69	56	50	82	67	129	

The similarity is then multiplied by 1000 for convenience, so that we can deal with whole numbers.

$$\text{similarity} = 0.053 \times 1000 = 53$$

We then complete the table of similarities (Table 11-8).

3. Next we search the table for the range of similarities. The highest is 129 for $H + G$, and the lowest is 27 for $E + B$. The vertical axis of our dendrogram should include this range, so for convenience we will set up a vertical scale of from 0 to 150; note that the scale increases from top to bottom (Fig. 11-4*A*).

4. We then begin linking individuals on the basis of similarities, starting with the largest and working to the smallest. Individuals $H + G$ provide our first association. The next similarity 122 is between G and B. This means we link B up with both G and H at the 122 level. This is an example of "chaining," which is an undesirable characteristic of the method, since it links through intermediates (Jardine and Sibson 1968) and is difficult to interpret visually. That is, the apparent association between B and H is really due to B's association with G. Morgan et al. (1976) discuss the problems of "chaining" in more detail.

The next association is $D + C$ at the 117 level of similarity. We then continue the similarities (Table 11-8) until all the individuals have been linked together. This happens when we make the association of similarity 75 between D and A. If there are ties in similarities between two or more pairs with a common individual, then they are all linked together at the same level. The association between E and G is difficult to graph, because E must be linked to the association between B and $G + H$ (Fig. 11-4*A*). This type of association becomes clearer when the individuals are rearranged on the abscissa.

5. It can be seen that we did not use over half of the ordinate in our dendrogram and that the ordering of individuals on the abscissa

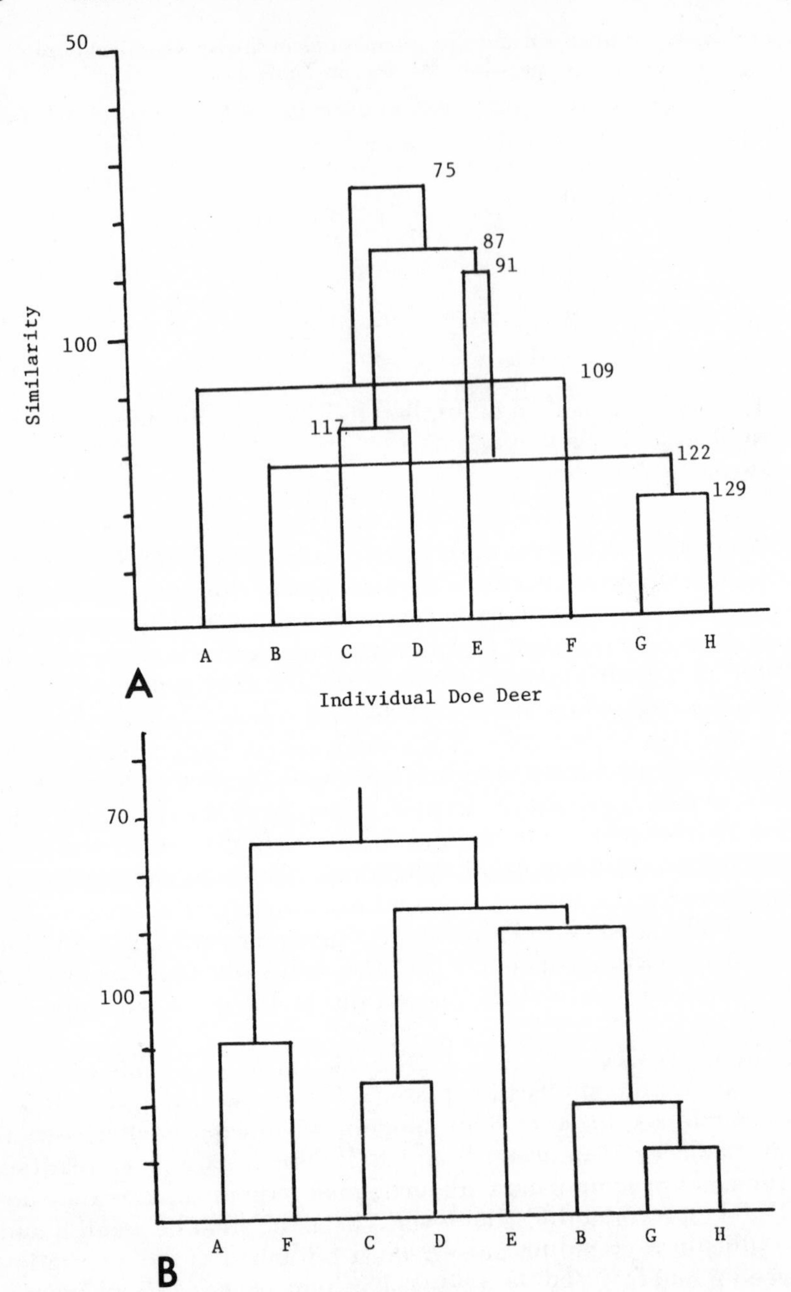

Figure 11-4 *A*, initial dendrogram for associations in a hypothetical herd of eight adult doe deer. *B*, final dendrogram for associations in a hypothetical herd of eight adult doe deer.

makes interpretation of the dendrogram difficult. Therefore we correct these shortcomings in Figure 11-4*B*.

6. The dendrogram can now be used in a visual inspection of the associations. We can easily see the relative "strength" of the various individual associations and the associations among three or more individuals.

Because of the chaining which occurs in this two-dimensional representation, distortion occurs and increases toward the lower levels of similarity. This distortion can be measured, but not tested statistically, by Sokal and Rohlf's (1962) *cophenetic correlation coefficient*, and Jardine and Sibson's (1971) *distortion measure*. DeGhett (1978) presents an excellent discussion of the use of cluster analysis in ethology.

4. Describing Differences Among Individuals

Once the individuals in Matrix *A* have been grouped by Q-factor analysis, we might want to know more about the differences among individuals and about the parameters (components) of their behavior that are most important in distinguishing differences.

Principal-components analysis separates individuals in a sample in terms of a few composite components. The *first principle component* is the one which accounts for the maximum individual difference, and the *second principle component* is that combination of variables (*e.g.*, behaviors), uncorrelated with the first principle component, that accounts for the largest proportion of the remaining individual differences. The analysis can be extended to additional components, if necessary, to explain the greatest proportion of individual differences.

Individuals may be represented on a multidimensional figure to provide a visual image of the results of analysis. Subgroups can also be delineated for a clearer presentation (Fig. 11-5).

5. Discriminating Among Groups of Individuals or Behaviors

Discriminant analysis 1) determines relationships among several identified groups (*e.g.*, predefined or resulting from Q-factor analysis or principal component analysis), 2) assesses the discriminability between the groups, and 3) places individuals or behaviors in the appropriate groups (Sparling and Williams, 1978).

Aspey (1977*c*) used *multiple stepwise discriminant analysis* to further discriminate between three groups of 40 spiders which Q-factor analysis had extracted and Aspey had labeled as "Dominant, Intermediate, and Subordinate." In his analysis ". . . a sequence of

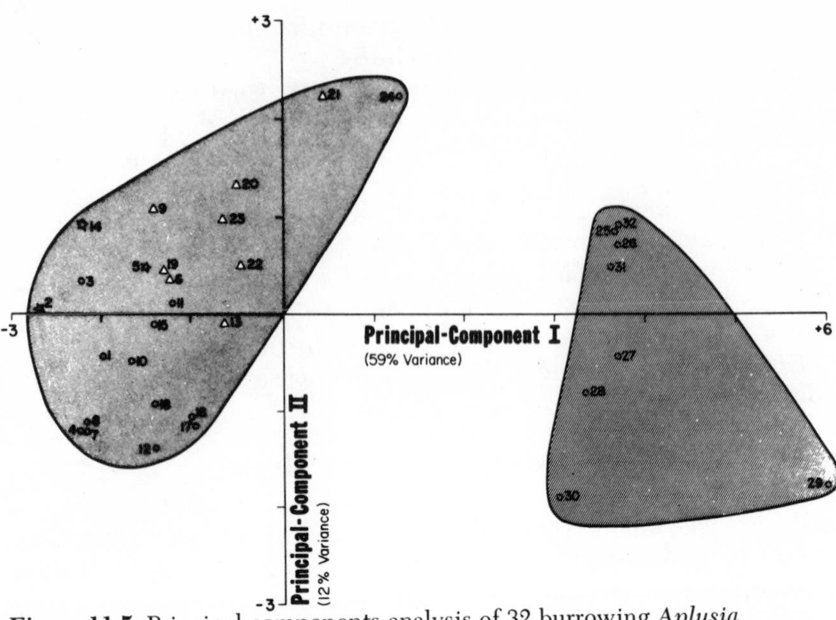

Figure 11-5 Principal-components analysis of 32 burrowing *Aplysia brasiliana* showing two extracted subgroups. The first principal component (shaded area on right) accounted for 59 percent of the variance and generally corresponded to individuals which had high Factor *I* (approach/signal) and Factor *II* (vigorous pursuit) values. The second principle component accounted for 12 percent of the variance and corresponded to individuals with low Factor *II* values and high Factor *III* (run/retreat) values. (From Aspey, W.P., and J.E. Blankenship, 1978.)

discriminant equations was computed in a stepwise manner so that one variable was added to the equation at each step, and a one-way analysis of variance F-statistic was then used to determine which variable (*i.e.*, behavior) should join the function next. The variable added is the one making the greatest reduction in the error sum of squares" (Aspey and Blankenship 1977:90). The results were then plotted (Fig. 11-6). The first discriminant function (abscissa) separated the "Dominant" and "Subordinate" spiders, but it took the second discriminant function (ordinate) to separate out the "Intermediate" group from the other two groups. The three groups were encircled in the figure for added clarity.

In contrast to Aspey's (1977*c*) use of multiple stepwise discriminant analysis, Bekoff et al. (1975) used *linear discriminant analysis* to assess the taxonomic relationships among infant wolves, coyotes, and "New England Canids" (Silver and Silver 1969). Linear discriminant analysis compares the means of variables (*e.g.*, behaviors) from two populations (*e.g.*, individuals from a canid taxon) and produces a *dis-*

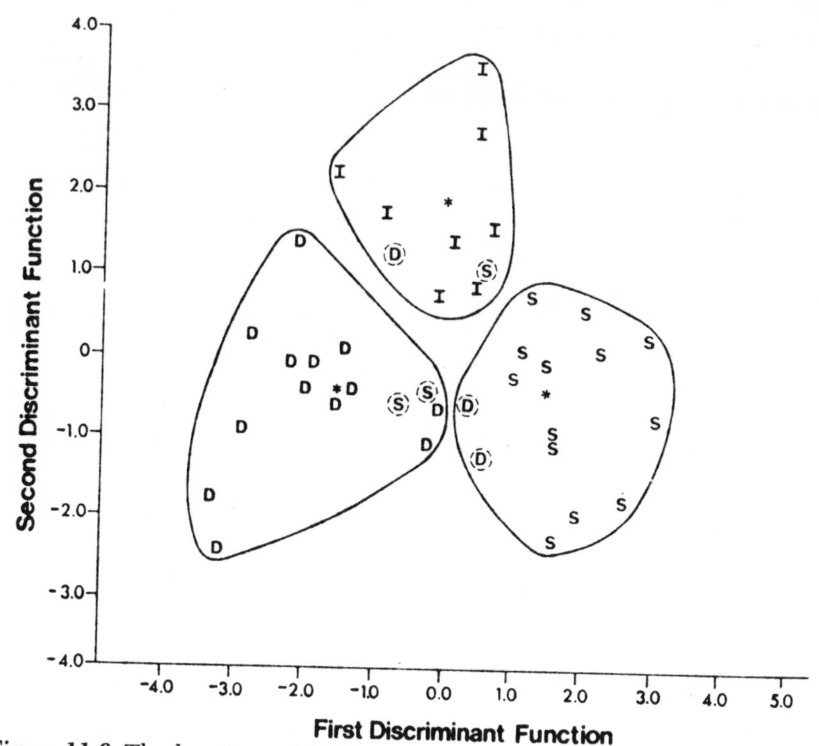

Figure 11-6 The locations of 16 Dominant (*D*), 8 Intermediate (*I*), and 16 Subordinate (*S*) adult male *Schizocosa crassipes* plotted in a geometric space of minimum dimensionality by multiple stepwise discriminant analysis on the basis of the frequency of 20 behaviors observed during agonistic encounters. The first discriminant function (abscissa) is plotted against the second discriminant function (ordinate) and * denotes group means. The spiders were initially grouped as Dominant, Intermediate, or Subordinate by a Dominance Index (*DI*). Dotted circles represent spiders "misclassed" by the *DI* relative to the discriminant analysis. (From Aspey, W.P., 1977*c*.)

criminant function which is a relative measure of the difference between the populations. Bekoff et al. (1975) used two behaviors (social play and agonistic behavior) in their analysis of differences among the three canid "types." Their results, when plotted on a *linear discriminant function axis* (Fig. 11-7), show that the wolves and coyotes were clearly separated, and that the "New England Canids" were intermediate, but fell closest to the coyotes. Interestingly, their results are in full agreement with the same type of analysis done on various anatomical measurements.

In contrast to the other multivariate analyses discussed in this section, discriminant analysis can be used to test hypotheses (Sparling and Williams, 1978).

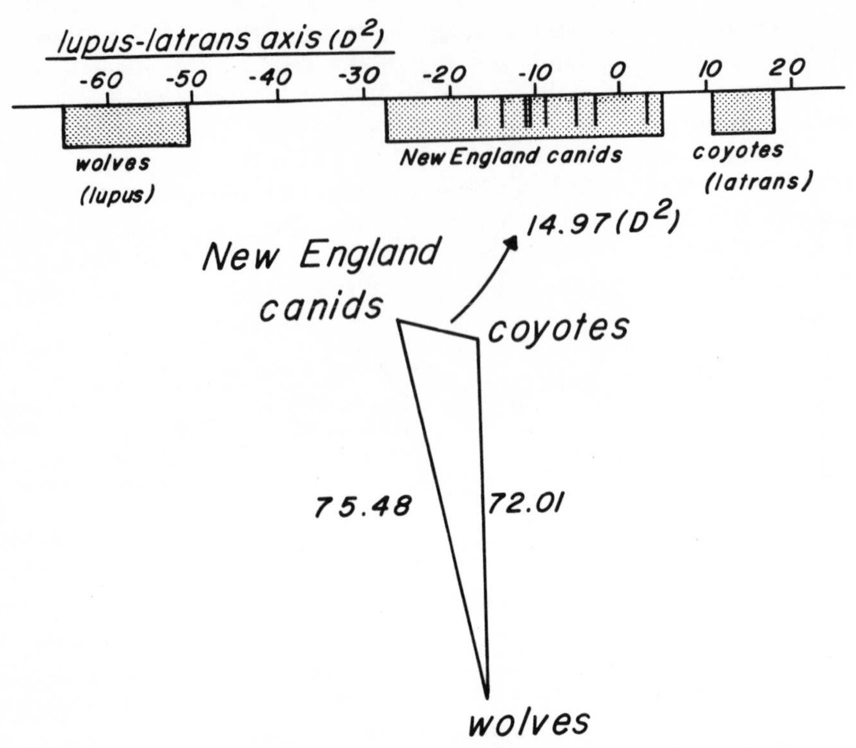

Figure 11-7 The results of a behavioral taxonomy study on infant coyotes, wolves, and New England canids (eastern coyotes). The relative frequencies of occurrence of social play and agonistic behavior were used as behavioral characters (Bekoff, et al., 1975). Top: Linear discriminant values of known *C. lupus* (wolf) and *C. Latrans* (coyote) litters cast on a *lupus-latrans* discriminant axis onto which New England canids are projected. Note that the New England canids fall between *lupus* and *latrans*, but closer to *latrans*. Bottom: Distances (D^2) in discriminant function units based on pairwise analyses of *lupus, latrans* and New England canids. Note the close relationship between coyotes and New England canids and that both fall approximately the same distances from wolves. This is an example of where a multivariate analysis of behavior has been applied to taxonomy, an area where these analyses are commonly used on morphological data (Jardine and Sibson 1971; Sneath and Sokal 1973).

6. Analyzing Sequences of Behavior

Multivariate analyses are only one method for analyzing sequences of behavior (see Chapter 11, Sec. D). Whereas Markov chain analysis tends to emphasize "sequential effects" (Slater 1973), *factor analysis* assumes that the measured variables (*e.g.*, behavior patterns) do not depend causally on each other (Blalock 1961), but rather that there are underlying "motivational processes" common to the behaviors grouped around the extracted factors (Andrew 1972; Hutt and Hutt 1974). Slater (1973) points out that both motivational changes and sequence effects are probably present in all behaviors so that neither analysis is perfect.

Factor analysis was used to analyze behavioral sequences (Matrix Type *C*, Fig. 11-2) by Wiepkema (1961) for bitterling (*Rhodeus amarus* Bloch) reproductive behavior, and Baerends and Van der Cingel (1962) for common-heron (*Ardea cinerea* L.) snap displays. Both started with matrices of Type *C*, and then generated transition frequencies on the basis of observed/expected frequencies.

$$\text{expected frequencies} = \frac{\text{row total} \times \text{column total}}{\text{grand total}}$$

Calculating transition frequencies on this basis provided a ratio which indicated relative frequency with which each of the behaviors pre-

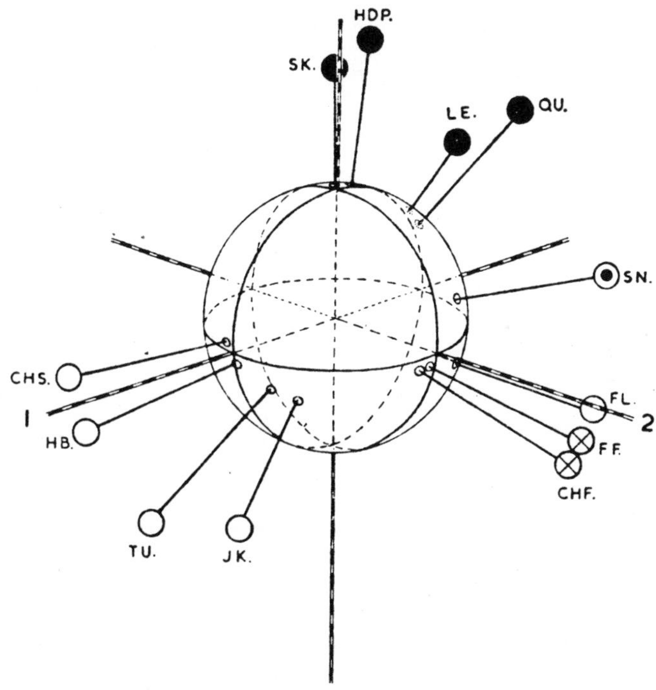

Figure 11-8 Vector diagram of the behavior of the male bitterling (*Rhodeus amarus*). *CHF* = chafing, *CHS* = chasing, *FF* = Fin flickering, *FL* = fleeing, *HB* = head butting, *HDP* = head-down posture, *JK* = jerking, *LE* = leading, *QU* = quivering, *SK* = skimming, *SN* = snapping, *TU* = turning beat. (From Wiepkema, P.R., 1961, An ethological analysis of the reproductive behaviour of the bitterling (*Rhodeus amarus bloch*), *Arch. Neerl. Zool.* 14:130, Fig. 13.)

ceded or followed the other, but it also included transitions between a given behavior and itself which may be impossible or difficult to interpret. Slater and Ollason (1972) discuss these difficulties and suggest another method provided by Goodman (1968). Wiepkema (1961) and Baerends and Van der Cingel (1962) then ranked the transition frequencies and generated correlation coefficients using Spearman's rank correlation, in contrast to the product-moment correlations used by Aspey (1977*c*) and Van Hooff (1970). Factor and analysis was then used to extract the minimum number of common causal factors necessary to explain the sequences observed. Since three factors explained the majority of the variability in the data, a three-dimensional vector model was used to illustrate the clustering of the behaviors around the three factors (Fig. 11-8).

The angle between any two vectors represents the extent of their correlation. For vectors of unit length the correlation is represented by an angle determined according to the following formula:

$$\text{the correlation coefficient} = \cos \phi_{AB}$$

Therefore, the vector angle = that angle whose cosine equals the correlation coefficient

If the correlation is perfect ($r = 1.00$) the two vectors will be the same (1 below). If there is no correlation ($r = 0.00$) the two vectors will diverge by 90° (2 below). If there is a positive correlation the angle will be between 0 and 90° (3 below), and if it is negative it will be between 90° and 180° (4 below).

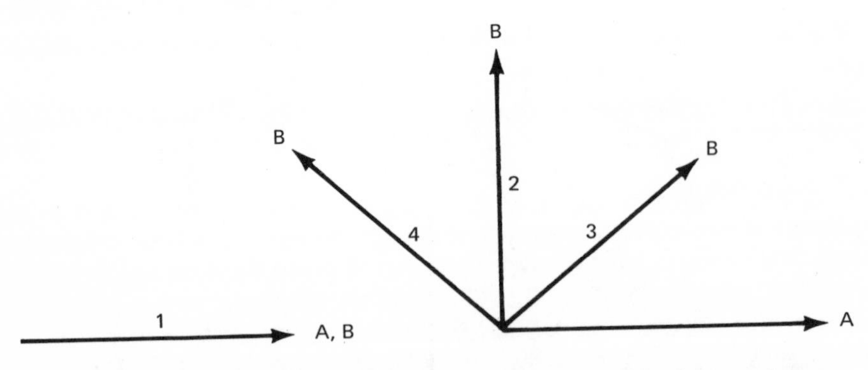

In the vector diagram (Fig. 11-8) it can be seen that the behaviors cluster around three axes. These three axes are the causal factors: 1 is the positive side of the aggressive factor; 2 is the positive side of the nonreproductive factor; 3 is the positive side of the sexual factor. The length of each vector represents the amount of the common variance of that behavior accounted for by the factors.

The use of factor analysis in ethological studies, in general, has been questioned by several investigators (Andrew 1972, Morgan et al. 1976, and Maurus and Pruscha 1973). Factor analysis often makes more assumptions than some other multivariate analyses (*e.g.*, cluster analysis) (Morgan et al. 1976; Sparling and Williams, 1978). Once causal factors have been extracted, some researchers question what they really mean. Slater (1973:145) concludes that "It is doubtful whether the extraction of factors which are themselves of complex causation advances understanding." However, it does allow us to visualize our data more clearly and generate hypotheses and experiments that truly will advance our understanding. For example, Aspey and Blankenship's (1976) experimental study of *Aplysia* grew out of factor analysis of groups and advanced our understanding of the function of burrowing and its relationship to reproduction behavior.

a. Multidimensional Scaling

Multidimensional scaling orders individuals or behaviors from a matrix and discriminates between them in terms of distances along coordinate axes. Golani (1973) used multidimensional scaling in his analysis of precopulatory behavioral sequences in two pairs of golden jackals *(Canis aureus)*. He utilized the Guttman-Lingoes multidimensional scalogram analysis (L. Guttman 1966), of which a computer program is available (Lingoes 1966). Guttman et al. (1969) had previously used the analysis to measure sequential behavior in mice.

Morgan et al. (1976) reanalyzed Wiepkema's (1961) data, using multidimensional scaling, and discussed their different results. They suggested that Wiepkema's grouping of behaviors around three causal factors may tend to oversimplify the situation.

This method is too complex to discuss in detail here; consult the references above.

7. Summary

Multivariate analyses provide a diversity of methods to treat ethological data in matrices. They should be used primarily 1) to aid in interpreting relationships in the data through grouping and visual representation and 2) to generate hypotheses for further testing.

To interpret, the researcher must have his objective clearly in mind and watch the analyses to the objectives and type of data.

Finally, Aspey and Blankenship (1977:78) "... caution that multivariate analyses are simply one research tool available to the animal behaviorist for determining inherent data structure ... misusing multivariate analyses in ethology could most certainly lead to a bewildering array of extraordinarily sterile papers."

Regardless of the mechanics, analysis must be accurate. Moving up the scale of sophistication from paper and pencil to calculator to large computer will speed up the analysis and may improve its accuracy, but it will not change the data. Computers are tremendous tools for the storage and analysis of data, and their use should be encouraged. However, poorly collected and invalid data cannot be laundered in a computer, although some researchers will try to sell it by wrapping it in complex analyses, tying it with a computer and presenting it with complicated diagrams.

Notterman cautions us to be on the lookout for "computeritis disease," which has the following symptoms:

> (1) *The diarrhea symptom*—this is manifested in the researcher's assumption that the more data, the better the experiment. (2) *The displacement symptom*—the more time spent in manipulating the computer, the more incisive the psychological research. (3) The *Lorenzian-territorial symptom*—the more expensive the computer, and the larger the laboratory in which it is housed, the greater the personal authority exerted either in the academic or in the institutional setting. [NOTTERMAN 1973:130]

In short, use the *best* methods and equipment available to both generate and test hypotheses; however, avoid *methodological overkill*. As Fagen and Young (1978:114) suggest "The future may well belong to those who can use simple methods effectively to test a theoretical hypothesis, or who are prepared to construct original models of behavior should no simple technique be available."

F. LONG-TERM SPATIAL PATTERNS

Ethologists generally measure the location of animals relative to 1) other animals (*e.g.*, mother-infant) and 2) the environment (*e.g.*, home range). Although animal-environment spatial relationships are measured relative to the physical environment, they often reflect the presence of other animals (intra- or inter-specific).

1. Animal-Animal Spatial Relationships

The researcher may be interested in intra- or inter-specific associations between individuals or groups. The use of data from sociometric matrices in determining interindividual spatial associations was discussed on page 283.

The relative observability of the animals will influence the type of sampling method employed. If the animals can be observed constantly (*e.g.*, dairy cattle) or for long periods of time, then focal-animal, all-occurrences, or instantaneous and scan sampling will be used. However, *ad libitum* or all-occurrences sampling will be necessary when the animals are visible for only short periods of time.

Knight (1970) used all-occurrences sampling to determine the dyadic (two-animal) associations in a herd of elk. He then employed Cole's coefficient of association (Cole 1949) as a relative measure.

$$\text{coefficient of association} = \frac{2ab}{a + b}$$

where a = total number of times individual a was seen
b = total number of times individual b was seen
ab = total number of times individuals a and b were seen together

This coefficient was then calculated for all combinations of two individuals in the herd. Fager (1957) used the same ratio measurement expressed as an "Index of Affinity" for different species of plants.

The criterion distance between individuals used to determine association depends on the species under study and the experience of the observer. However, final choice is often arbitrary, based primarily on intuition. Grant (1973) used as his "measure of association" between grey kangaroos *(Macropus giganteus)* the number of times each animal occurred within 120 cm of another at set 15-minute intervals (instantaneous samples).

When animals are maintained in an enclosure, each animal's position in the enclosure can be designated according to a predetermined grid. The grid positions can then be compared to obtain a relative measure of association between individuals (see Weeden's study (1965) of wild tree sparrows using a grid system). Aspey (1977*a, b*) has written computer programs in BASIC for computing interindividual distances within a gridded rectangular area (RECDIS) and a gridded circular area (CIRDIS). Stricklin et al. (1977) reported on a FORTRAN computer program which analyzes relative distances between individual animals within a gridded enclosure (square, circular, or rectangular) and also considers each animal's angle of orientation on a 360° basis. This allows the calculation of angular relationships and a determination of the angles any two individuals would have to turn in order to be facing each other.

The resolution of association data taken from animals in a gridded enclosure is proportional to the size of the grids, since an animal is recorded only as being present in a particular grid, regardless of its actual position in that grid. Figure 11-9 illustrates how a large grid system may not clearly reflect the actual distances between animals.

In Grid *A*, Figure 11-9, individuals *A* and *B* are in adjoining

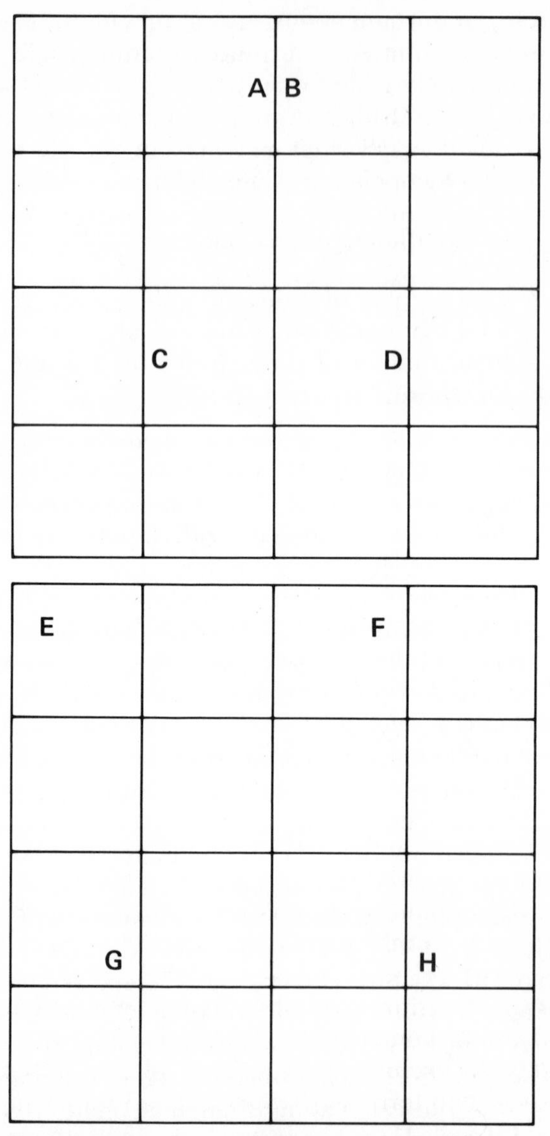

Figure 11-9 Hypothetical positions of eight individuals (*A–H*) in two 4×4 gridded enclosures. (See text for explanation.)

quadrats as are individuals *C* and *D*. Although automatic data analysis would treat these pairs of individuals as being at equal distances from each other, in fact *A* and *B* are much closer together than *C* and *D*. If the same enclosure was gridded into an 8 × 8 format, then grid columns 2 and 3 would be halved, separating individuals *C* and *D* by two quadrats, but leaving *A* and *B* in adjoining quadrats. Hence, smaller quadrats improve resolution and accuracy. However, it may be difficult to locate animals in small quadrats with precision; therefore reliability may suffer. Grid *B*, Figure 11-9, illustrates how pairs of individuals (*E* and *F*; *G* and *H*) may actually be the same distance apart but separated by one and two quadrats, respectively.

Bekoff and Corcoran (1975) stress the importance of knowing not only the spatial relationships between individuals, but also the behavior of each. For example, copulation, fighting, and nursing all require close spatial relationships; but their interpretation is quite different.

2. Animal-Environment Spatial Relationships

The grid systems discussed above can also be used to measure animal-environment spatial relationships in the laboratory, enclosures, or over large study areas where the grid becomes quite large.

A grid system can be automated to collect and process positional data directly. For example, Kleerekoper (1969) used photosensors in a 50 × 50 *X*-*Y* grid to detect the position of fish in a large tank relative to olfactory gradients. An on-line computer then analyzed the *X* and *Y* coordinate positions of the fish, the time of occurrence of a change in position, velocity of movement, distance covered, angle of turns, left- and right-handedness, and the statistical significance of changes in these parameters. Pitcher (1973) used a similar system (Fig. 11-10) in his research on minnow schooling.

Long-term spatial patterns in the environment are generally reflected as home ranges and/or territories. The *home range* of an individual is that area covered in normal daily activities (Blair 1953). It may overlap the home ranges of other individuals (or groups, Fig. 11-16) and may or may not contain a *territory*, an area defended against members of the same species and occasionally other species. The designation of territories is often the result of finding contiguous and nonoverlapping home ranges, suggesting that the entire home ranges are mutually exclusive and are thus assumed to be territories.

Home ranges and territories are calculated from successive locations obtained through 1) direct observations (continuous or sampled) or 2) indirect methods, including a) location of natural signs, b) capture-recapture, c) radioactive material, d) dyes for urine and feces, e) photographic devices, and f) radiotelemetry. See page 199; see also Sanderson (1966) and Giles (1969) for reviews.

Computer programs, available from the University of Minnesota Cedar Creek Software Library, calculate home ranges based on radiotelemetry fixes in a grid system. The horizontal and vertical axes of the

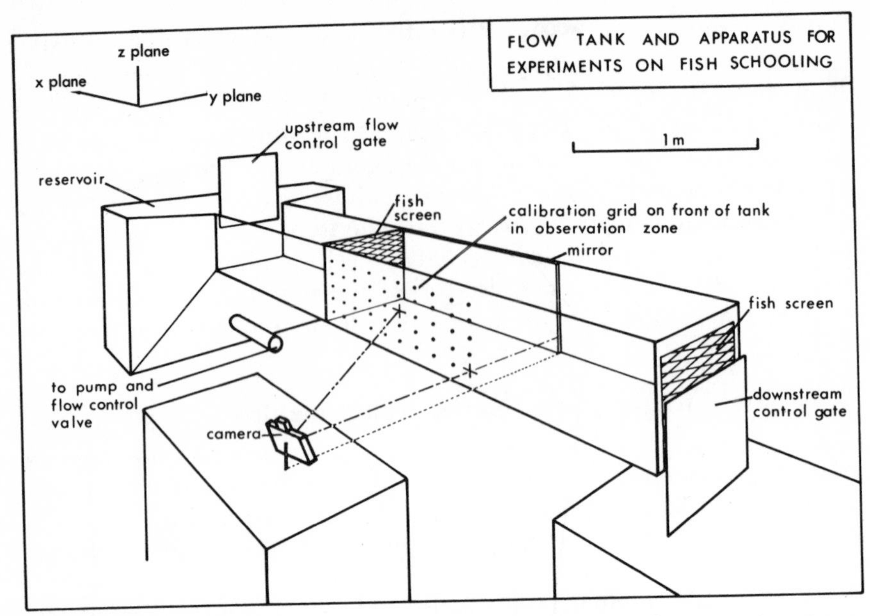

Figure 11-10 Diagram of the flow tank and apparatus used in experiments on minnow schooling. The *X* axis was taken as running along the length of the tank; the *Y* axis across it; and the Z axis was taken as vertical. Therefore the photographs showed fish and their reflections in the *XZ* plane only. (From Pitcher, T.J., 1973, The three-dimensional structure of schools in the minnow, *Phoxinus phoxinus* (L.), *Anim. Behav.* 21(4):674, Fig. 1.)

grid are searched by the computer according to some criterion such as that used by Rongstad and Tester (1969:367) in their study of white-tailed deer movements; "Squares with fixes that were separated along either axis by not more than two vacant squares were considered to be within the boundary of the home range." The Cedar Creek programs also test for randomness of movement (observed vs. expected fixes per grid unit) and for spatial relationships between individuals and generate an intensity of use coefficient determined by the fixes per grid unit as a percent of the total number of fixes for that period (Gull 1977).

The *observation-area curve* (Fig. 11-11) was developed by Odum and Kuenzler (1955) to assist in determining the number of observations (locations) necessary in order to determine the territorial extent of several bird species.

The observation-area curve is based on the same concept as the curves used to assess repertoire size (p. 50), *i.e.*, as the number of observations increases the rate of increase in area utilized by the bird decreases, so that an asymptotic curve results (Fig. 11-11). Odum and

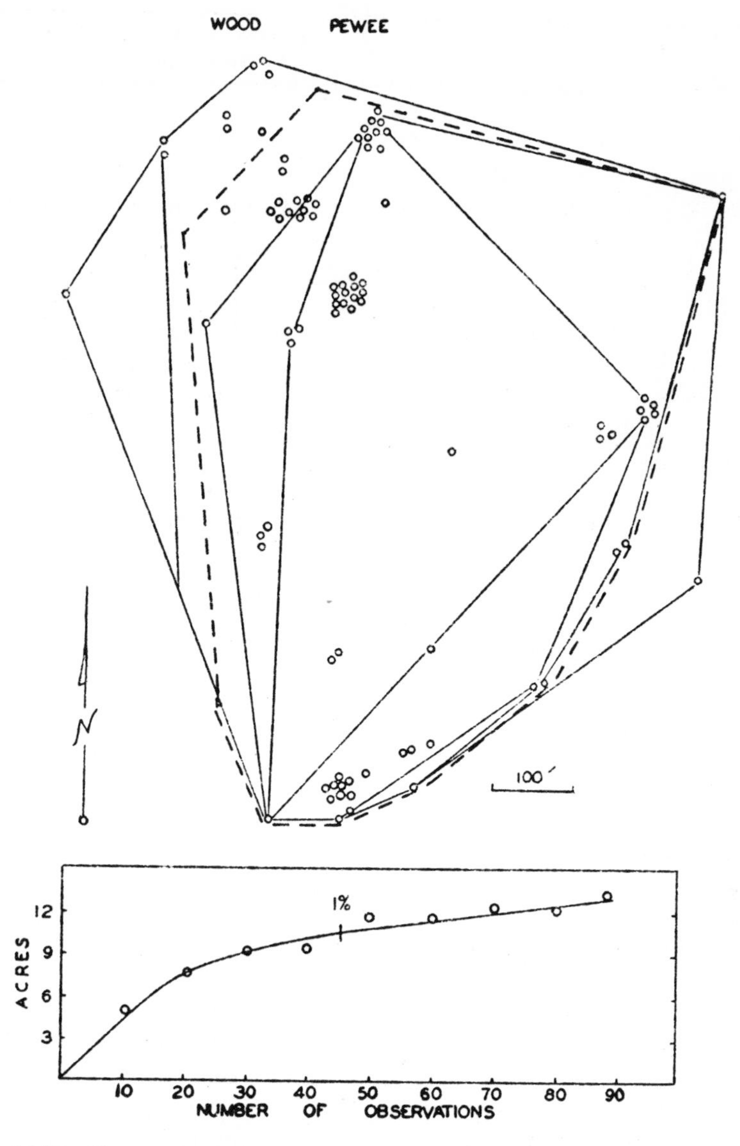

Figure 11-11 Observed positions of a male wood pewee at five-minute intervals (small circles), with maximum observed area enclosed in solid lines after successive tens of observations. The broken line in the upper diagram encloses the calculated maximum territory size (10.8 acres) at the 1-percent level, (see text for explanation) as shown on the observation-area curve below. (From Odum, E.P., and E.J. Kuenzler, 1955, Measurement of territory and home range size in birds, *Auk* 72:131, Fig. 1.)

Kuenzler (1955) arbitrarily selected the one-percent level on the curve as the point at which the territory size was calculated. The one-percent level is that point on the curve where each additional observation produces less than a one-percent increase in the area calculated as the territory. The number of observations necessary to reach the one percent level varied with the species and the stage of the nesting cycle. Sanderson (1966) suggested that live-trapping mammals would probably provide insufficient data to apply the observation-area curve technique, but that radiotelemetry probably would.

The *accuracy* (resolution) and *precision* of the calculated home range, or territory, will depend on the technique employed. When *direct observations* are made, the animal's location can be considered along an essentially *continuous distribution*. That is, it may be found at any point within its true home range, which is affected by sociological and physiographic variables. Many *indirect measures* can also provide continuous data (*e.g.*, tracking in snow and radiotelemetry); however, other methods, such as the commonly used capture-recapture method for small mammals, provides a *discontinuous distribution* of observations of animals restricted to specified trap locations. Therefore, the validity of this method is affected by trap spacing (Stickel 1954) and the trap response of the animals (Balph 1968).

Most small mammal trapping designs used to reveal home ranges are grids similar to that in Figure 11-12, but on a larger scale. However, Lockie (1966) placed his traps near features in the environment likely to be frequented by weasels *(Mustela nivalis)* and stoats *(M. erminea)*.

With capture-recapture methods, the animal is individually marked (see p. 141) on the first capture, and its location recorded on subsequent captures. The following assumptions are inherent in the use of trapping data to calculate home ranges:

1. The animal will be trapped over all of the ecologically significant area of its home range, *i.e.*, it will be trapped wherever it goes under the following conditions:
   a. The grid being as large as or larger than its home range.
   b. On encountering a trap the probability of capture being high.
   c. The probability of capture on encountering a trap being the same throughout its home range.
2. The frequency of capture at a particular trap site reflects the frequency of visits by the animal.
3. Each animal has an equal chance of being captured upon encountering a trap.

These assumptions are probably never completely met and perhaps seldom approximated.

The following are the most common methods of calculating home ranges based on data from trap grids (Stickel 1954):

1. *Minimum area method.* The outermost capture sites are connected by straight lines (*e.g.*, Fig. 11-12*A*).

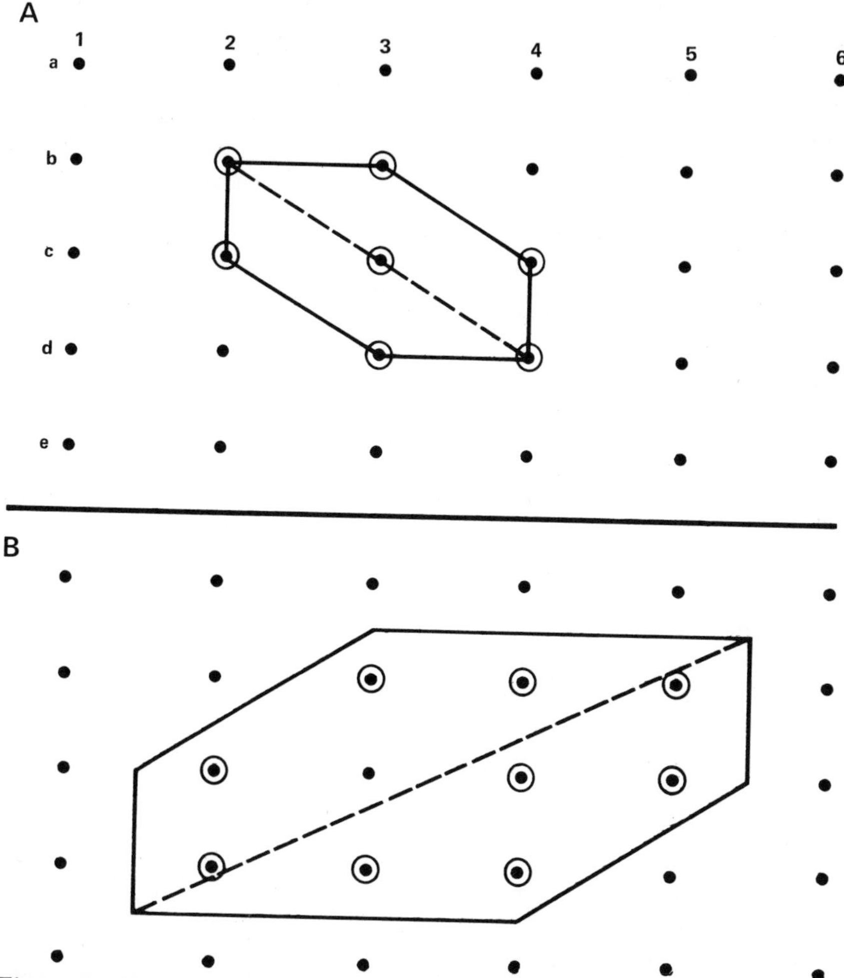

Figure 11-12 Hypothetical captures of an individual small mammal in a trap grid. Circled dots indicate sites of captures. Home-range measurement of: *A*, minimum area and observed range length; *B*, boundary strip and adjusted range length.

2. *Boundary strip method.* Points halfway between the outermost capture sites and the next closest trap sites are connected by straight lines (*e.g.*, Fig. 11-12*B*). The rationale behind this method is that on the average the animal will have traveled halfway to the next trap site during its movements (see Stickel 1954 for the *inclusive* and *exclusive variations* on this method).
3. *Observed range length.* The distance between the two most widely separated capture sites is measured (*e.g.*, Fig. 11-12*A*).
4. *Adjusted range length.* The furthest distance across the home range calculated by the boundary strip method is measured (*e.g.*, Fig. 11-12*B*).

Stickel (1954) concluded that the boundary-strip method and the adjusted range length provide closer estimates of the true home range than the other methods. Stickel also found that trap spacing altered the apparent size of the home range even when trap visitations are random and biological factors are excluded.

The two-dimensional methods can be extended to a third dimension, and home range volume can then be calculated for arboreal species (*e.g.*, cricetid rodents, Meserve 1977).

An individual will occasionally be caught at a great distance from the cluster of other captures. These may reflect occasional excursions out of the animal's usual home range and are often disregarded in calculating the home range.

Balph (1968) observed the behavioral responses of uinta ground squirrels *(Citellus armatus)* to live traps. He found that the trap was initially an attractant which could be enhanced by baiting; however, there was an equal probability of capture on the first encounter whether the trap was baited or not. Capture appeared to be punishing while the bait served as a reward (see Model p. 7); this produced a conflict between the tendencies to approach and avoid the trap on subsequent encounters, and recaptures were influenced by the relative strengths of these tendencies (Fig. 11-13).

Getz (1972) used multiple captures (more than one individual in the same trap) to infer associations between sex and age groups in a population of *Microtus pennsylvanicus*. He compared the number of multiple captures of the different sex and age categories to that expected from their relative frequency in the population. He used a chi-square test to determine whether they were found together more frequently or less frequently than expected and inferred attraction and avoidance, respectively. Slade (1976) elaborated further on Getz's procedure.

Individuals are often captured more than once in the same trap, perhaps reflecting a disproportionate use of those portions of its home range. Hayne (1949) calculated the *center of activity*, the mathemati-

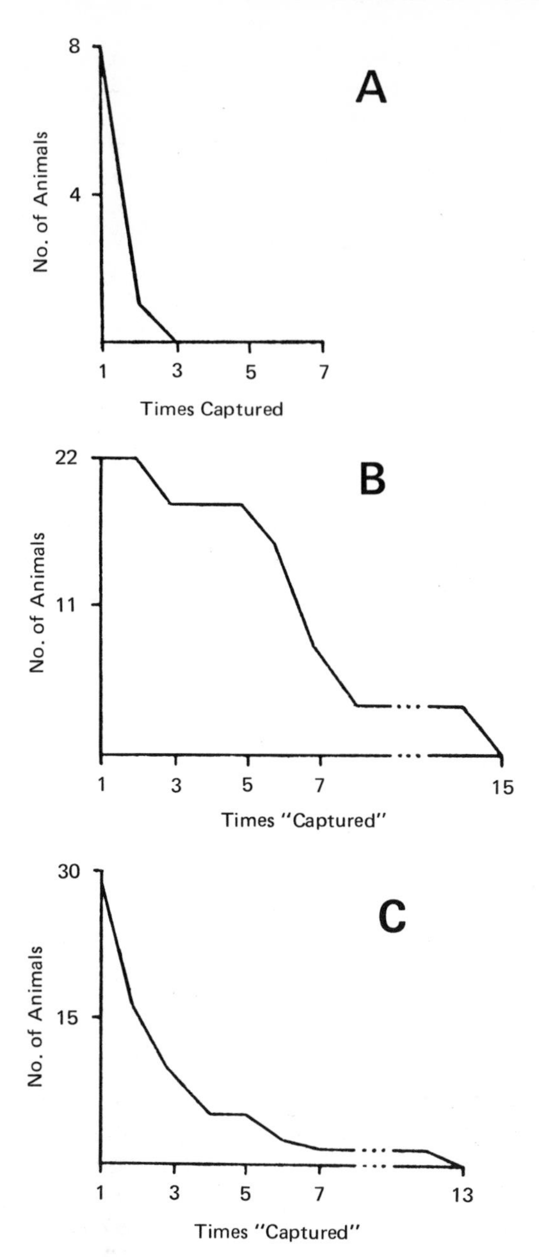

Figure 11-13 Distribution of number of times uinta ground squirrels were captured in an unbaited and functional trap (*A*), a baited and nonfunctional trap (*B*), and a baited and functional trap (*C*). (From Balph, D.F., 1968, Behavioral responses of unconfined uinta ground squirrels to trapping, *J. Wildl. Mange.* 32(4):789, Fig. 2.)

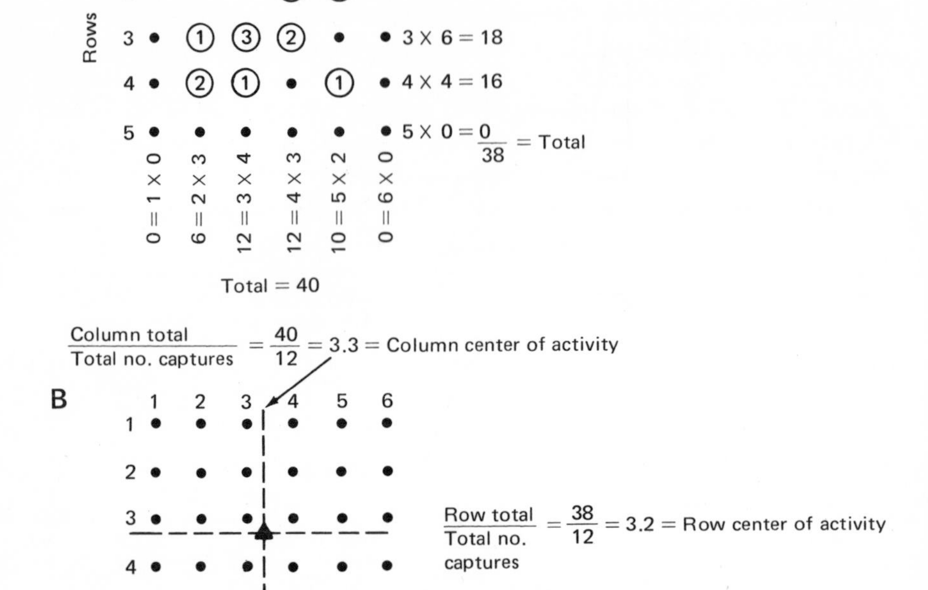

Figure 11-14 Calculation of *center of activity* (triangle, grid *B*) based on hypothetical recaptures (circled numbers, grid *A*). The numbers in the circles indicate the number of captures at each trap site (see text for explanation).

cal center of the distribution of total captures, taking into account the number of captures at each trap site (Fig. 11-14). This can be obtained by weighting the rows and columns of the trap grid and multiplying the number of captures in each row and column by its respective weight. The totals for all rows is then divided by the total number of captures in the grid to determine the center of activity for the rows; the same procedure is repeated for the columns. The point at which the centers of activity for the rows and columns intersect is the center of activity for the animal's home range (Fig. 11-14). However, this does not necessarily reflect the location of the animal's nest, burrow, etc.

Koeppl et al. (1975) suggested that this point might better be called the "center of familiarity" based on Ruff's (1969) correlation of heart rate with position in the home range for uinta ground squirrels. Koeppl and his colleagues also showed how confidence ellipses can be calculated around the center of activity for elliptical home ranges and then used to determine the probability of finding the resident at any given location. The distance between centers of activity for residents of adjacent territories, or home ranges, can be used to infer their relative avoidance throughout the year (see Clark and Evans 1954).

Weeden (1965) used a gridded area to study territories and their utilization by tree sparrows *(Spizella arborea)*. An observer spent an entire four-hour observation period with a single pair of sparrows (focal pair) plotting every location visited on a map. Weeden then used the observation-area curve method to determine the number of observations necessary to calculate a reasonably accurate territorial

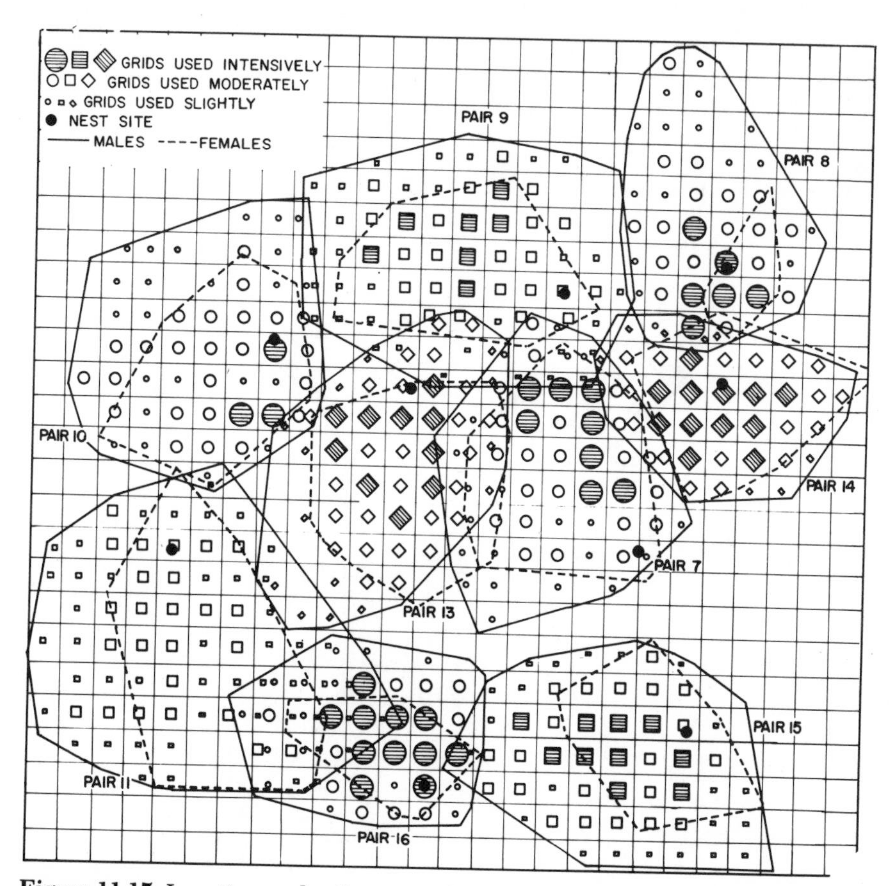

Figure 11-15 Location and utilization of total activity spaces of males, location of total activity spaces of females, and location of nest sites of tree sparrows. (From Weeden, 1965, Territorial behavior of the tree sparrow, *Condor* 67(3):197, Fig. 1.)

area. The relative number of visitations to the different quadrats in the territory revealed "central cores of more concentrated use" (Fig. 11-15).

Continual tracking of individual animals (or groups) for several days by direct observation or radiotelemetry (*e.g.*, Sargent 1972) reveals disproportionate use of the home range. Primatologists often observe groups of primates for ten-day periods in order to construct ten-day ranges (Fig. 11-16). Within these ranges are generally found areas of heavy usage designated "core areas" (*e.g.*, DeVore and Hall 1965).

The method(s) selected for determining the home range, or territory, should be based on a knowledge of the animal's behavior. The method will be limited by constraints on your time and equipment; therefore, a constant assessment of its validity is mandatory.

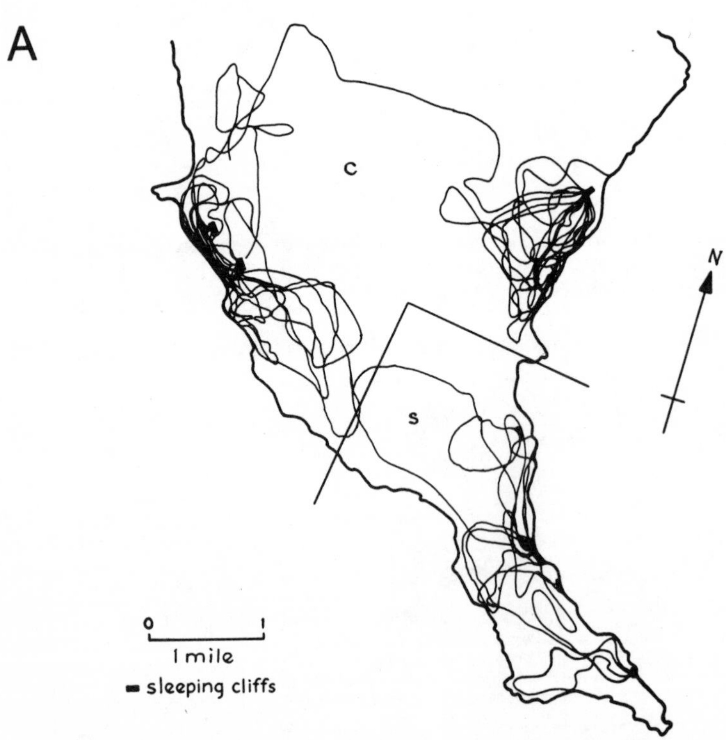

Figure 11-16 *A*, 10-day ranges for group *S* and 21-day ranges for group *C*, Cape Reserve. The sharp line indicates the approximate limit of each group's home range. *B*, areas occupied by *C* and *S* groups and southern limit of *N* group's range, indicating amount of overlap in home ranges and location of

B

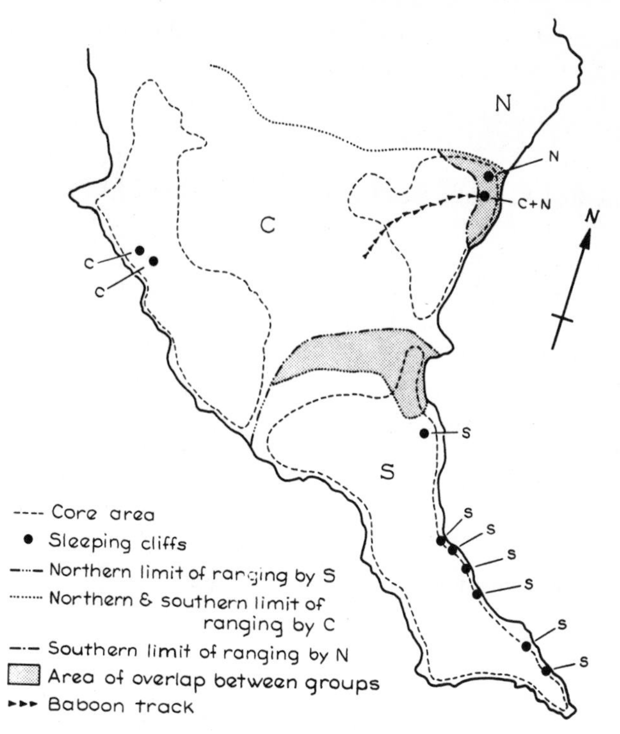

core areas. (From Primate Behavior: Field Studies of Monkeys and Apes edited by Irven DeVore. Copyright © 1965 by Holt, Rinehart and Winston, Inc. Reprinted by permission of Holt Rinehart and Winston and Irven DeVore.)

12
Presentation and Interpretation of Results

Discovery consists in seeing what everybody else has seen and thinking what nobody else has thought.
[ALBERT SZENT-GYORGYI]

A. WHAT DO YOUR RESULTS MEAN?

The end point (temporary pause in the ethological approach cycle) in your research is not the results of your data analyses, but their *interpretation*.

Have you been able to reject your null hypothesis (H_0)? If so, you can accept the alternate hypothesis (H_A) and take temporary pleasure in your accomplishment. If you were not able to reject your H_0, then you have *negative results* which are difficult to interpret (Kerlinger 1967). Most importantly, note that failure to reject the H_0 does not mean that you automatically accept it. Rather, it means that several factors could have contributed to your results, only one of which was that H_0 was true. Also consider your techniques—were they reliable and valid? Did you overlook an important parameter in your original design? Are you now convinced that the H_0 is true or should you design another experiment to test the H_0 in what you consider to be a more valid approach?

If you were merely making reconnaissance observations or using analyses (*e.g.*, cluster analysis) in hypothesis seeking, can you now generate a meaningful and testable hypothesis?

Once you have your results it is often useful to prepare a visual representation for further inspection.

B. VISUAL REPRESENTATIONS

Most researchers can learn more about the results of a research project by studying visual presentations (other than tables) of results. Figures and graphs are not only useful in presenting your results to other people, but also help you interpret the results and see relationships that were not apparent in the tabular data. Increased insight into interpretation and new hypotheses are often the result of careful contemplation of visual representations. Multiple presentations of the same results sometimes allow you to recognize subtleties in relationships which were previously hidden.

The *dendrogram*, described previously for cluster analysis (p. 283), is obviously a valuable tool. Others which are useful are described below.

1. Graphs and Figures

There are several types of graphic formats which have proven valuable in interpreting results. The *simple scatter diagram* (scattergram) is generally used to graph correlation data. The interpretation of the graph is dependent on the location and distribution of the points relative to the axes (p. 257).

For example, Goss-Custard (1977) examined the hypothesis that the redshank *(Tringa totanus)* selects the sizes of polychaete worms (prey) that maximize the biomass ingested per unit time. He plotted the number of large worms taken relative to their density in the mud (Fig. 12-1). The graph revealed an apparent positive correlation that was later shown to be statistically significant.

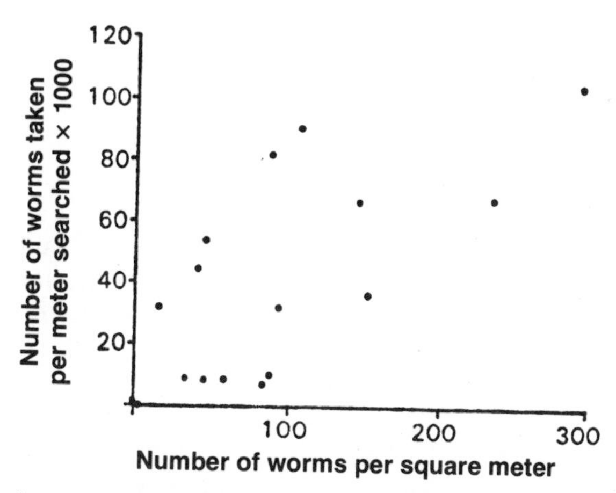

Figure 12-1 The numbers of worms above 30 mg dry weight taken per meter searched in relation to their density in the mud. (From Goss-Custard, J.D., 1977, Optimal foraging and the size selection of worms by redshank, *Tringa totanus*, in the field, *Anim. Behav.* (25)1:15, Fig. 4.)

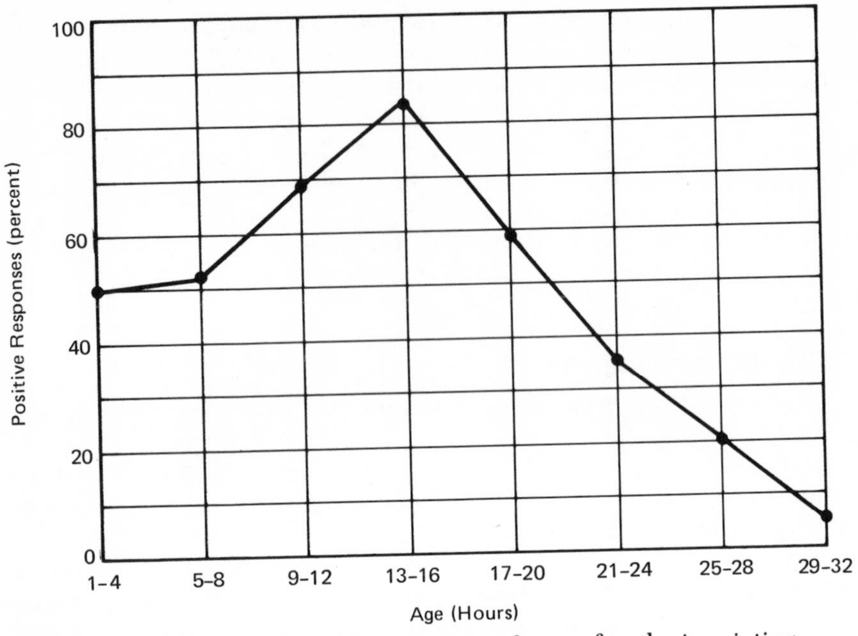

Figure 12-2 Mean scores made in testing, 24 hours after the imprinting experience, by animals which had been given the imprinting experience at different ages. A score of 100 percent is regarded as perfect. (From Hess, 1962.)

Frequency distributions are plotted in two common formats—frequency polygons and histograms.

Frequency polygons are usually produced by connecting points along a *continuous* frequency distribution. Points should generally not be connected if the distribution is not continuous or if the sample points are distantly separated. Connection implies that the line between the points is a reasonable representation of the missing data.

Hess (1962), in his studies of imprinting, exposed ducklings to an adult model (decoy) for a limited period early in life and later tested them for the imprinting response (Fig. 12-2). The frequency polygon shows that the highest percentage of positive responses was given by ducklings exposed to the model at 13–16 hours after hatching, the "critical period" or "sensitive period" for this species.

Frequency polygons are sometimes used with *discontinuous* data to show relative changes from one condition to another, while the identity of several individuals or levels of another variable are maintained. Figure 12-3 shows the mean number of courtship displays for 11 male guppies *(Poecilia reticulata)* at different population densities.

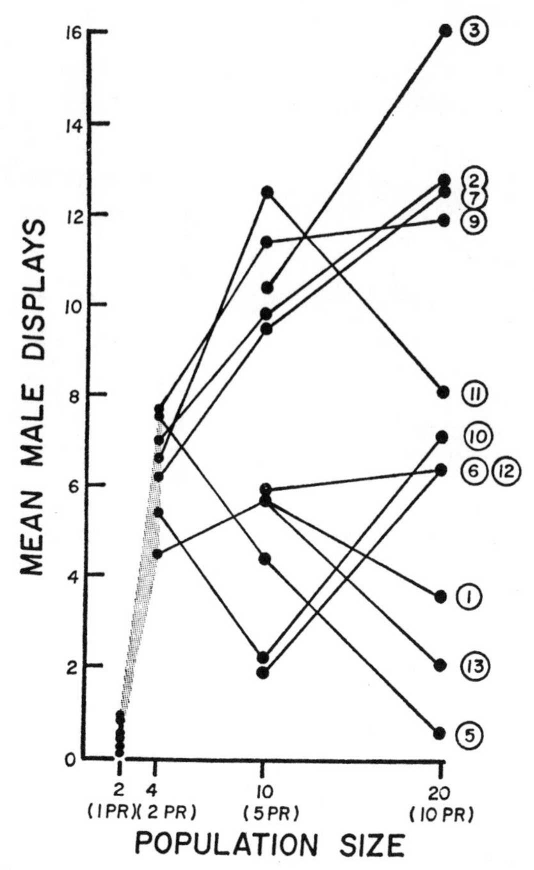

Figure 12-3 The mean number of courtship displays per observation period for specific male guppies. Each line represents a single male. (From Farr, J. and W. E. Herrnkind, 1974, A quantitative analysis of social interaction of the guppy, *Poecilia reticulata* (Pisces:Poeciliidae) as a function of population density, *Anim. Behav.* 22(3):586, Fig. 4, 5.)

As the population density increased beyond two pairs, the mean number of displays for most males increased or decreased in an unpredictable manner (Farr and Herrnkind 1974).

Several frequency polygons can be combined into a single graph to show changes in several dependent variables relative to an independent variable. Figure 12-4 illustrates the change in relative occurrence of five behaviors in female woodchucks *(Marmota monax)* as their infants grew older (Barash 1974). See Figure 12-7 for another approach to illustrating similar data.

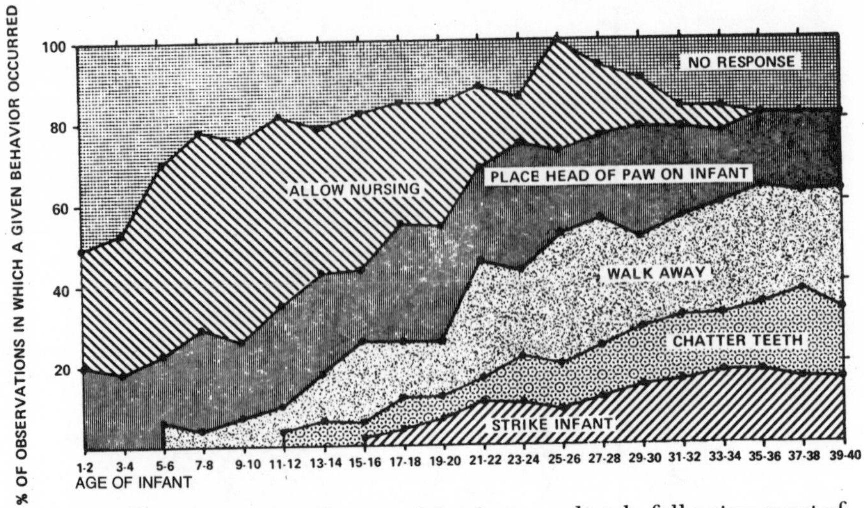

Figure 12-4 Behavior of mother woodchucks immediately following onset of squeaking by infants (cumulative data for three litters, based on a total of 387 incidents, with a minimum of fourteen per two-day interval). (From Barash, D.P., 1974, Mother-infant relations in captive woodchuck *(Marmota monax)*, *Anim. Behav.* (22):448, Fig. 2.)

Some ethologists have found it illustrative to present results in a three-dimensional frequency polygon. For example, Figure 12-5 demonstrates that the probability of a stickleback eating a food item is due, in part, to a complex relationship between the cumulative number of eats and rejects which precede the encounter with the food item (Thomas 1977). It can be seen that, in general, the probability of an eat occurring is greatest when the number of prior rejects is low irrespective of the number of prior eats.

Frequency distributions within categories of a discontinuous (*i.e.*, discrete) independent variable are often illustrated with *histograms*. When the means of a single variable are illustrated, the standard error of the mean (Fig. 12-6*A*) or the standard deviation is each informative relative to the statistical significance of the difference between the means. For example, Nyby et al. (1977) showed that male mice made significantly more ultrasounds in response to facial chemicals from females than from either males or controls (*i.e.*, clean surgical cotton swabs), and their responses to male and control facial chemicals were not significantly different (Fig. 12-6*A*).

It is often illustrative to incorporate two or more groups within an independent variable (*e.g.*, dominant or subordinate within sex) into the same histogram (Fig. 12-6*B*).

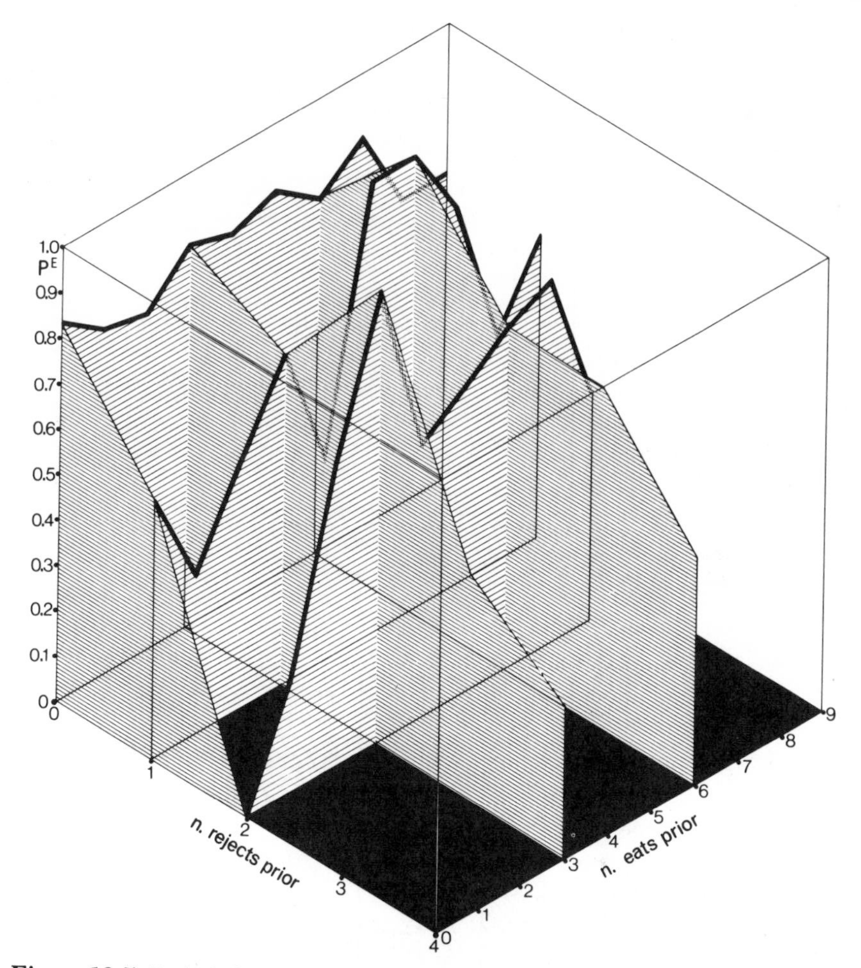

Figure 12-5 Probabilities of an eat occurring (P^E) in any encounter correlated against the accumulative numbers of foregoing eats on one plane and foregoing rejects on the other. (From Thomas, G., 1977, The influence of eating and rejecting prey items upon feeding and food searching behavior in *Gasterosteus aculeatus* L., (Anim. Behav.) 25(1):52–66, Fig. 11.)

A type of histogram can be employed to show the emergence and disappearance of behavior over time. For example, Figure 12-7 illustrates the timing of emergence of postural, locomotor, and related skills in the laboratory rat (Altman and Sudarshan 1975). Note that the results present in this figure are of the same type presented in the frequency polygon (Fig. 12-4).

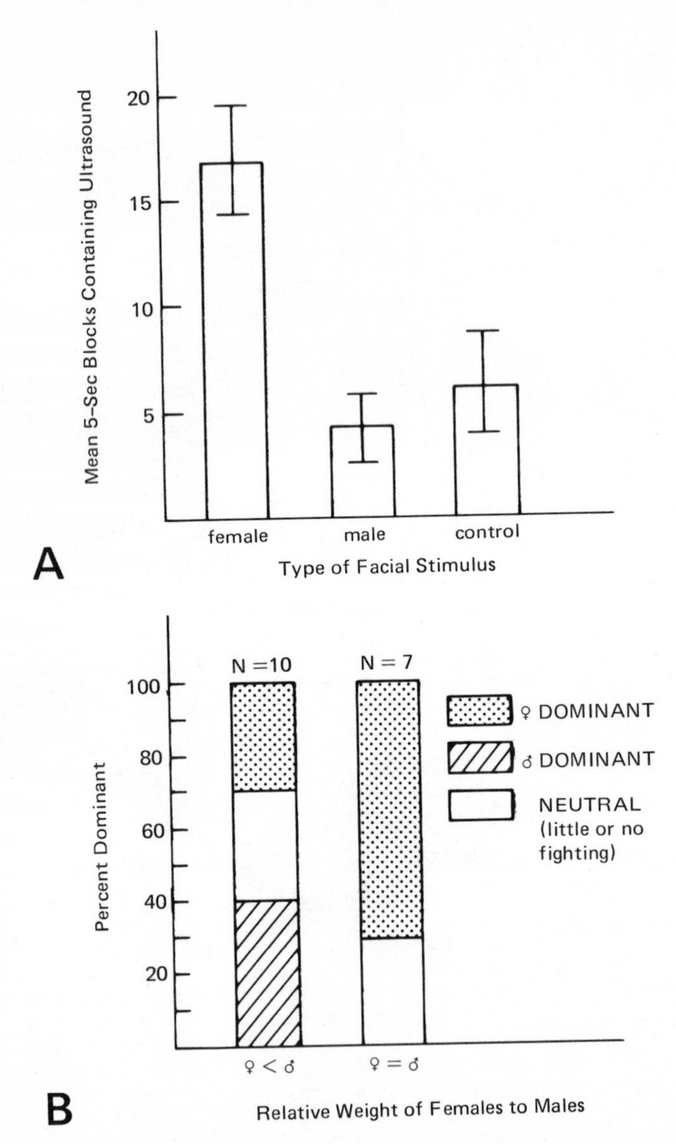

Figure 12-6 *A*, mean number of 5-second blocks containing ultrasound from DBA/J2 inbred male house mice in response to facial chemicals on a cotton swab (measurements = mean ± SE). (From Nyby, J., *et al.*, 1977, Pheromonal regulation of male mouse ultrasonic courtship *(Mus musculus). Anim. Behav.* 25(2):336, Fig. 2.) *B*, male-female dominance record in hamsters as a function of relative body weight. A female was considered lighter if its weight was 20 g or more below that of the male. (From Marques, D.M., and E.S. Valenstein, 1977, Individual differences in aggressiveness of female hamsters: response to intact and castrated males and to females, *Anim. Behav.* 25(1):134, Fig. 1.)

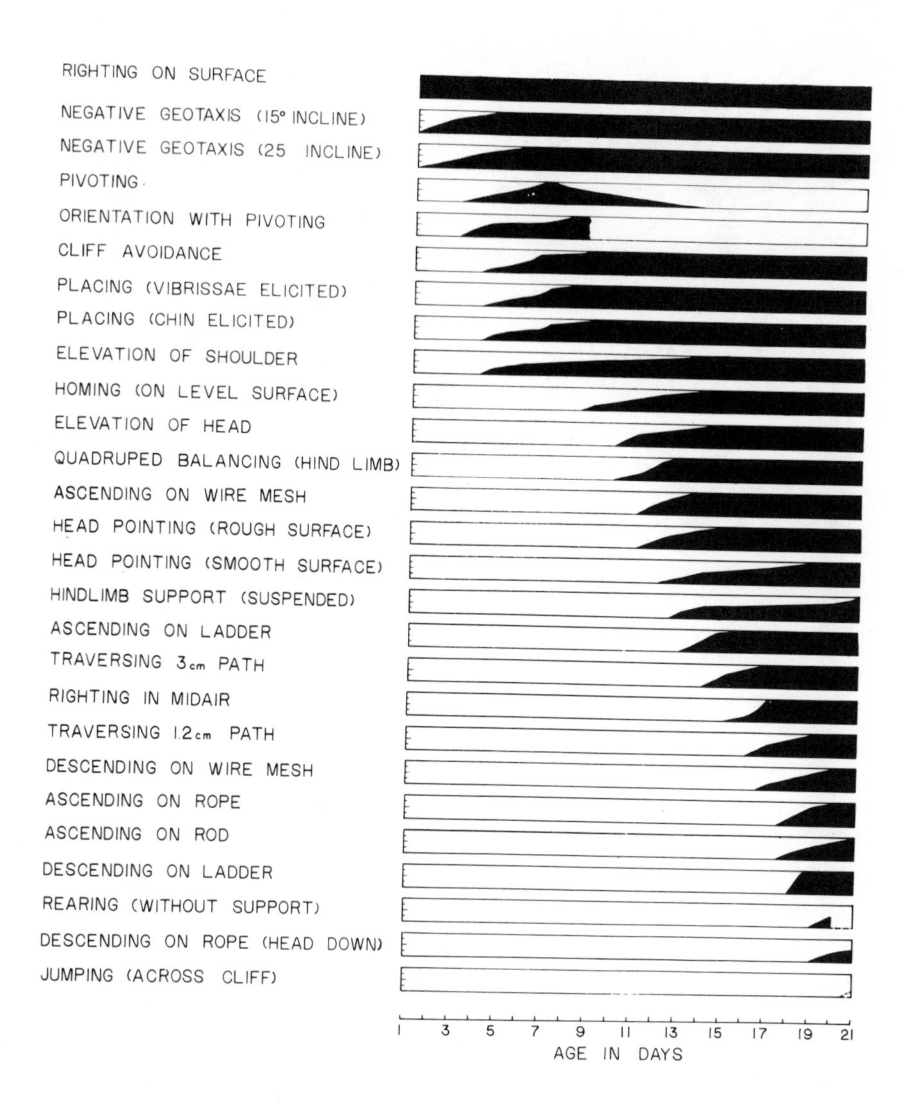

Figure 12-7 Summary diagram of the emergence of different postural, locomotor, and related skills in the laboratory rat. In the majority of instances performance level (vertical axis of each graph; 0, 25, 50, 75, and 100 percent) refers to the percentage of animals successful in the full display of the response. In a few instances the reference is to level of performance with respect to asymptotic response frequency. (From Altman, J., and K. Sudarshan, 1975, Postnatal development of locomotion in the laboratory rat, *Anim. Behav.* 23(4)916, Fig. 22.)

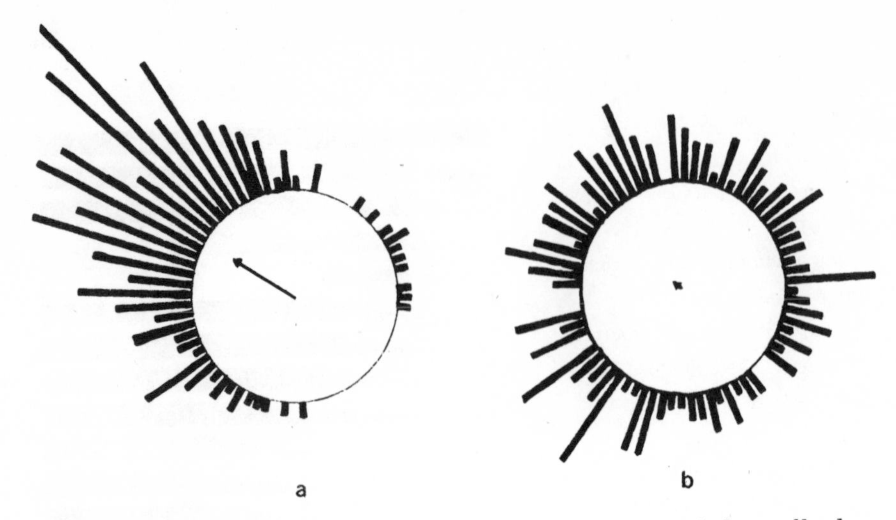

Figure 12-8 The superimposed vanishing bearings of borough fen mallards released under (*a*) sunny and (*b*) overcast conditions. The centrifugal arrow indicates the mean vector (*m*) whose length (*r*) increases the tighter the bearings cluster about the mean. The shortest bar represents one bird, the longest 21. (From Matthews, G.V.T., and W.A. Cook, 1977, The role of landscape features in the 'nonsense' orientation of mallards, *Anim. Behav.* 25(2):511, Fig. 2.)

2. Vector Diagrams

Vector diagrams are used to illustrate the distribution of data relative to two or more coordinate axes. Recall that three-dimensional vector diagrams were previously discussed as a useful method for presenting and interpreting the results of factor analysis (p. 291).

Two-dimensional vector diagrams are often used to illustrate the directional responses of individual animals in orientation studies. For example, Figure 12-8 shows the compass direction in which individual hen mallards vanished from sight after being released (Matthews and Cook 1977). The effect of overcast on the mean direction vector can easily be seen.

McKinney (1975) used vector diagrams to illustrate the orientations, distances, and positions of the male green-winged teal relative to the female during selected courtship displays (Fig. 12-9).

3. Kinematic Graphs

Kinematic graphs (often called flow diagrams) are useful to illustrate transitions between behaviors (see p. 118). Sustare (1978) discusses, in detail, the use of various systems diagrams including information

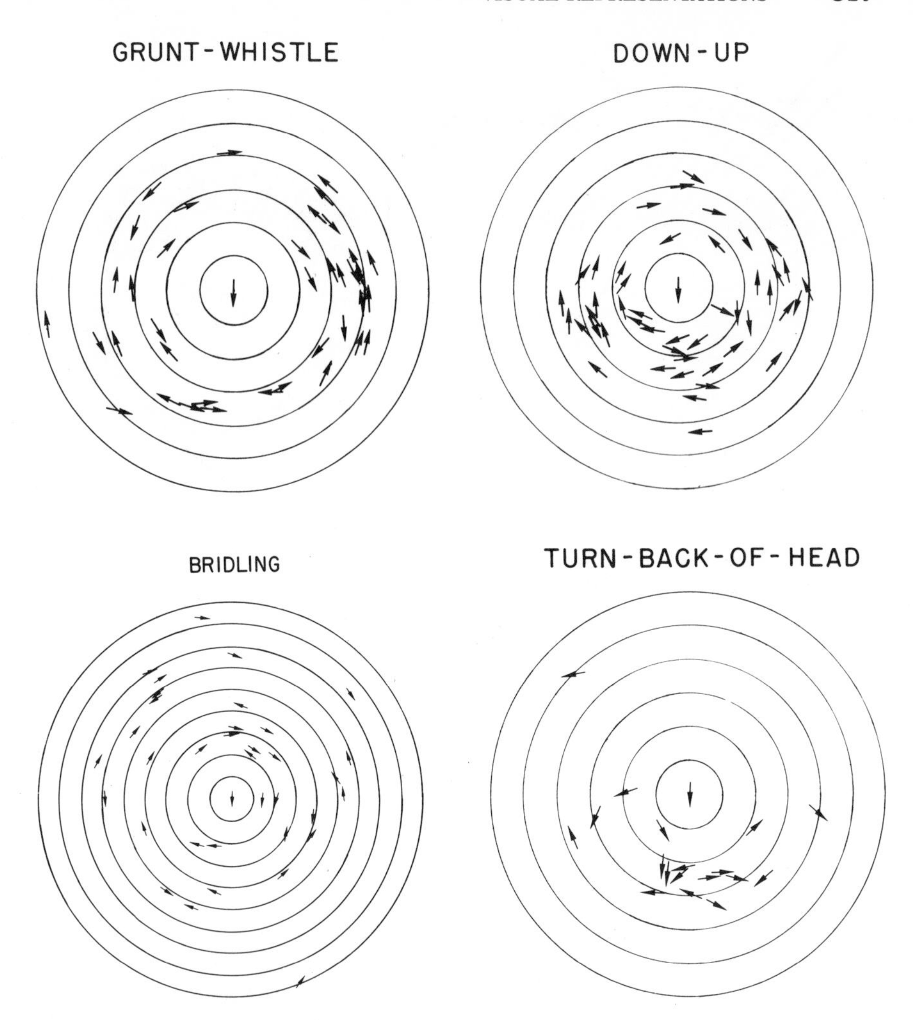

Figure 12-9 Orientations, distances, and positions of male green-winged teal in relation to the female (center arrow) during performances of grunt-whistle, down-up, bridling, and turn-back-of-head displays (see Figure 7-2). Note the precise lateral body orientation of males when performing the grunt-whistle, shorter distance from the female in the case of the down-up, variability in distance for bridling, and positioning in front of the female for turn-back-of-head. Grunt-whistle, bridling, and turn-back-of-head can occur when only one male is present, but down-up is performed only when a second male is present also. The distance between concentric circles is one foot; a swimming teal measures slightly less from bill-tip to tail-tip. (From McKinney, F., 1975, The evolution of duck displays, p. 339, Fig. 16.1 *in* Baerends, G., *et al.*, (eds.), Function and evolution in behaviour, Oxford Univ. Press, London.)

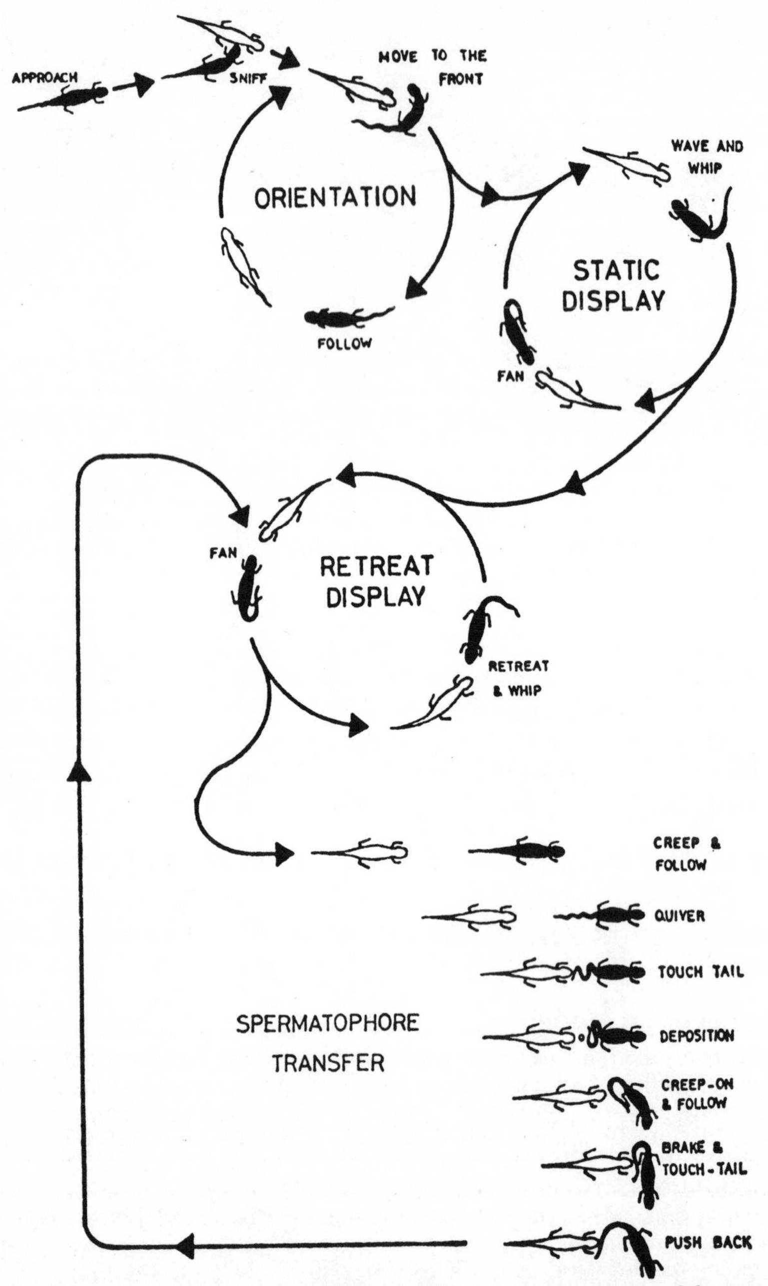

Figure 12-10 Kinematic graph of the sexual-behavior sequence of the smooth newt. The male is in black. (From Halliday, T.R., 1975, An observational and experimental study of sexual behaviour in the smooth newt, *Trifurus vulgaris* (Amphibia : Salamandridae), *Anim. Behav.* 23(2):295, Fig. 3.)

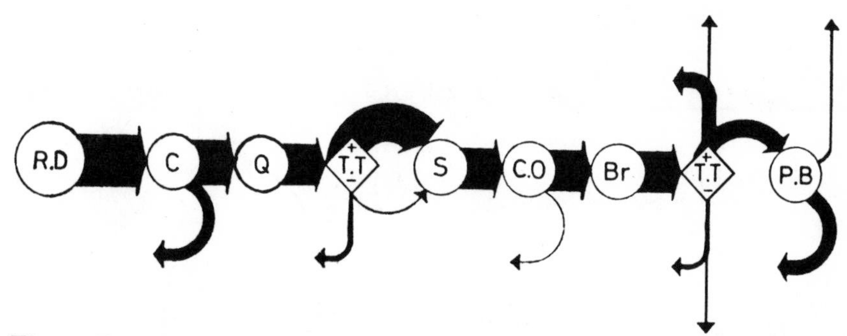

Figure 12-11 Kinematic graph of the spermatophore-transfer phase of the smooth newt's sexual behavior sequence (Fig. 12-10). Width of arrows is proportional to the frequency of transition. Arrows pointing to the left are returning to retreat display; arrows pointing outwards are leaving sexual behavior, for example, to breathe. *Br:* Brake; *C:* Creep; *C.O.:* Creep-on; *P.B.:* Push-back; *Q:* Quiver; *R.D.:* Retreat Display; *S:* Spermatophore deposition; *T.T.:* Touch-tail. (From Halliday, 1975, An observational and experimental study of sexual behaviour in the smooth newt, *Troturus vulgaris* (Amphibia: Salamandridae), *Anim. Behav.* 23(2):312, Fig. 15.)

networks (*e.g.*, sociogram Fig. 9-12), association diagrams (Fig. 12-12), state-space diagrams, and kinematic graphs. Halliday (1975) used two types of kinematic graph to show the sexual behavior sequence in the smooth newt *(Triturus vulgaris)*. Figure 12-10 includes drawings of the male and female, which increases the ability of the reader to "visualize" the sequence through the orientation of the two sexes.

Figure 12-11 provides increased information on the probability of particular transitions occurring (width of arrows). Altering the size of the arrows can be supplemented or replaced with the actual probabilities of transitions in percent.

4. Conceptual Models

Conceptual models are a means of maintaining perspective. They allow you to fit together pieces of information about a behavioral complex (*e.g.*, reproductive behavior) in an attempt to better understand and illustrate the interrelationships. Models are generally hypothetical and temporary, being changed as new results come forth. For example, Baerends (1976) proposed a model (Fig. 12-12) to explain the occurrence of interruptive behavior during the incubation of the herring gull. Baerends broke the model into "systems," "subsystems," and "acts" which he relates to Tinbergen's (1950) earlier conceptual model of the hierarchical organization of behavior (see also Dawkins 1976).

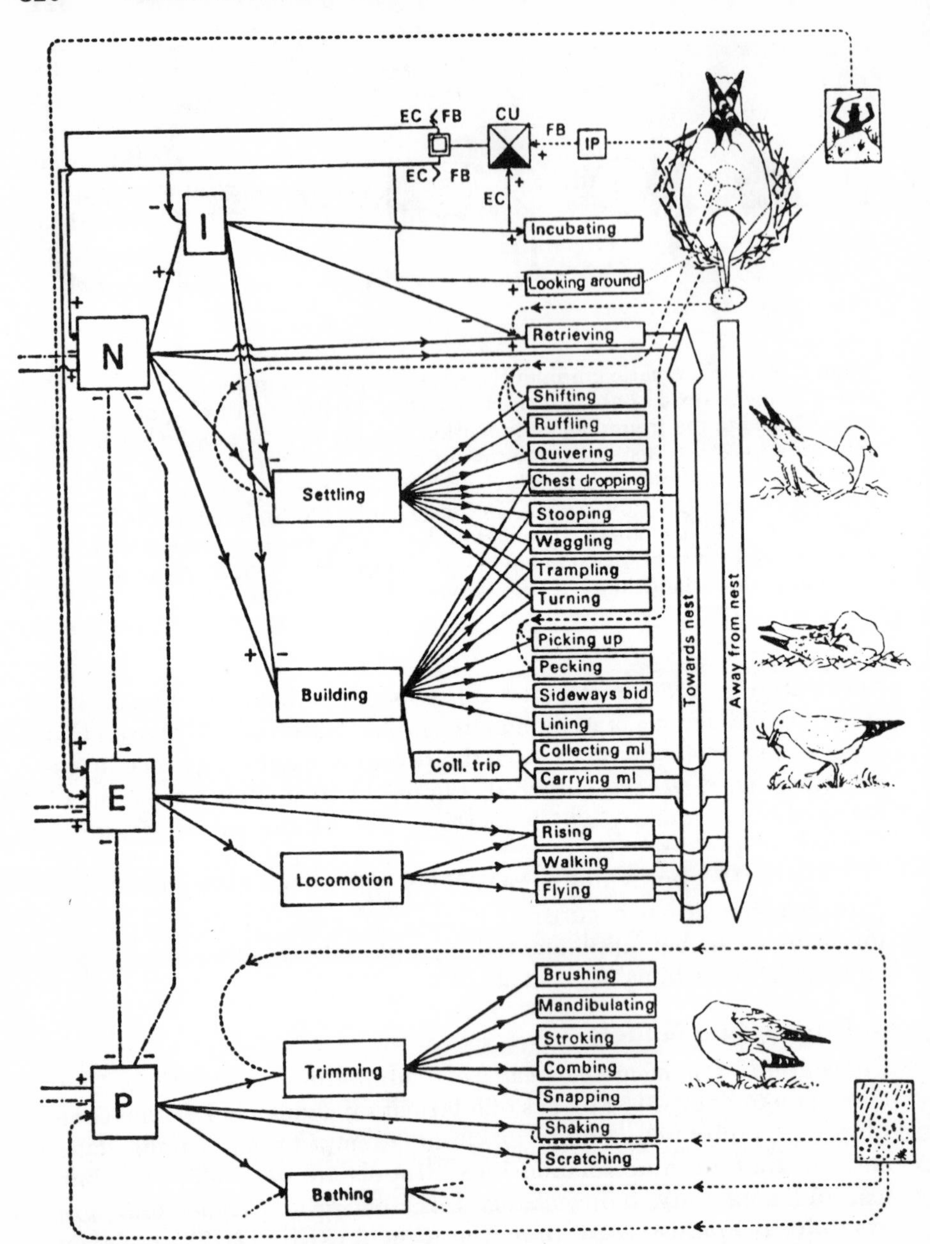

Figure 12-12 Model for the explanation of the occurrence of interruptive behavior during the incubation of a herring gull (after Baerends 1970). The fixed action patterns are in the right column and superimposed control systems of first and second order are represented left of them (N = incubation system, E = escape system, P = preening system). The large vertical arrows

Further perspective is provided by *generalized conceptual models* which help the researcher envision the complex of variables which impinge on behavior (*e.g.*, p. 87). Some models aid the researcher in recognizing how his research fits within the "big picture" and assists in identifying important variables to investigate in future studies. A model of this general type was described and discussed in detail by Crook et al. (1976). The universality of their model is illustrated in Figure 12-13. Colgan (1978) discusses the role of modeling in ethological research.

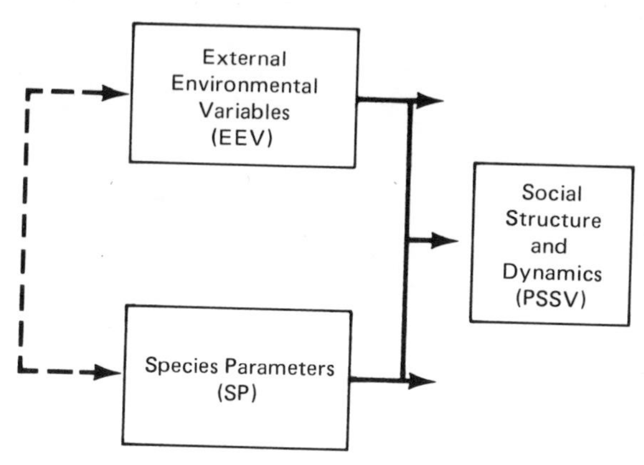

Figure 12-13 A conceptual model showing how external environmental variables (*EEV*) are expected to interact with species parameters (*SP*, for example, morphological and physiological characteristics) to determine social structure [measured as the principal social system variables (PSSV) and social dynamics (changes in *PSSV* over time)]. The dotted arrow takes note of the fact that *EEV*'s also affect *SP*, but on a slower (evolutionary) time scale than the effects on *PSSV*'s, which may change within the lifespan of an individual through learning. (From Crook et al., 1976.)

Figure 12-12 (Continued)
represent orientation components with regard to the nest. Incubating is the consummatory act. Feedback stimulation from the clutch, after being processed in *IP*, flows to a unit (*CU*) where it is compared with expectancy, an efference copy or corollary on the input for incubation. This input is fed through a unit (*I*), necessary to explain the inhibition of settling and building when feedback matches expectancy. The effect of feedback discrepancy on *N* (and *I*), *E*, and *P* can be read from the arrows. The main systems mutually suppress one another; *P* is thought to occur as interruptive behaviour through disinhibition of *N* and *E*. *P* can be activated directly by external stimuli like dust, rain, or parasites; *E* can also be stimulated by disturbances other than deficient feedback from the clutch. (From Baerends, G.P., 1976, The functional organisation of behaviour, *Anim. Behav.* 24(2):732, Fig. 5.)

Conceptual models often lead to *predictive models*, which are generally expressed in mathematical terms to enable tests of their validity. These models are beyond the scope of this book, but an interesting example can be found in Hazlett and Bach (1977).

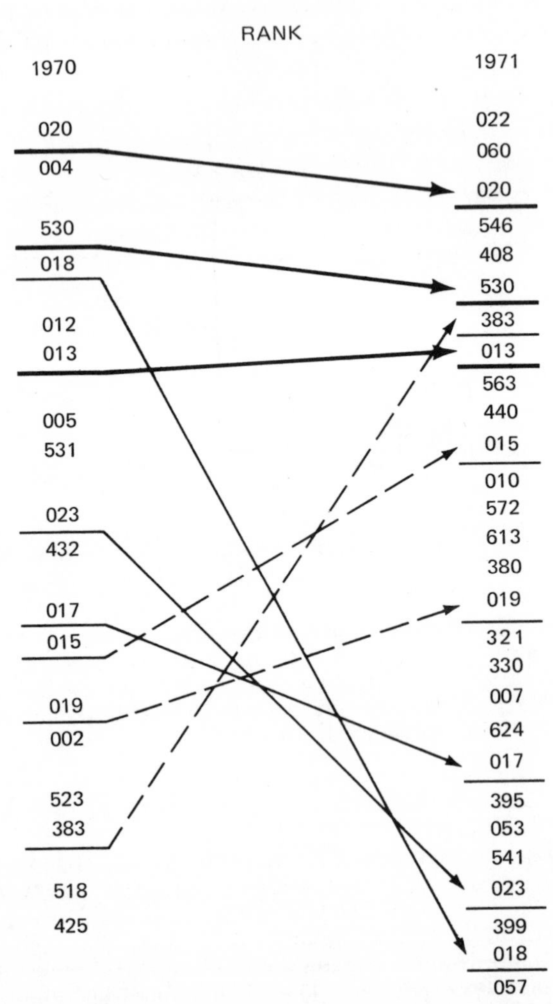

Figure 12-14 Changes in rank order between years. The figures are the serial numbers of individual marked males arranged in rank order. Birds which were ranked in both years are joined by arrows. (From Patterson, I.J., 1977, Aggression and dominance in winter flocks of shelduck *Tadorna tudorna* (L.), *Anim. Behav.* 25(2):454, Fig. 7.)

5. Other Illustrations

The type of visual representation employed and its value in interpreting results are limited only by the ingenuity of the researcher. Simplicity in illustrations is often a virtue worth pursuing. For example, Patterson (1977) used a simple diagram which clearly demonstrates the rank-order changes of male shelducks *(Tadorna tadorna)* over a two-year observation period (Fig. 12-14). The positional and relative extent of the changes in rank order are obvious and conducive to further interpretation.

Hutt and Hutt (1974) followed up on a suggestion by Altmann (1965) and described the application of a *phase structure grammar model* to the analysis of behavioral sequences. The model was first developed by Chomsky (1957) for the study of psycholinguistics.

The model consists of the sequential partitioning of a sentence into its constituent parts based on its explicit meaning. The result is a

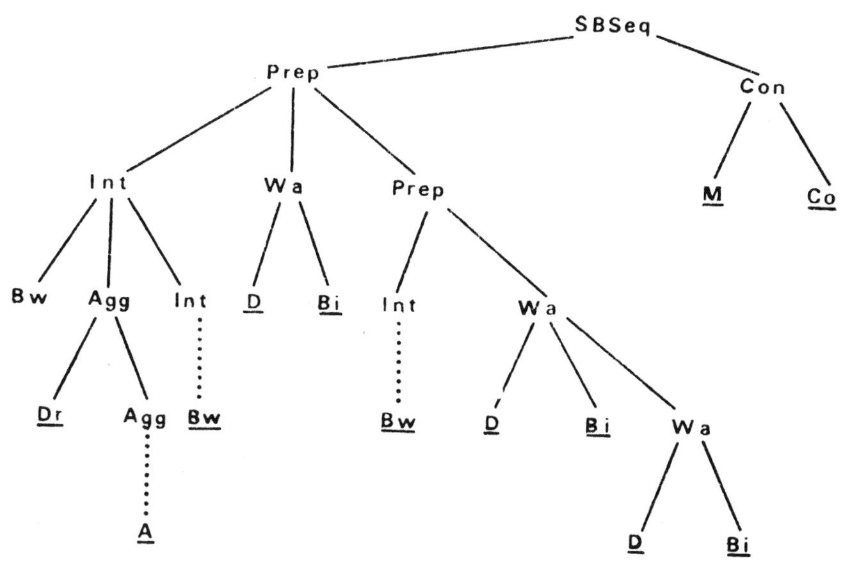

Figure 12-15 Tree diagram showing application of generative grammar in its recursive form to reproductive behavior of the male pigeon. *SBSeq*, sexual Behavior Sequence; *Prep*, preparatory behavior; *Con*, consummatory behavior; *Int*, introduce; *Wa*, warm up; *Agg*, aggressive behavior; *Bw*, bowing; *Dr*, driving; *A*, attacking; *D*, displacement preening; *Bi*, billing; *M*, mounting; *Co*, copulation. (From Hutt, S.J., and C. Hutt, 1970, Direct observation and measurement of behavior, courtesy of Charles C. Thomas, Springfield, ILL., p. 183, Fig. 63.)

tree diagram of sequentially smaller clusters of words that together carry the meaning of the sentence. This hierarchical model, discussed by R. Dawkins (1976) and Westman (1977), was used by Marshall (1970) in his study of syntax in the reproductive behavior of the pigeon (Fig. 12-15).

The "Catch-22" of this method for the ethologist is that to apply the model to gain understanding of the message in communication, we must first understand the message.

> Quite aside from the fact that the word units in Chomsky's linguistic analysis are not strictly comparable with the behavioural units in our study, there is the more difficult problem that Chomsky's analysis depends upon being able to distinguish that class of word chains which constitute the grammatically correct statements within the language. [S.A. ALTMANN 1964:519]

This difficulty is illustrated by applying the analysis to the sentence "We fed her dog bones," which can have two meanings; hence it can be diagrammed in two ways (Fig. 12-16).

As Dale (1976) states, the ambiguity does not arise from a difference in words or in their order, but rather from a difference in their constituent structure. Do we have this level of resolution in analyzing sequences of animal behavior? Altmann (1965) suggests that with sufficient experience it can be done.

> If one's goal is to draw up an exclusive and exhaustive classification of the animals' repertoire of socially significant behaviour patterns, then these units of behaviour are not arbitrarily chosen. On the contrary, they can be empirically determined. One divides up the continuum of action wherever the ani-

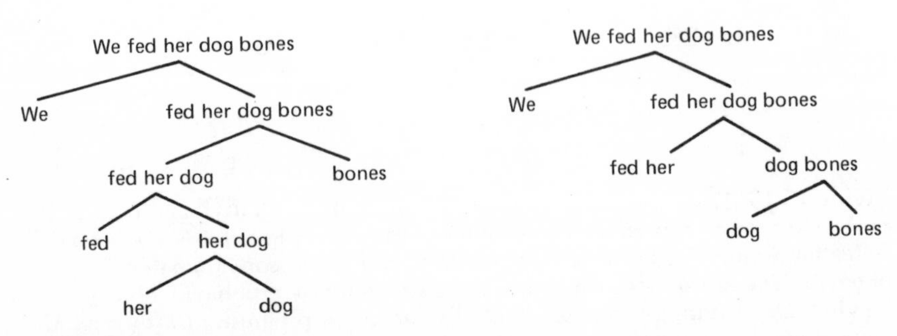

Figure 12-16 Two-phase-structure grammar models of a single sentence to illustrate the different meanings (adapted from Dale 1976).

mals do. If the resulting recombination units are themselves communicative, that is, if they affect the behaviour of other members of the social group, then they are social messages. Thus, the splitting and lumping that one does is, ideally, a reflection of the splitting and lumping that the animals do. [ALTMANN 1964:492]

To summarize, all of these techniques of visual representation (and others not discussed) can aid in the interpretation of results. They should be examined not only to better understand the particular behavior studied, but also to put it in perspective relative to the various levels of behavior (p. 2). Are the results similar to those seen for other behaviors, and species and under other environmental conditions? Are the results consistent with a conceptual model or valuable for use in a predictive model? Does your interpretation help you to develop new models and generate new hypotheses? At this point it is again important to consider what other research has shown.

C. COMPARISONS WITH PREVIOUS RESULTS

How do your results compare with those of other researchers? Discuss your results with the same researchers you consulted before beginning your study (p. 54). Even though you reviewed the literature before beginning your study, it is wise to again search for relevant material in the light of your results. You may want to know more about similar behavior in other species or related behavior in the same species. You may discover that your results have a bearing on a general concept or current theoretical issues. The importance of results are often unforeseen when a study begins, but become apparent as the study proceeds and finally come to light as the results are carefully interpreted.

D. RE-EVALUATION

You've now reached the point where you can re-evaluate the entire study. You know where you began, and you think you know what your results mean. Even though your results were conclusive, your study could have been better. Re-evaluate the economics, efficiency, and validity of your methods. Did you select the proper species, study area, behavioral units, data-collection method, analytical tests, etc.? Re-evaluate your study at each phase of the ethological approach (p. 11). You should improve with each study, but this can only come through a critical re-evaluation of each study as it is completed.

E. REVISING AND RESTATING HYPOTHESES

You may want to revise or restate your hypotheses whether your results were positive or negative. Testing revised hypotheses can help reinforce positive results and isolate the source of negative results. You might choose to isolate additional variables or test the external validity of your results on other species.

Whether you revise, restate, or generate new hypotheses, you are now back at the beginning of the ethological approach cycle, ready to begin again. This time you are more experienced and, hopefully, wiser.

Bibliography

Adams, R.M. and R.P. Markley. 1978. Assessment of the accuracy of point and one-zero sampling techniques by computer simulation (unpublished manuscript). Paper presented at 1978 Animal Behavior Society Meeting, June 19–23, Seattle, Washington.

Alcock, J. 1970. Punishment levels and the response of black-capped chickadees *(Parus atricapillus)* to three kinds of artificial seeds. Anim. Behav. 18(3):592–599.

Alcock, J. 1973. Cues used in searching for food by red-winged blackbirds *(Agelaius phoeniceus)*. Behaviour 46:174–188.

Alcock, J. 1975. Animal behavior: an evolutionary approach. Sinauer Associates, Sunderland, Mass. 547 pp.

Alexander, R.D. 1975. The search for a general theory of behavior. Behav. Sci. 20(2):77–100.

Allee, W.C. 1938. The social life of animals. Beacon Press, Boston. 233 pp.

Altman, J. and K. Sudarshan. 1975. Positioned development of locomotion in the laboratory rat. Anim. Behav. 23(4):896–920.

Altmann, J. 1974. Observational study of behavior: sampling methods. Behaviour 49(3,4):227–265.

Altmann, S.A. 1965. Sociobiology of rhesus monkeys. II. Stochastics of social communication. J. Theor. Biol. 8:490–522.

Altmann, S.A. 1968. Sociobiology of rhesus monkeys IV: Testing Mason's hypothesis of sex differences in affective behaviour. Behaviour 32(1–3): 49–69.

Altmann, S.A. and J. Altmann. 1977. On the analysis of rates of behaviour. Anim. Behav. 25(2):364–372.

Altmann, S.A. and S.S. Wagner. 1970. Estimating rates of behaviour from Hansen frequencies. Primates 2:181–183.

Andrew, R.J. 1972. The information potentially available in mammalian displays. Pages 179–204 *in* R.A. Hinde, ed. Non-verbal communication. Cambridge Univ. Press, Cambridge, England.

Anonymous. 1970. Basic scientific photography. Eastman Kodak Scientific Data Book N-9, Rochester, New York. 40 pp.

Ansell, A.D. 1967. Leaping and other movements in some cardiid bivalves. Anim. Behav. 15(4):421–426.

Ashby, W.R. 1963. An introduction to cybernetics. John Wiley Inc., New York. 295 pp.

Aspey, W.P. 1977*a*. RECDIS: A BASIC program for computing interindividual distances within a rectangular area. Behav. Res. Methods Instrum. 9(1):26–27.

Aspey, W.P. 1977*b*. CIRCIS: A BASIC program for computing interindividual distances within a circular area. Behav. Res. Methods Instrum. 9(1): 50–51.

Aspey, W.P. 1977*c*. Wolf spider sociobiology: I. Agonistic display and dominance-subordinance relations in adult male *Schizocosa crassipes*. Behaviour 62(1–2):103–141.

Aspey, W.P. and J.E. Blankenship. 1976. *Aplysia* behavioral biology: I. A multivariate analysis of burrowing in *A. brasiliana*. Behav. Biol. 17:279–299.

Aspey, W.P. and J.E. Blankenship. 1977. Spiders and snails and statistical tales: application of multivariate analyses to diverse ethological data. Pages 75–120 *in* B.A. Hazlett, ed. Quantitative methods in the study of behavior. Academic Press, New York. 222 pp.

Aspey, W.P. and J.E. Blankenship. 1978. Comparative ethometrics: congruence of different multivariate analyses applied to the same ethological data. Behav. Proc. 3:173–195.

Baerends, G.P. 1976. The functional organization of behaviour. Anim. Behav. 24(4):726–738.

Baerends, G.P., R. Brouwer and H.T. Waterbolk. 1955. Ethological studies on *Lebistes reticulatus* (Peters): I. An analysis of the male courtship pattern. Behaviour 8:249–334.

Baerends, G.P. and N.A. Van der Cingel. 1962. On the phylogenetic origin of the snap display in the common heron *(Ardea cinerea L.)*. Symp. Zool. Soc. London. 8:7–24.

Baerends, G.P., R.H. Drent, P. Glas and H. Groenewold. 1970. An ethological analysis of incubation behaviour in the herring gull. Behaviour suppl. 17:135-235.

Baerends, G.P. and J.P. Kruijt. 1973. Stimulus selection. Pages 23–49 *in* R.A. Hinde and J. Hinde, eds. Constraints on learning. Academic Press, London.

Bakeman, R. 1978. Untangling streams of behavior. Pages 63–78 *in* G.P. Sackett, ed. Observing behavior. Vol. 2. Data collection and analysis methods. University Park Press, Baltimore.

Balgooyen, T.G. 1976. Behavior and ecology of the American kestrel *(Falco sparverius L.)* in the Sierra Nevada of California. Univ. Calif., Publ. Zool. vol. 103. Berkeley. 83 pp.

Balph, D.F. 1968. Behavioral responses of unconfined uinta ground squirrels to trapping. J. Wildl. Manage. 32(4):778–794.

Barash, D.P. 1974. Mother-infant relations in captive woodchucks *(Marmota monax)*. Anim. Behav. 22(2)446–448.

Barlow, G.W. 1977. Modal action patterns. Pages 94–125 *in* T.A. Sebeok, ed. How animals communicate. Univ. Indiana Press, Bloomington.

Bateson, P.P.G. 1977. Testing an observer's ability to identify individual animals. Anim. Behav. 25(1):247–248.

Baufle, J.M. and J.P. Varin. 1972. Photographing wildlife. Oxford Univ. Press, New York. 157 pp.

Beach, F.A. 1950. The snark was a boojum. Am. Psychol. 5:115–124.

Beer, C.G. 1977. What is a display? Am. Zool. 17:155–165.

Bekoff, M. 1976. Animal play: problems and perspectives. Pages 165–188 *in* P.P.G. Bateson and P. Klopfer, eds. Perspectives in ethology, vol. 2, Plenum, New York.

Bekoff, M. 1977*a*. Social communication in canids: evidence for the evolution of a stereotyped mammalian display. Science 197:1097–1099.

Bekoff, M. 1977*b*. Quantitative studies of three areas of classical ethology: social dominance, behavioral taxonomy, and behavioral variability. Pages 1–46 *in* B. Hazlett, ed. Quantitative methods in the study of animal behavior. Academic Press, New York. 222 pp.

Bekoff, M. 1978*a*. A field study of the development of behavior in Adelie penguins; univariate and numerical taxonomic approaches. *In* G.M. Burghart and M. Bekoff, eds. The development of behavior. Garland STPM Press, New York.

Bekoff, M. 1978*b*. Behavioral development in coyotes and eastern coyotes. Pages 97–126 *in* M. Bekoff, ed. Coyotes: biology, behavior and management. Academic Press, New York.

Bekoff, M. 1979. Scent marking by free-ranging domestic dogs: olfactory and visual components. Biol. of Behav. (in press)

Bekoff, M. and J. Corcoran. 1975. A method for the analysis of activity and spatial relations in animal groups. Behav. Res. Methods Instrum. 7(6):569.

Bekoff, M., H.L. Hill, and J.B. Mitton. 1975. Behavioral taxonomy in canids by discriminant function analyses. Science 190:1223–1225.

Bentley, D.R. and R.R. Hoy. 1972. Genetic control of the neuronal network generating cricket (Teleogryllus gryllus) song patterns. Anim. Behav. 20(3): 478–492.

Berger, J. 1977. Organizational systems and dominance in feral horses in the Grand Canyon. Behav. Ecol. Sociobiol. 2:131–146.

Bernstein, I.S. 1970. Primate status hierarchies. Pages 71–109 *in* L.A. Rosenblum, ed. Primate behavior, vol. 1. Academic Press, New York.

Beveridge, W.I.B. 1950. The art of scientific investigation. Vintage Books, New York. 239 pp.

Blair, W.F. 1953. Population dynamics of rodents and other small mammals. Adv. Genet. 5:1–41.

Blaker, A.A. 1976. Field photography. W.H. Freeman and Co., San Francisco. 451 pp.

Blalock, H.M. 1961. Causal inferences in nonexperimental research. Univ. North Carolina Press, Chapel Hill. 200 pp.

Bouissou, M.F. 1972. Influence of body weight and presence of horns on social rank in domestic cattle. Anim. Behav. 20(3):474–477.

Bradbury, J.W. and F. Nottebohm. 1969. The use of vision by the little brown bat, *Myotis lucifugus*, under controlled conditions. Anim. Behav. 17(3):480–485.

Bradley, R. 1977. Making animal sound recordings. Am. Birds 31(3):279–285.

Brander, R.B. and W.M. Cochran. 1969. Radio-location telemetry. Pages 95–103 *in* R.H. Giles, ed. Wildlife management techniques, 3rd ed. The Wildlife Society, Washington, D.C.

Brockway, B.F. 1964. Ethological studies of the budgerigar *(Melopsittacus unclulatus)*:non-reproductive behaviour. Behaviour 22:193–222.

Bronowski, J. 1973. The ascent of man. Little, Brown, Boston. 448 pp.

Brown, J.L. 1975. The evolution of behavior. W.W. Norton, New York. 761 pp.

Buchler, E.R. 1976. The use of echolocation by the wandering shrew *(Sorex vagrans)*. Anim. Behav. 24(4):858–873.

Buckley, P.A. and J.T. Hancock, Jr. 1968. Equations for estimating and a simple computer program for generating unique color- and aluminum-band sequences. Bird-Banding 39:123–129.

Bullock, R.E. 1974. Functional analysis of locomotion in pronghorn antelope. Pages 274–305 *in* V. Geist, and F. Walther, eds. The behaviour of ungulates and its relation to management. Int. Union Conserv. Nat. Resour. Publ. New Ser. 24.

Burghardt, G.M. 1973. Instinct and innate behaviour: toward an ethological psychology. *In* J.A. Nevin and G.S. Reynolds, eds. The study of behavior: learning, motivation, emotion and instinct. Scott, Foresman and Co., Glenview, Ill.

Cade, W. 1975. Acoustically orienting parasitoids: fly phontaxis to cricket song. Science 190:1312–1313.

Carey, M. and V. Nolan Jr. 1975. Polygyny in indigo buntings: a hypothesis tested. Science 190:1296–1297.

Carlier, C. and E. Noirot. 1965. Effects of previous experience on maternal retrieving by rats. Anim. Behav. 13(4):423–426.

Carpenter, C.C. and G. Grubitz. 1961. Time-motion study of a lizard. Ecology 42:199–200.

Carthy, J.D. 1966. The study of behaviour. Arnold, London. 57 pp.

Cates, R.G. and G.H. Orians. 1975. Successional status and the palatability of plants of generalized herbivores. Ecology 56:410–418.

Celhoffer, L., C. Boukyois and K. Minde. 1977. The DCR-II event recorder; a portable high-speed digital cassette system with direct computer access. Behav. Res. Methods Instrum. 9(5):442–446.

Chalmers, N.R. 1968. The social behaviour of free living Mangabeys in Uganda. Folia Primat. 8:263-281.

Chase, I.D. 1974. Models of hierarchy formation in animal societies. Behav. Sci. 19:374–382.

Chatfield, C. 1973. Statistical inference regarding Markov chain models. Appl. Stat. 22:7–20.

Chomsky, N. 1957. Syntactic structures. Mouton, The Hague. 116 pp.

Clark, P.J. and F.C. Evans. 1954. Distance to nerest neighbor as a measure of spatial relationships in populations. Ecology 35:445–453.

Clayton, D. 1976. The effects of pre-test conditions on social facilitation of drinking in ducks. Anim. Behav. 24:125–134.

Cochran, W.W., D.W. Warner, J.R. Tester and V.B. Kuechle. 1965. Automatic radio-tracking system for monitoring animal movements. BioScience 15(2):98–100.

Cohen, J.A. 1960. A coefficient of agreement for nominal scales. Educ. Psychol. Meas. 20:37–46.

Cole, L.C. 1949. The measurement of interspecific association. Ecology 30(4):411–424.

Colgan, P. 1978. Modeling. Pages 313–326 *in* P. Colgan, ed. Quantitative Ethology. John Wiley, New York.

Comrey, A.L. 1973. A first course in factor analysis. Academic Press., New York. 316 pp.

Conger, R.D. and D. McLeod. 1977. Describing behavior in small groups with the Datamyte event recorder. Behav. Res. Methods Instrum. 9(5):418–424.

Conner, W.E. and W.M. Masters. 1978. Infrared video viewing. Science 199:1004.

Cooley, W.W. and P.R. Lohnes. 1971. Multivariate data analysis. John Wiley, New York. 364 pp.

Crews, D. 1977. The annotated anole: studies on the control of lizard reproduction. Amer. Sci. 65:428–434.

Crook, J.H., J.E. Ellis and J.D. Goss-Custard. 1976. Mammalian social systems: structure and function. Anim. Behav. 24(2):261–274.

Croze, H. 1970. Searching image in carrion crows. Z. Tierpsychol. Beih. 5:85 pp.

Curio, E. 1975. The functional organization of anti-predator behaviour in the pied flycatcher: A study of avian visual perception. Anim. Behav. 23(1):1–115.

Dale, P.S. 1976. Language development: Structure and function, 2nd ed. Holt, Rinehart and Winston, New York. 358 pp.

Dane, B. and W.G. Van Der Kloot. 1964. An analysis of the display of the goldeneye duck *(Bucephala clangula L.).* Behaviour 22:282–328.

Darling, F. 1937. A herd of red deer. Oxford Univ. Press, New York. 215 pp.

Davis, D.E. 1964. The hormonal control of aggressive behavior. Pages 994–1003 *in* C.G. Sibley, ed. Proc. 13th Int. Ornithol. Congr. The Am. Ornithologists' Union.

Davis, G.J. and J.F. Lussenhop. 1970. Roosting of starlings *(Sturnus vulgaris)*: a function of light and time. Anim. Behav. 18(2):362–365.

Davis, J.M. 1975. Socially induced flight reaction in pigeons. Anim. Behav. 23(3):597–601.

Dawkins, M. 1971. Perceptual changes in chicks: another look at the 'search image' concept. Anim. Behav. 19(3):566–574.

Dawkins, R. 1971. A cheap method of recording behavioral events, for direct computer-access. Behaviour 40(1–2):162-173.

Dawkins, R. 1976. Hierarchical organization: a candidate principle for ethology. Pages 7–54 *in* P.P.G. Bateson and R.A. Hinde, eds. Growing points in ethology. Cambridge Univ. Press, London.

Dawkins, R. and M. Dawkins. 1973. Decisions and the uncertainty of behaviour. Behaviour 45(1–2):83–103.

Dawkins, R. and M. Dawkins. 1976. Hierarchical organization and postural facilitation: rules for grooming in flies. Anim. Behav. 24(4):739–755.

DeGhett, V.J. 1978. Hierarchical cluster analysis. Pages 115–144 *in* P. Colgan, ed. Quantitative ethology. John Wiley, New York.

Delgado, R.R. and J.M.R. Delgado. 1962. An objective approach to measurement of behavior. Philos. of Sci. 29:253–268.

Denenberg, V.H. 1976. Statistics and experimental design for behavioral and biological researchers. Hemisphere Publ. Corp., Washington, D.C. 344 pp.

Denes, P.B. and E.N. Pinson. 1973. The speech chain: The physics and biology of spoken language. Anchor Press, Garden City, New York. 217 pp.

Dethier, V.G. 1962. To know a fly. Holden-Day, San Francisco. 119 pp.

Dethier, V.G. and D. Bodenstein. 1958. Hunger in the blowfly. Z. Tierpsychol. Beih. 15:129–140.

DeVore, I. and K.R.L. Hall. 1965. Baboon ecology. Pages 20–52 *in* I. DeVore, ed. Primate behavior. Holt, Rinehart and Winston, New York.

Dewsbury, D.A. 1975. Filming animal behavior. Pages 13–15 *in* E.D. Price, and A.W. Stokes, eds. Animal behavior in laboratory and field, 2nd ed. W.H. Freeman Co., San Francisco.

Diakow, C. 1975. Motion picture analysis of rat mating behavior. J. Comp. Physiol. Psychol. 88(2):704–712.

Dice, L.R. 1945. Measures of the amount of ecologic association between species. Ecology 26:297–302.

Dilger, W.C. 1962. The behavior of lovebirds. Sci. Am. 206(1):88–98.

Dingle, H. 1972. Aggressive behavior in somatopods and the use of information theory in the analysis of animal communication *in* H.E. Winn and B.L. Olla, eds. Behavior of marine animals, vol. 1. Invertebrates. Plenum Press, New York.

Dixon, W.J. 1954. Power under normality of several nonparametric tests. Ann. Math. Stat. 25:610–614.

Drori, D. and Y. Folman. 1967. The sexual behaviour of male rats unmated to 16 months of age. Anim. Behav. 15(1):20–24.

Drummond, H. (in press). The nature and description of behavior patterns. *In* P.P.G. Bateson and P.H. Klopfer, eds. Perspectives in Ethology, Vol. IV. Plenum Press, New York.

Dudziński, M.L. and J.M. Norris. 1970. Principal component analysis as an aid for studying animal behaviour. Forma et Functio 2:101–109.

Dunbar, R.I.M. 1976. Some aspects of research design and their implications in the observational study of behaviour. Behaviour 58(1–2):78–98.

Duncan, I.J.H. and D.G.M. Wood-Gush. 1972. An analysis of displacement preening in the domestic fowl. Anim. Behav. 20:68–71.

Dwyer, T.J. 1975. Time budget of breeding gadwalls. Wilson Bull. 87(3):335–343.

Ehrenfeld, D.W. and A. Carr. 1967. The role of vision in the sea-finding orientation of the green turtle *(Chelonia mydas)*. Anim. Behav. 15(1): 25–36.

Eibl-Eibesfeldt, I. 1972. Similarities and differences between cultures in ex-

pressive movements. Pages 297–314 *in* R.E. Hinde, ed. Non-verbal communication. Cambridge Univ. Press, London.
Eibl-Eibesfeldt, I. 1975. Ethology: The biology of behavior, 2nd ed. Holt, Rinehart and Winston, New York. 625 pp.
Eiseley, L. 1964. The unexpected universe. Harcourt, Brace, Jovanovich, New York. 239 pp.
Eisenberg, J.F. 1963. The behavior of heteromyid rodents. Univ. California Publ. Zool. vol. 69:1–100.
Eisenberg, J.F. 1966. The social organization of mammals. Handb. Zool. 10(7):1–92.
Eisenberg, J.F. 1967. A comparative study in rodent ethology with an emphasis on evolution of social behavior. Proc. U.S. Nat. Mus. 122:1–51.
Eisner, T. and E.O. Wilson. 1975. Animal behavior: Readings from Scientific American. W. H. Freeman and Co., San Francisco. 339 pp.
Emlen, J.T., Jr. 1958. The art of making field notes. Jack-Pine Warbler 36(4):178–181.
Emlen, S.T. 1967. Migratory orientation in the indigo bunting, *Passerina cyanea*. I. Evidence for use of celestial cues. Auk 84:309–342.
Emlen, S.T. 1970. Celestial rotation: its importance in the development of migratory orientation. Science 170:1198–1201.
Emlen, S.T. 1972. An experimental analysis of the parameters of bird song eliciting species recognition. Behaviour 41(1–2):130–171.
Emlen, S.T., W. Wiltschko, N. Demong, R. Wiltschko, and S. Bergman. 1976. Magnetic direction finding: evidence for its use in migratory indigo buntings. Science 193:505–508.
Estes, R.D. 1967. Predators and scavengers. Nat. Hist. 76(2, 3):20–29, 38–47.
Ettlinger, D.M.T., ed. 1974. Natural history photography. Academic Press, New York. 389 pp.
Evans, H.E. 1957. Studies on the comparative ethology of digger wasps of the genus *Bembix*. Comstock Publ. Assoc., Ithaca, N.Y. 248 pp.
Fagen, R.M. 1977. Quantitative ethology: new results in catalog and sequence analysis (unpubl.). Paper presented at Animal Behavior Society Meeting, University Park, Pennsylvania, June 5–10, 1977.
Fagen, R.M. and R.N. Goldman. 1977. Behavioural catalogue analysis methods. Anim. Behav. 25:261–274.
Fagen, R.M. and D.Y. Young. 1978. Temporal patterns of behaviors: durations, intervals, latencies, and sequences. Pages 79–114 *in* P.W. Colgan, ed. Quantitative Ethology, John Wiley and Sons, New York. 364 pp.
Fager, E.W. 1957. Determination and analysis of recurrent groups. Ecology 38:586–595.
Falls, J.B. and R.J. Brooks. 1975. Individual recognition by song in white-throated sparrows. II. Effects of location. Can. J. Zool. 53:1412–1420.
Farr, J. and W.F. Herrnkind. 1974. A quantitative analysis of social interaction of the guppy, *Poecilia reticulata* (Pisces: Poecilidiae) as a function of population density. Anim. Behav. 22(3):582–591.
Federer, W.T. 1955. Experimental design: Theory and application. MacMillan Co., New York. 544 pp.

Fenton, M.B. 1970. A technique for monitoring bout activity with results obtained from different environments in southern Ontario. Can. J. Zool. 48:847–851.

Fenton, M.B., S.L. Jacobson and R.N. Stone. 1973. An automatic ultrasonic sensing system for monitoring activity of some bats. Can. J. Zool. 51:291–299.

Fentress, J.C., ed. 1977. Simpler networks and behavior. Sinauer, Sunderland, Mass. 403 pp.

Fernald, R.D. 1977. Quantitative behavioural observations of *Haplochromis burtoni* under semi-natural conditions. Anim. Behav. 25(3). 643–653.

Fernald, R.D. and P. Heinecke. 1974. A computer compatible multi-purpose event recorder. Behaviour 58(3–4):268–275.

Field, R. 1976. Application of a digitizer for measuring sound spectrograms. Behav. Biol. 17:579–583.

Fienberg, S.E. 1972. On the use of Hansen frequencies for estimating rates of behavior. Primates 13:323–326.

Finney, D.J. 1960. An introduction to the theory of experimental design. Univ. Chicago Press, Chicago. 222 pp.

Fitzpatrick, L.J. 1977. Automated data collection for observed events. Behav. Res. Methods Instrum. 9(5):477–451.

Free, J.B. 1967. Factors determining the collection of pollen by honeybee foragers. Anim. Behav. 15(1):134–144.

Frey, D.F. and R.A. Pimentel. 1978. Principle component analysis and factor analysis. Pages 219–245 *in* P. Colgan, ed. Quantitative ethology. John Wiley, New York.

Frisch, K. von 1953. The dancing bees. Harcourt, Brace and Co., New York. 182 pp.

Fruchter, B. 1954. Introduction to factor analysis. D. van Nostrand Co., New York. 280 pp.

Fryer, T.B., H.A. Miller and H. Sandler, eds. 1976. Biotelemetry III. Third International Symposium, Pacific Grove, California. Academic Press, New York. 381 pp.

Fuller, J.L. 1967. Effects of the albino gene upon behaviour of mice. Anim. Behav. 15(4):467–470.

Gans, C. 1978. All animals are interesting! Am. Zool. 18(1):3–9.

Gass, C.L. 1977. A digital encoder for field recording of behavioral, temporal, and spatial information in directly computer-accessible form. Behav. Res. Methods Instrum. 9(1):5–11.

Geist, V. 1971. Mountain sheep: A study in behaviour and evolution. Univ. Chicago Press, Chicago. 383 pp.

Getz, L.L. 1972. Social structure and aggressive behavior in a population of *Microtus pennsylvanicus*. J. Mammal. 53:310–317.

Giles, R.H., ed. 1969. Wildlife management techniques. The Wildlife Society. Washington, D.C. 623 pp.

Glickman, S.E. 1973. Responses and reinforcement. Pages 207–241 *in* R.A. Hinde and J. Stevenson-Hinde, eds. Constraints on learning. Academic Press, London.

Golani, I. 1973. Non-metric analysis of behavioral interaction sequences in captive jackals *(Canis aureus L.)*. Behaviour 44(1–2):89–112.

Golani, I. 1976. Homeostatic motor processes in mammalian interactions: a choreography of display. Pages 69–134 *in* P.P.G. Bateson and P.H. Klopfer, eds. Perspectives in ethology, vol. 2. Plenum Press, N.Y.

Goodman, L.A. 1968. The analysis of crossclassified data: independence, quasi-independence and interactions in contingency tables with or without missing entries. J. Am. Stat. Assoc. 63:1091–1131.

Goss-Custard, J.D. 1977. Optimal foraging and the size selection of worms by redshank, *Tringa totanus,* in the field. Anim. Behav. 25(1):10–29.

Gottman, J.M. 1978. Nonsequential data analysis techniques in observational research. Pages 45–61 *in* G.P. Sackett, ed. Observing behavior. Vol. 2. Data collection and analysis methods. University Park Press, Baltimore.

Grant, E.C. and J.H. Mackintosh. 1963. A comparison of the social postures of some common laboratory rodents. Behaviour 21:246–259.

Grant, T.R. 1973. Dominance and association among members of a captive and a free-ranging group of grey kangaroos *(Macropus giganteus)*. Anim. Behav. 21(3):449–456.

Gull, J. 1977. Movement and dispersal patterns of immature gray squirrels *(Sciurus carolinensis)* in East-Central Minnesota. Unpublished M.S. Thesis. Univ. Minnesota. 117 pp.

Guttman, B.S. 1976. Is "levels of organization" a useful biological concept? BioScience 26(2):112–113.

Guttman, L. 1966. The nonmetric breakthrough for the behavioral sciences. Proceedings of the Second National Conference on Data Processing. Rehovot, Israel.

Guttman, R., I. Lieblich and G. Naftali. 1969. Variation in activity scores and sequences in two inbred mouse strains, their hybrids, and back crosses. Anim. Behav. 17(2):335–374.

Hailman, J.P. 1967. The ontogeny of an instinct. The pecking response in chicks of the laughing gull *(Larus africilla L.)* and related species. Behaviour (suppl. 15). 159 pp.

Hailman, J.P. 1969. How an instinct is learned. Sci. Am. 221(6):98–106.

Hailman, J.P. 1971. The role of stimulus-orientation in eliciting the begging response from newly-hatched chicks of the laughing gull *(Larus atricilla)*. Anim. Behav. 19(2):328–335.

Hailman, J.P. 1973. Fieldism. BioScience 23(3):149.

Hailman, J.P. 1975. The scientific method: *modus operandi* or supreme court? Am. Biol. Teach. 37:309–310.

Hailman, J.P. 1977. Optical signals: Animal communication and light. Indiana Univ. Press, Bloomington. 362 pp.

Hailman, J.P. and B.D. Sustare. 1973. What a stuffed toy tells a stuffed shirt. BioScience 23(11):644–651.

Halliday, T.R. 1975. An observational and experimental study of sexual behavior in the smooth newt, *Triturus vulgaris* (Amphibia: Salamandridae). Anim. Behav. 23(2):291–322.

Hamilton, W.J., III, 1966. Social aspects of bird orientation mechanisms.

Pages 57–71 *in* R.L. Storm, ed. Animal orientation and navigation. Proc. 27th Ann. Biol. Colloquium, May 6–7, 1966. Oregon State Univ. Press, Corvallis. 125 pp.

Hanenkrat, F.T. 1977. Wildlife watcher's handbook. Winchester Press, New York, 241 pp.

Harthoorn, A.M. 1976. The Chemical Capture of Animals: A Guide to the Chemical Restraint of Wild and Captured Animals. Baillière Tindall. London. 416 pp.

Hartmann, D.P. 1972. Notes on methodology: 1. On choosing an interobserver reliability measurement. Unpublished manuscript. Univ. Utah.

Hausfater, G. 1977. Tail carriage in baboons *(Papio cynocephalus)*: relationship to dominance rank and age. Folia Primatol. 27(1):41–59.

Hayne, D.W. 1949. Calculation of size of home range. J. Mammal 39(2): 190–206.

Hazlett, B.A. and C.E. Bach. 1977. Predicting behavioral relationships. Pages 121–144 *in* B.A. Hazlett, ed. Quantitative methods in the study of animal behavior. Academic Press, New York. 222 pp.

Hazlett, B.A. and W.H. Bossert. 1965. A statistical analysis of the aggressive communications systems of some hermit crabs. Anim. Behav. 13(2,3): 357–373.

Heiligenberg, W. 1965. A quantitative analysis of digging movements and their relationship to aggressive behaviour in Cichlids. Anim. Behav. 13:163–170.

Heimstra, N.W. and R.T. Davis. 1962. A simple recording system for the direct observation technique. Anim. Behav. 10:202–210.

Heinrich, B. 1971. The effect of leaf geometry on the feeding behaviour of the caterpillar of *Munduca sexta* (Sphingidae). Anim. Behav. 19(1):119–124.

Heinroth, O. 1911. Beitrage zur Biologie, namentlich Ethologie und Psychologie der Anatiden. Pages 598–702 *in* Proc. V. Int. Ornith. Congr. Berlin, 1910.

Heppner, F. 1965. Sensory mechanisms and environmental clues used by the American robin in locating earthworms. Condor 67(3):247–256.

Hess, E.H. 1962. Imprinting and the "critical period" concept. Pages 254–263 *in* E.L. Bliss, ed. Roots of behavior. Hafner, New York.

Hess, E.H. 1972. "Imprinting" in a natural laboratory. Sci. Amer. 227(2): 24–31.

Hildebrand, M. 1965. Symmetrical gaits of horses. Science 150:701–709.

Hildebrand, M. 1977. Analysis of asymmetrical gaits. J. Mammal. 58(2): 131–156.

Hinde, R.A. 1954. Changes in responsiveness to a constant stimulus. Brit. J. Anim. Behav. 2(1):41–55.

Hinde, R.A. 1957. Consequences and goals. Brit. Jour. Anim. Behav. 5(3):116–118.

Hinde, R.A. 1970. Animal behaviour: a synthesis of ethology and comparative psychology. McGraw-Hill Book Co., New York. 876 pp.

Hinde, R.A. 1973. On the design of checksheets. Primates 14:393–406.

Hinde, R.A. 1975. The concept of function. Pages 3–15 *in* G. Baerends, C. Beer and A. Manning, eds. Function and evolution in behaviour. Clarendon Press, Oxford.

Hinde, R.A. and Y. Spencer-Booth. 1967. Behaviour of socially living rhesus monkeys in their first two and a half years. Anim. Behav. 15(1):169–196.

Hinde, R.A. and J. Stevenson-Hinde. 1976. Towards understanding relationships: dynamic stability. Pages 451–479 *in* P.P.G. Bateson end R.A. Hinde, eds. Growing points in ethology. Cambridge Univ. Press, Cambridge.

Hoffman, H.S. and A.M. Ratner. 1973. A reinforcement model of imprinting: implications for socialization in monkeys and men. Psych. Rev. 80(6):527–544.

Hollenbeck, A.R. 1978. Problems of reliability in observational research. Pages 79–98 *in* G.P. Sackett, ed. Observing behavior. vol. 2. Data collection and analysis methods. University Park Press, Baltimore.

Hollenbeck, A.R., L. Smythe, G.P. Sackett and C. Boulais. 1977. A manual and computer programs for analyzing digitally coded observational data. Behav. Res. Methods Instrum. 9(1):34.

Holst, E. von and V. von St. Paul. 1963. On the functional organization of drives. Anim. Behav. 11:1–20.

Hopkins, C.D., M. Rossetto and A. Lutjen. 1974. A continuous sound spectrum analyzer for animal sounds. Z. Tierpsychol. Beih. 34:313–320.

Horii, Y. 1974. Digital sound spectrograms with simultaneous plotting of intensity and fundamental frequency for speech study. Behav. Res. Methods Instrum. 6(1):55.

Hotelling, H. 1958. The statistical method and the philosophy of science. Am. Statistician 12(5):9–14.

Huck, U.W. and E.O. Price. 1970. Effect of the post-weaning environment on the climbing behavior of wild and domestic Norway rats. Anim. Behav. 24(2):364–371.

Huntingford, F.A. 1976. The relationship between anti-predator behaviour and aggression among conspecifics in the three-spined stickleback, *Gasterosteus aculeatus*. Anim. Behav. 24(2):245-260.

Hutt, S.J. and C. Hutt. 1974. Direct observation and measurement of behavior. Charles C. Thomas, Springfield, Ill. 224 pp.

Jardine, N. and R. Sibson. 1968. The construction of hierarchic and nonhierarchic classifications. Comput. J. 11:177–184.

Kalinoski, R. 1975. Intra- and interspecific aggression in house finches and house sparrows. Condor 77:375–384.

Kandel, E.R. 1977. Cellular basis of behavior. W.H. Freeman & Co., San Francisco. 725 pp.

Kaufman, C. and L.A. Rosenblum. 1966. A behavioral taxonomy for *Macaca nemestrina* and *Macaca radiata:* based on longitudinal observation of family groups in the laboratory. Primates 8(2):205–252.

Keeton, W.T. 1974. The mystery of pigeon homing. Sci. Amer. 231(6):96–98, 101–107.

Kelly, J.G. 1967. Naturalistic observations and theory confirmation: an example. Hum. Dev. 10:212–222.

Kelly, J.G. 1969. Naturalistic observations in contrasting social environments. Pages 183–199 *in* E.P. Willems and H.L. Rausch, eds. Naturalistic viewpoints in psychological research. Holt, Rinehart and Winston, New York.

Kendall, M.G. 1948. Rank correlation methods. Charles Griffin and Co., Ltd., London. 199 pp.

Kerfoot, W.B. 1967. The lunar periodicity of *Sphecodogastra texana*, a nocturnal bee (Hymenoptera: Halictidae). Anim. Behav. 15(4):479–486.

Kerlinger, F.N. 1967. Foundations of behavioral research. Holt, Rinehart and Winston, New York. 739 pp.

Kessel, E.L. 1955. The mating activities of balloon flies. Syst. Zool. 4:97–104.

Kikkawa, J. and M.J. Thorne. 1971. The behaviour of animals. The Jacaranda Press, Queensland, Australia. 223 pp.

King, M.G. 1965. The effect of social context on dominance capacity of domestic hens. Anim. Behav. 13(1):132–133.

Kinsey, K.P. 1976. Social behaviour in confined populations of the Allegheny woodrat, *Neotoma floridana magister.* Anim. Behav. 24(1):181–187.

Kirk, R.E. 1968. Experimental design: Procedures for the behavioral sciences. Brooks/Cole Publ. Co., Belmont, Cal. 577 pp.

Kleerekoper, H. 1969. Olfaction in fishes. Indiana Univ. Press, Bloomington. 222 pp.

Kleiman, D.G. 1974. Activity rhythms in the giant panda *Ailuropoda melanoleucca:* An example of the use of checksheets for recording behaviour data in zoos. Int. Zoo Yearbk. 14:165–169.

Klein, M.S. 1955. Ionophone on haut-parleur ionique. Pages 46–49 *in* R.G. Busnel, ed. Collogue sur l'acoustigue des orthopteres. Institut National de la Recherche Agronomique, Paris.

Klingel, H. 1965. Notes on the biology of the plains zebra *Equus guagga boehmi* Matschie. East Afr. Wildl. J. 3:86–88.

Klopfer, P. 1963. Behavioral aspects of habitat selection: The role of early experience. Wilson Bull. 75:15–22.

Knight, R.R. 1970. The Sun River elk herd. Wildl. Monogr. No. 23, 66 pp.

Koeppl, J.W., N.A. Slade and R.S. Hoffman. 1975. A bivariate home range model with possible application to ethological data analysis. J. Mammal. 56(1):81–90.

Kramer, G. 1952. Experiments on bird orientation. Ibis 94:265–285.

Krebs, H.A. 1975. The August Krogh principle: "For many problems there is an animal on which it can be most conveniently studied." J. Exp. Zool. 194:221–226.

Krebs, J.R. 1971. Territory and breeding density in the great tit, *Parus major* L. Ecology 52(1):2–22.

Krogh, A. 1929. Progress in physiology. Am. J. Physiol. 90:243–251.

Kroodsma, D.E. 1977. A re-evaluation of song development in the song sparrow. Anim. Behav. 25(2):390–399.

Kruijt, J.P. 1964. Ontogeny of social behaviour in Burmese Red Junglefowl *(Gallus gallus spadiceous)*. Brill, Leiden. 201 pp.

Kruskal, J.B. 1964. Nonmetric multidimensional scaling: a numerical method. Psychometrika 29:115–129.

Kruuk, H. 1972. The spotted hyena: A study of predation and social behavior. Univ. Chicago Press, Chicago. 335 pp.

Kucera, T.E. 1978. Social behavior and breeding system of the desert mule deer. J. Mammal. 59(3):463–476.

Kummer, H. 1968. Social organization of hamadryas baboons. A field study. Univ. Chicago Press, Chicago; Basil, S. Karger, Bibl. Primatol. No. 6. 189 pp.

Kunz, T.H. and C.E. Brock. 1975. A comparison of mist nets and ultrasonic detectors for monitoring flight activity of bats. J. Mammal. 56(4):907–911.

Larkin, R.P. 1977. Reactions of migrating birds to sounds broadcast from the ground (unpubl.). Paper presented Animal Behavior Society Meeting, University Park, Penn., June 5–10, 1977.

Larkin, R.P. and P.J. Sutherland. 1977. Migrating birds respond to Project Seafarer's electromagnetic field. Science 195:777–779.

Layne, J.N. 1967. Evidence for the use of vision in diurnal orientation of the bat *Myotis austroriparius*. Anim. Behav. 15(4):409–415.

Ledley, R.S. 1965. Use of computers in biology and medicine. McGraw-Hill, New York. 965 pp.

Lehner, P.N. 1976. Coyote howls and Wynne-Edward's hypothesis revisited (unpubl.). Paper presented at Animal Behavior Society Meeting, Boulder, Colo. June 20–25, 1976.

Lehrman, D.S. 1955. The perception of animal behavior. Pages 259–267 *in* B. Schaffner, ed. Group processes: Transactions of the first conference. Josiah Macy, Jr., Foundation, New York.

Lemon, R.E. and C. Chatfield. 1971. Organization of song in cardinals. Anim. Behav. 19(1):1–17.

Leuthold, W. 1977. African ungulates: A comparative review of their ethology and behavioral ecology. Zoophysiology and Ecology, vol. 8. Springer-Verlag, New York. 307 pp.

Lindzey, G., D.D. Thiessen and A. Tucker. 1968. Development and hormonal control of territorial marking in the male mongolian gerbil *(Meriones unguiculatus)*. Dev. Psychobiol. 1(2):97–99.

Lingoes, J.C. 1966. An IBM-7090 program for Guttman-Lingoes multidimensional scalogram analysis-I. Behav. Sci. 11:76–78.

Lissaman, P.B.S. and C.A. Shollenberger. 1970. Formation flight of birds. Science 168:1003–1005.

Littlejohn, M.J. and A.A. Martin. 1969. Acoustic interaction between two species of Leptodactylid frogs. Anim. Behav. 17(4):785–791.

Lockard, J.S. 1976. Small interval timer for observational studies. Behav. Res. Methods Instrum. 8(5):478.

Lockie, J.D. 1966. Territory in small carnivores. Pages 143–165 *in* P.A. Jewell and C. Loizos, eds. Play, exploration and territory in mammals. Symp. Zool. Soc. London no. 18. Academic Press, London.

Long, F.M., ed. 1977. Proceedings of the First International Conference on Wildlife Biotelemetry. Laramie, Wyoming, July 27–29, 1977. 159 pp.

Lorenz, K. 1935. Companionship in bird life: Fellow members of the species as releasers of social behavior. Pages 83–128 *in* C.E. Schiller, ed. Instinctive behavior. International Universities Press, New York. 328 pp.

Lorenz, K. 1941. Verliechends Bewegungsstudien an Anatinen. J. Ornithol. 89 (suppl.):194–294.

Lorenz, K. 1951–1953. Comparative studies on the behavior of Anatinae. Avic. Mag. 57:157–182; 58:8–17, 61–72, 86–94, 172–184; 59:24–34, 80–91.

Lorenz, K. 1960*a*. Methods of approach to the problems of behavior. Pages 60–103 *in* The Harvey Lectures 1958–59. Academic Press. New York.

Lorenz, K. 1960*b*. Foreword. Pages xi-xii *in* N. Tinbergen, The herring gull's world, rev. ed. Harper and Row, New York.

Lorenz, K. 1973. The fashionable fallacy of dispensing with description. Naturwissenschaften 60:1–9.

Lorenz, K. 1974. Analogy as a source of knowledge. Science 185:229–234.

Losey, G.S., Jr. 1977. The validity of animal models: A test for cleaning symbiosis. Biol. Behav. 2(3):223–238.

Losey, G.S., Jr. 1978. Information theory and communication. Pages 43–78 *in* P. Colgan, ed. Quantitative ethology. John Wiley, New York.

Lott, D.F. 1975. Protestations of a field person. BioScience 25(5):328.

MacArthur, R.H. 1958. Population ecology of some warblers of northeastern coniferous forests. Ecology 39:599–619.

McBride, G. 1976. The study of social organizations. Behaviour 59(1–2): 96–115.

Machlis, L. 1977. An analysis of the temporal patterning of pecking in chicks. Behaviour 63(1–2):1–70.

McKinney, F. 1975. The evolution of duck displays. Pages 331–357 *in* G. Baerends, C. Beer and A. Manning, eds. Function and evolution in behavior. Clarendon Press, Oxford. 393 pp.

Marchington, J. and A. Clay. 1974. An Introduction to bird and wildlife photography. Faber & Faber Ltd., London. 149 pp.

Marion, W.R. and J.D. Shamis. 1977. An annotated bibliography of bird marking techniques. Bird-Banding 48(1):42-61.

Marler, P. 1975. Observation and description of behavior. Pages 2–4 *in* E.O. Price and A.W. Stokes, eds. Animal behavior in laboratory and field, 2nd ed. W.H. Freeman and Co., San Francisco.

Marler, P. and W.J. Hamilton III. 1966. Mechanisms of animal behavior. John Wiley, New York. 771 pp.

Marler, P. and D. Isaac. 1960. Physical analysis of a simpler bird song as exemplified by the chipping sparrow. Condor 62:124–135.

Marsden, H.M. 1968. Agonistic behaviour of young rhesus monkeys after changes induced in social rank of their mothers. Anim. Behav. 16(1): 38–44.

Marshall, J.C. 1965. The syntax of reproductive behaviour in the male pigeon. Medical Research Council Psycholinquistics Unit Report, Oxford.

Mason, W.A. 1960. The effects of social restriction on the behaviour of rhesus monkeys. J. Comp. Physiol. Psychol. 53:582–589.

Mason, W.A. 1968. Naturalistic and experimental investigations of the social behavior of monkeys and apes. Pages 398–419 *in* P.C. Jay, ed. Primates: Studies in adaptation and variability. Holt, Rinehart and Winston, New York. 529 pp.

Matthews, G.U.T. 1951. The experimental investigation of navigation in homing pigeons. J. Exp. Biol. 28:508–536.

Matthews, G.U.T. and W.A. Cook. 1977. The role of landscape features in the 'nonsense' orientation of the mallard. Anim. Behav. 25(2):508–517.

Matzkin, M.A. 1975. Super 8mm movie making. Amphoto, Garden City, New York. 96 pp.

Maurus, M. and H. Pruscha. 1973. Classification of social signals in squirrel monkeys by means of cluster analysis. Behaviour 47:106–128.

Mech, L.D., K.L. Heezen and D.B. Siniff. 1966. Onset and cessation of activity in cottontail rabbits and snowshoe hares in relation to sunset and sunrise. Anim. Behav. 14(4):410–413.

Menzel, E.W. Jr. 1969. Naturalistic and experimental approaches to primate behavior. Pages 78–121 *in* E.P. Willems and H.L. Rausch, eds. Naturalistic viewpoints in psychological research. Holt, Rinehart and Winston, New York.

Meserve, P.L. 1977. Three-dimensional home ranges of cricetid rodents. J. Mammal. 58(4):549–558.

Metzgar, L. 1967. An experimental comparison of screech owl predation on resident and transient white-footed mice *(Peromyscus leucopus)*. J. Mammal. 48(3):387–391.

Meyer, M.E. 1964. Discriminative basis for astronavigation in birds. J. Comp. Physiol. Psychol. 58(3):403–406.

Meyer, M.E. 1976. A statistical analysis of behavior. Wadsworth Publ. Co., Belmont, Cal. 408 pp.

Mitchell, G. and D.L. Clark. 1968. Long term effects of social isolation in nonsocially adapted rhesus monkeys. J. Gen. Psychol. 113:117–128.

Moore, F.R. 1977. Geomagnetic disturbance and the orientation of nocturnally migrating birds. Science 196:682–684.

Morgan, B.J.T., M.J.A. Simpson, J.P. Hanby and J. Hall-Craggs. 1976. Visualizing interaction and sequential data in animal behaviour: Theory and application of cluster-analysis methods. Behaviour 56(1–2):1–43.

Moss, C. 1975. Portraits in the wild. Houghton Mifflin Co., Boston. 363 pp.

Mrosovsky, N. and S.J. Shettleworth. 1975. On the orientation circle of the leatherback turtle, *Dermochelys coriacea*. Anim. Behav. 23(3):568–591.

Müller-Schwarze, D. 1968. Locomotion in animals. Pages 13–19 *in* A.W. Stokes, ed. Animal behavior in laboratory and field. W.H. Freeman and Co., San Francisco.

Mulligan, J.A. 1963. A description of song sparrow song based on instrumental analysis. Proc. 13th International Congr. Ornithol. 272–284.

Murchison, C. 1935. The experimental measurement of a social hierarchy in

Gallus domesticus: IV. Loss of body weight under conditions of mild starvation as a function of social dominance. J. Gen. Psychol. 12:296–312.

Myrberg, A.A., Jr. and S.H. Gruber. 1974. The behavior of the bonnethead shark, *Sphyrna tiburo.* Copeia 1974:358–374.

Narins, P.M. and R.R. Capranica. 1977. An automated technique for analysis of temporal features in animal vocalizations. Anim. Behav. 25(3):615–621.

Nice, M.M. 1937. Studies in the life history of the song sparrow. I. Trans. Linn. Soc. New York. 4:1–247.

Nice, M.M. 1943. Studies in the life history of the song sparrow. Trans. Linn. Soc. New York 6:1–328.)

Nie, N.I., H. Hull, J. Jenkins, K. Steinbrenner and D. Bent. 1975. Statistical package for the social sciences. McGraw-Hill Book Co., New York. 675 pp.

Nielsen, E.T. 1958. The method of ethology. Proc. 10th Int. Congr. Entomol. 2:563–565.

Nisbet, I.C.T. and W.H. Drury. 1968. Short-term effects of weather on bird migration: A field study using multivariate statistics. Anim. Behav. 16(4):496–530.

Nisbett, A. 1977. Konrad Lorenz: A biography. Harcourt Brace Jovanovich, New York. 240 pp.

Notterman, J.M. 1973. Discussion of "On-line computers in the animal laboratory." Behav. Res. Methods Instrum. 5(2):129–131.

Nyby, J., C. Wysocki, G. Whitney and G. Dizinno. 1977. Phenomenal regulation of male mouse ultrasonic courtship *(Mus musculus).* Anim. Behav. 25:333–341.

Odum, E.P. and E.J. Kuenzler. 1955. Measurement of territory and home range size in birds. Auk 72:128–137.

Orcutt, F.S. 1967. Oestrogen stimulation of nest material preparation in the peach-faced lovebird *(Agapornis roseicollis).* Anim. Behav. 15(4): 471–478.

Overall, J.E. and S.M. Free. 1972. Multidimensional scaling based on a subset of objects or variables. Psychometric Laboratory Report, No. 30. Univ. Texas Medical Branch of Galveston.

Overall, J.E. and C.J. Klett. 1972. Applied multivariate analysis. McGraw-Hill Book Co., New York. 500 pp.

Owen-Smith, R.N. 1974. The social system of the white rhinoceros. Pages 341–351 *in* V. Geist and F. Walther, eds. The behaviour of ungulates and its relation to management, vol. 1. IUCN Publ. 24, Morges, Switzerland.

Parker, J.W. 1972. A mirror and pole device for examining high nests. Bird Banding 43(3):216–218.

Patterson, I.J. 1977. Aggression and dominance in winter flocks of shelduck *Tadorna tadorna* (L.). Anim. Behav. 25(2):447–459.

Pengelley, E.T. and S.T. Asmundson. 1971. Annual biological clocks. Sci. Amer. 224(4):72–79.

Pennycuick, C.J. and J.A. Rudnai. 1970. A method of identifying individual lions *Panthera leo* with an analysis of the reliability of identification. J. Zool. Lond. 160:497–508.

Perdeck, A.C. 1958. Two types of orientation in migrating starlings, *Sturnus vulgaris* L. and chaffinches, *Fringilla coelebs* L., as revealed by displacement experiments. Ardea 46:1–37.

Pettingill, O.S. 1970. Ornithology in laboratory and field, 4th ed. Burgess Publ. Co., Minneapolis. 524 pp.

Pimentel, R.A. and D.F. Frey. 1978. Multivariate analysis of variance and discriminant analysis. Pages 247–274 *in* P. Colgan, ed. Quantitative ethology. John Wiley, New York.

Pitcher, T.J. 1973. The three-dimensional structure of schools in the minnow, *Phoxinus phoxinus* (L.). Anim. Behav. 21(4):673–685.

Potash, L.M. 1972. A signal detection problem and possible solution in Japaneze quail *(Coturnix coturnix japonica).* Anim. Behav. 20(1):192–195.

Ralston, S.L. 1977. The social organization of two herds of domestic horses. Unpubl. M.S. Thesis. Colorado State Univ. 100 pp.

Reichert, R.J. and E. Reichert. 1961. Binoculars and scopes: How to choose, use and photograph through them. Chilton Co. Book Div., Philadelphia. 128 pp.

Remsen, J.V. 1977. On taking field notes. Am. Birds 31:946–953.

Reynierse, J.H. 1968. Effects of temperature and temperature change on earthworm locomotor behaviour. Anim. Behav. 16(4):480–484.

Reynierse, J.H. and J.N. Toeus. 1973. An ideal signal generator for time-sampling observation procedures. Behav. Res. Methods Instrum. 5(1): 57–58.

Rice, J.O. and W.L. Thompson. 1968. Song development in the indigo bunting. Anim. Behav. 16(4):462–469.

Richards, S.M. 1974. The concept of dominance and methods of assessment. Anim. Behav. 22(4):914–930.

Rioch, D.M. 1967. Discussion of agonistic behavior. Pages 115–122 *in* S.A. Altmann, ed. Social communication among primates. Univ. Chicago Press, Chicago.

Robson, C. 1973. Experiment, design and statistics in psychology. Penguin Books, Baltimore. 174 pp.

Roeder, K. 1961. Summary of the VII ethological conference. Z. Tierpsychol. Beih. 18:491–494.

Rohlf, F.J. 1968. Stereograms in numerical taxonomy. Syst. Zool. 17:246–255.

Rohwer, S. 1977. Status signaling in Harris sparrows: Some experiments in deception. Behaviour 61(1–2):107–129.

Rongstad, O.J. and J.R. Tester. 1969. Movements and habitat use of white-tailed deer in Minnesota. J. Wildl. Manage. 33(2):366–379.

Rosenblum, L.A. 1978. The creation of a behavioral taxonomy. Pages 15–24 in G.P. Sackett, ed. Observing Behavior, vol. 2. Data collection and analysis methods. University Park Press, Baltimore. 110 pp.

Rosenblum, L.A., I.C. Kaufman and A.J. Stynes. 1964. Individual distance in two species of macaque. Anim. Behav. 12(2–3):338–342.

Rosenthal, R. 1976. Experimenter effects in behavioral research. Halsted Press. New York. 500 pp.

Royama, T. 1970. Factors governing the hunting behaviour and selection of seed by the great tit *(Parus major* L.). J. Anim. Ecol. 39:619–668.

Rudnai, J. 1973. The social life of the lion. Washington Square East Publ., Wallingford, Penn. 122 pp.

Ruff, R.L. 1969. Telemetered heart rates of free-living Uinta ground squirrels in response to social interactions. Unpublished Ph.D. thesis. Utah State Univ., Logan. 71 pp.

Ryden, H. 1975. God's dog. Coward, McCann and Geoghegan. New York. 288 pp.

Sackett, G.P. 1978. Measurement in observational research. Pages 25–45 *in* G.P. Sackett, ed. Observing behavior, vol. 2. Data collection and analysis methods. University Park Press, Baltimore. 110 pp.

Sackett, G.P., G.C. Ruppenthal and J. Clark. 1978. Introduction: An overview of methodological and statistical problems in observational research. Pages 1–14 *in* G.P. Sackett, ed. Observing behaviour, vol. 2. Data collection and analysis methods. University Park Press, Baltimore. 110 pp.

Sade, D.S. 1966. Ontogeny of social relations in a group of free-ranging rhesus monkeys *(Macaca mulatta).* Zimmerman. Unpubl. Ph.D. dissertation. Univ. of California, Berkeley.

Saiz, R.B. 1975. Ecology and behavior of the gemsbok at White Sands Missile Range, New Mexico. Unpublished M.S. thesis, Colorado State Univ., Ft. Collins. 145 pp.

Sales, G. and D. Pye. 1974. Ultrasonic communication by animals. Halsted Press, New York. 281 pp.

Sanderson, G.C. 1966. The study of mammal movements—a review. J. Wildl. Manage. 30:215–235.

Sargent, A.B. 1972. Red fox spatial characteristics in relation to waterfowl predation. J. Wildl. Manage. 36:225–236.

Sauer, E.G.F. 1957. Die Sternenorientierung nachtlich ziehender Grasmucken *(Sylvia africapilla, borin,* and *curruca* L.). Z. Tierpsychol. Beih. 14:29–70.

Sawin, D.B., J.H. Langlois and E.F. Leitner. 1977. What do you do after you say hello? Observing, coding, and analyzing parent-infant interactions. Behav. Res. Methods Instrum. 9(5):425–428.

Schaffner, B. (ed.). 1955. Group processes: Transactions of the First Conference. Josiah Macy, Jr., Foundation, New York. 334 pp.

Schaller, G.B. 1972. The Serengeti lion: A study of predator-prey relations. Univ. Chicago Press, Chicago. 480 pp.

Schaller, G.B. 1973. Golden shadows, flying hooves. Alfred A. Knopf, New York, 287 pp.

Schladweiler, J.L. and I.J. Ball, Jr. 1968. Telemetry bibliography emphasizing studies of wild animals under natural conditions. Bell Museum of Natural History Technical Report No. 15, 31 pp. (Mimeo.)

Schleidt, W. 1973. Tonic communication: Continual effects of discrete signs in animal communication systems. J. Theor. Biol. 42:359–386.

Schmidt-Koenig, K. 1961. Die sonne als Kompass im Heim-orientierungsytem der Brieftauben. Z. Tierpsychol. Beih. 18:221–224.
Schmitt, J.C. 1977. Factor analysis in BASIC for minicomputers. Behav. Res. Methods Instrum. 9(3):302–304.
Scott, J.P., ed. 1950. Methodology and techniques for the study of animal societies. Ann. N.Y. Acad. Sci. 51(6):1001–1122.
Scott, J.P. 1963. Animal Behavior. Anchor Books. Doubleday and Co., Garden City, New York. 331 pp.
Scott, J.P. and J.L. Fuller. 1965. Genetics and the social behavior of the dog. Univ. of Chicago Press, Chicago. 468 pp.
Scott, K.G. and W.S. Masi. 1977. Use of the Datamyte in analyzing duration of infant visual behaviors. Behav. Res. Methods Instrum. 9(5):429–433.
Seal, H.L. 1964. Multivariate statistical analysis for biologists. Methuen, London. 209 pp.
Shepard, R.N., A.K. Romney and S. Nerlove. 1972. Multidimensional scaling, vol. 1. Seminar Press, London. 261 pp.
Short, L.L., Jr. 1970. Bird listing and the field observer. Calif. Birds 1:143–145.
Sibson, R. 1973. SLINK: An optimally efficient algorithm for the single-link cluster method. Comput. J. 16:30–34.
Siegel, S. 1956. Nonparametric statistics for the behavioral sciences. McGraw-Hill, Book Co., New York. 312 pp.
Silver, H. and W.T. Silver. 1969. Growth and behavior of the coyote-like canid of Northern New England with observations on canid hybrids. Wildl. Monogr. no. 17, 41 pp.
Simmons, J.A. 1971. The sonar receiver of the bat. Ann. N.Y. Acad. Sci. 188:161–174.
Simpson, M.J.A. and A.E. Simpson. 1977. One-zero and scan methods for sampling behaviour. Anim. Behav. 25(3):726–731.
Slade, N.A. 1976. Analysis of social structure from multiple capture data. J. Mammal. 57(4):790–795.
Slater, L.E., ed. 1965. Biotelemetry. BioScience 15(2):79–121.
Slater, P.J.B. 1973. Describing sequences of behavior. Pages 131–154 *in* P.P.G. Bateson and P.H. Klopfer, eds. Perspectives in ethology. Plenum Press, New York. 336 pp.
Slater, P.J.B. 1974. Bouts and gaps in behaviour of zebra finches, with special reference to preening. Rev. Comp. Animal 8:47–61.
Slater, P.J.B. 1978. Data collection. Pages 7–24 *in* P. Colgan, ed. Quantitative ethology. John Wiley, New York.
Slater, P.J.B. and J.C. Ollason 1972. The temporal pattern of behaviour in isolated male zebra finches: Transition analysis. Behaviour 42:248–269.
Smith, D.G. 1972. The role of the epaulets in the red-winged blackbird *(Agelaius phoeniceus)* social system. Behaviour 41(3–4):251–268.
Smith, D.G. and D.A. Spencer. 1976. A simple pole and mirror device. N. Amer. Bird Bander 1(4):175.
Smith, N.G. 1967. Visual isolation in gulls. Sci. Amer. 217(4):94–102.
Smith, R.J.F. and W.S. Hoar. 1967. The effects of prolactin and testosterone on

the parental behaviour of the male stickleback *Gasterosteus aculeatus*. Anim. Behav. 15(2–3):342–352.

Smith, W.J. 1968. Message-meaning analyses. Pages 44–60 *in* T.A. Sebeok, ed. Animal communication. Univ. Indiana Press, Bloomington.

Sneath, P.H. and R. Sokal. 1973. Numerical taxonomy. W. H. Freeman and Co., San Francisco. 573 pp.

Sokal, R.R. and F.J. Rohlf. 1962. The comparison of dendrograms by objective methods. Taxon 11:33–40.

Sokal, R.R. and F.J. Rohlf. 1969. Biometry: The principles and practice of statistics in biological research. W.H. Freeman and Co., San Francisco. 776 pp.

Southern, W.E. 1975. Orientation of gull chicks exposed to Project Sanguine's electromagnetic field. Science 189:143–145.

Sparling, D.W. and J.D. Williams. 1978. Multivariate analyses of avian vocalizations. J. Theor. Biol. 74(4):83–107.

Spence, I. 1978. Multidimensional scaling. Pages 175–217 *in* P. Colgan, ed. Quantitative ethology. John Wiley, New York. 364 pp.

Staddon, J.E.R. 1972. A note on the analysis of behavioural sequences in *Columba livia*. Anim. Behav. 20(2):284–292.

Stefanski, R.A. 1967. Utilization of the breeding territory in the black-capped chickadee. Condor 69(3):259–267.

Steinberg, J.B. 1977. Information theory as an ethological tool. Pages 47–74 *in* B. Hazlett, ed. Quantitative methods in the study of animal behavior. Academic Press, New York.

Stephenson, G.R. and T.W. Roberts. 1977. The SSR System 7: A general encoding system with computerized transcription. Behav. Res. Methods Instrum. 9(3):434–441.

Stephenson, G.R., D.P.B. Smith and T.W. Roberts. 1975. The SSR system: An open format event recording system with computerized transcription. Behav. Res. Methods Instrum. 7(6):497–515.

Stevens, S.S. 1951. Handbook of experimental psychology. John Wiley, New York. 1436 pp.

Stewart, R.E. and J.W. Aldrich. 1951. Removal and repopulation of breeding birds in a spruce-fir forest community. Auk 68:471–482.

Stickel, L.F. 1954. A comparison of certain methods of measuring ranges of small mammals. J. Mammal 35(1):1–15.

Stokes, A.W. 1962. Agonistic behavior among blue tits at a winter feeding station. Behaviour 19:118–138.

Stonehouse, B., ed. 1978. Animal marking: Recognition marking of animals in research. University Park Press. Baltimore. 224 pp.

Stout, J.F. and M.E. Brass. 1969. Aggressive communication by *Larus glaucescens*. II. Visual communication. Behaviour 34(1–2):42–52.

Stricklin, W.R., H.B. Graves and L.L. Wilson. 1977. DISTANGLE: A Fortran program to analyze and simulate spacing behavior of animals. Behav. Res. Methods Instrum. 9(4):367–370.

Sustare, B.D. 1978. Systems diagrams. Pages 275–311 *in* P. Colgan, ed. Quantitative ethology. John Wiley, New York. 364 pp.

Svendsen, G.E. and K.B. Armitage. 1973. Mirror image stimulation applied to field behavior studies. Ecology 54:623–627.

Syme. G.J. 1974. Competitive orders as measures of social dominance. Anim. Behav. 22(4):931–940.

Thomas, G. 1977. The influence of eating and rejecting prey items upon feeding and food searching behaviour in *Gasterosteus aculeatus* L. Anim. Behav. 25(1):52–66.

Tinbergen, L. 1960. The natural control of insects in pinewoods. I. Factors influencing the intensity of predation by songbirds. Arch. Neerl. Zool. 13:265–343.

Tinbergen, N. 1950. The hierarchical organization of nervous mechanisms underlying instinctive behaviour. Symp. Soc. Exp. Biol. 4:305–312.

Tinbergen, N. 1951. The study of instinct. Oxford Univ. Press., N.Y. 228 pp.

Tinbergen, N. 1953. Social behaviour in animals. John Wiley, New York. 150 pp.

Tinbergen, N. 1958. Curious naturalists. Basic Books, New York. 301 pp.

Tinbergen, N. 1959. Comparative studies of the behaviour of gulls (Laridae): A progress report. Behaviour 15:1–70.

Tinbergen, N. 1960. The evolution of behaviour in gulls. Sci. Amer. 203(6):118–126, 128, 130.

Tinbergen, N. 1963. On aims and methods of ethology. Z. Tierpsychol. Beih. 20:410–433.

Tinbergen, N. 1965. Animal behavior. Time, New York. 199 pp.

Tinbergen, N. 1972. The animal in its world. Harvard Univ. Press, Cambridge, Mass. 2 vols. 343 pp., 231 pp.

Tinbergen, N. and W. Kruyt. 1938. Uber die Orientierung des Bienenwolfes (*Philanthus triangulum* Fabr.) III. Die Bevorzugung bestimmter Wegmarken. Z. V. Physiol. 25:292–334. Pages 146–196 *in* N. Tinbergen, 1972. The Animal in its World. Harvard Univ. Press, Cambridge, Mass.

Tinbergen, N. and D.J. Kuenen. 1939. Uber die ausloesenden und die richtunggebenden Resizsituationen der Sperrbewegung von jungen Drosseln (*Turdus m. merula* L. und *T. e. ericetorium* Turton). Z. Tierpsychol. Beih. 3:37–60.

Tinbergen, N. and A.C. Perdeck. 1950. On the stimulus situation releasing the begging response in the newly hatched herring gull chick (*Larus argentatus* Pont.) Behaviour 3(1):1–39.

Tobach, E., T.C. Schneirla, L.R. Aronson and R. Laupheimer. 1962. The ASTL: An observer-to-computer system for a multivariate approach to behavioral study. Nature 194:257–258.

Tolman, E. 1958. Behavior and psychological man. Univ. Calif. Press, Berkeley. 269 pp.

Torgerson, L. 1977. Datamyte 900. Behav. Res. Methods Instrum. 9(5): 405–406.

Trochim, W.M.K. 1976. The three-dimensional graphic method for quantifying body position. Behav. Res. Methods Instrum. 8(1):1–4.

Trotter, J.R. 1959. An aid to field observation. Anim. Behav. 7(1–2):107.

Turner, E.R.A. 1964. Social feeding in birds. Behaviour 24(1–2):1–46.

Ulinski, P.S. 1972. Tongue movements in the common boa *(Constrictor constrictor)*. Anim. Behav. 20(2):373–382.

Van Abeelen, J.H.F. 1966. Effects of genotype on mouse behaviour. Anim. Behav. 14(2–3):218–225.

Van Der Kloot, N. and M.J. Morse. 1975. A stochastic analysis of the display behavior of the red-breasted merganser *(Mergus serrator)*. Behaviour 54(3–4):181–216.

Van Hoof, J.A.R.A.M. 1970. A component analysis of the structure of the social behaviour of a semi-captive chimpanzee group. Experientia 26:549–550.

Van Tets, G.G. 1965. A comparative study of some social communication patterns in the Pelecaniformes. Ornithol. Monogr. 2:1–88.

Verberne, G. and P. Leyhausen. 1976. Marking behaviour of some Viverridae and Felidae: Time-interval analysis of the marking pattern. Behaviour 58(3–4):192–253.

Walcott, C. and R.P. Green. 1974. Orientation of homing pigeons altered by a change in the direction of an applied magnetic field. Science 184: 180–182.

Wallace, R.A. 1973. The ecology and evolution of animal behavior. Goodyear Publ. Co., Pacific Palisades, Calif. 342 pp.

Walther, F. 1978. Behavioral observations on oryx antelope *(Oryx beisa)* invading Serengeti National Park, Tanzania, J. Mammal. 59(2):243–260.

Waser, P.M. 1975*a*. Diurnal and nocturnal strategies of the bushbuck *Tragelaphus scriptus* (Pallas). East Afr. Wildl. 13:49–63.

Waser, P.M. 1975*b*. Experimental playbacks show vocal mediation of intergroup avoidance in a forest monkey. Nature 255(5503):56–58.

Watt, K.E.F. 1966. Ecology in the future. Pages 253–267 *in* K.E.F. Watt, ed. Systems analysis in ecology. Academic Press, New York.

Wecker, S.C. 1964. Habitat selection. Sci. Amer. 221(4):109–116.

Weeden, J.S. 1965. Territorial behavior of the tree sparrow. Condor 67: 193–209.

Weidmann, V. and J. Darley. 1971. The role of the female in the social display of mallards. Anim. Behav. 19(2):287–298.

Wells, M.C. 1977. The relative importance of the senses in coyote predatory behavior. Unpublished Ph.D. thesis. Colorado State Univ. 118 pp.

Wells, M.C. and P.N. Lehner. 1978. The relative importance of the distance senses in coyote predatory behaviour. Anim. Behav. 26(1):251–258.

Westman, R.S. 1977. Environmental languages and the functional bases of animal behavior. Pages 145–201 *in* B.A. Hazlett, ed. Quantitative methods in the study of animal behavior. Academic Press, New York.

White, R.E.C. 1971. WRATS: A computer compatible system for automatically recording and transcribing behavioral data. Behaviour 40(1–2):135–161.

Wiens, J.A., S.G. Martin, W.R. Holthaus and F.A. Iwen. 1970. Metronome timing in behavioral ecology studies. Ecology 51(2):350–352.

Wiepkema, P.R. 1961. An ethological analysis of the reproductive behaviour of the bitterling (*Rhodeus amarus* Bloch). Arch. Neerl. Zool. 14:103–199.

Wiepkema, P.R. 1968. Behaviour changes in CBA mice as a result of one gold thioglucose injection. Behaviour 32:179–210.

Wildi, E. 1973. 16mm movie making. Petersen Publishing Co., Los Angeles. 80 pp.

Will, Gary B. and E.F. Patric. 1972. A contribution toward a bibliography on wildlife telemetry and radio tracking. 56 pp. (mimeo.).

Willems, E.P. and H.L. Raush. 1969. Introduction. Pages 1–10 *in* E.P. Willems and H.L. Raush, eds. Naturalistic viewpoints in psychological research. Holt, Rinehart and Winston, New York.

Wilson, E.O. 1975. Sociobiology, Harvard Univ. Press, Cambridge, Mass. 697 pp.

Wiltschko, W. and R. Wiltschko. 1972. Magnetic compass of European robins. Science 176:62–64.

Wolach, A.H., P. Roccaforte, S.N. Van Berschot and M.A. McHale. 1975. Converting an electronic calculator into a combination stopwatch-calculator. Behav. Res. Method. Instrum. 7(6):549–551.

Wood-Gush, D.G.M. 1972. Strain differences in response to sub-optimal stimuli in the fowl. Anim. Behav. 20(1):72–76.

Würsig, B. and M. Würsig. 1977. The photographic determination of group size, composition, and stability of coastal porpoises *(Tursiops truncatus).* Science 198:755–756.

Young, E., ed. 1975. The capture and care of wild animals. Curtis Books. Hollywood, Fla. 224 pp.

Appendix A
Statistical Tables

Table A1. Values of the t-statistic for degrees of freedom v and two-tailed probability level P percent.

P	.90	.80	.70	.60	.50	.40	.30	.20	.10	.05	.02	.01	0.1
ν = 1	0·158	0·325	0·510	0·727	1·000	1·376	1·936	3·078	6·314	12·706	31·821	63·657	636·619
2	0·142	0·289	0·445	0·617	0·816	1·061	1·386	1·886	2·920	4·303	6·965	9·925	31·598
3	0·137	0·277	0·424	0·584	0·765	0·978	1·250	1·638	2·353	3·182	4·541	5·841	12·924
4	0·134	0·271	0·414	0·569	0·741	0·941	1·190	1·533	2·132	2·776	3·747	4·604	8·610
5	0·132	0·267	0·408	0·559	0·727	0·920	1·156	1·476	2·015	2·571	3·365	4·032	6·869
6	0·131	0·265	0·404	0·553	0·718	0·906	1·134	1·440	1·943	2·447	3·143	3·707	5·959
7	0·130	0·263	0·402	0·549	0·711	0·896	1·119	1·415	1·895	2·365	2·998	3·499	5·408
8	0·130	0·262	0·399	0·546	0·706	0·889	1·108	1·397	1·860	2·306	2·896	3·355	5·041
9	0·129	0·261	0·398	0·543	0·703	0·883	1·100	1·383	1·833	2·262	2·821	3·250	4·781
10	0·129	0·260	0·397	0·542	0·700	0·879	1·093	1·372	1·812	2·228	2·764	3·169	4·587
11	0·129	0·260	0·396	0·540	0·697	0·876	1·088	1·363	1·796	2·201	2·718	3·106	4·437
12	0·128	0·259	0·395	0·539	0·695	0·873	1·083	1·356	1·782	2·179	2·681	3·055	4·318
13	0·128	0·259	0·394	0·538	0·694	0·870	1·079	1·350	1·771	2·160	2·650	3·012	4·221
14	0·128	0·258	0·393	0·537	0·692	0·868	1·076	1·345	1·761	2·145	2·624	2·977	4·140
15	0·128	0·258	0·393	0·536	0·691	0·866	1·074	1·341	1·753	2·131	2·602	2·947	4·073
16	0·128	0·258	0·392	0·535	0·690	0·865	1·071	1·337	1·746	2·120	2·583	2·921	4·015
17	0·128	0·257	0·392	0·534	0·689	0·863	1·069	1·333	1·740	2·110	2·567	2·898	3·965
18	0·127	0·257	0·392	0·534	0·688	0·862	1·067	1·330	1·734	2·101	2·552	2·878	3·922
19	0·127	0·257	0·391	0·533	0·688	0·861	1·066	1·328	1·729	2·093	2·539	2·861	3·883
20	0·127	0·257	0·391	0·533	0·687	0·860	1·064	1·325	1·725	2·086	2·528	2·845	3·850

Table A1 *(cont.)*

P	.90	.80	.70	.60	.50	.40	.30	.20	.10	.05	.02	.01	0.1
$\nu = 21$	0·127	0·257	0·391	0·532	0·686	0·859	1·063	1·323	1·721	2·080	2·518	2·831	3·819
22	0·127	0·256	0·390	0·532	0·686	0·858	1·061	1·321	1·717	2·074	2·508	2·819	3·792
23	0·127	0·256	0·390	0·532	0·685	0·858	1·060	1·319	1·714	2·069	2·500	2·807	3·767
24	0·127	0·256	0·390	0·531	0·685	0·857	1·059	1·318	1·711	2·064	2·492	2·797	3·745
25	0·127	0·256	0·390	0·531	0·684	0·856	1·058	1·316	1·708	2·060	2·485	2·787	3·725
26	0·127	0·256	0·390	0·531	0·684	0·856	1·058	1·315	1·706	2·056	2·479	2·779	3·707
27	0·127	0·256	0·389	0·531	0·684	0·855	1·057	1·314	1·703	2·052	2·473	2·771	3·690
28	0·127	0·256	0·389	0·530	0·683	0·855	1·056	1·313	1·701	2·048	2·467	2·763	3·674
29	0·127	0·256	0·389	0·530	0·683	0·854	1·055	1·311	1·699	2·045	2·462	2·756	3·659
30	0·127	0·256	0·389	0·530	0·683	0·854	1·055	1·310	1·697	2·042	2·457	2·750	3·646
40	0·126	0·255	0·388	0·529	0·681	0·851	1·050	1·303	1·684	2·021	2·423	2·704	3·551
60	0·126	0·254	0·387	0·527	0·679	0·848	1·046	1·296	1·671	2·000	2·390	2·660	3·460
120	0·126	0·254	0·386	0·526	0·677	0·845	1·041	1·289	1·658	1·980	2·358	2·617	3·373
∞	0·126	0·253	0·385	0·524	0·674	0·842	1·036	1·282	1·645	1·960	2·326	2·576	3·291

From Campbell, R.C. 1974. Statistics for Biologists, 2nd Edition. Cambridge University Press, Cambridge, 385 pp.

Table A2. The variance ratio (F).

5% significance level for two-tailed test.
N_1 are the degrees of freedom for greater variance. N_2 are the degrees of freedom for smaller variance.

	$N_1 = 1$	2	3	4	5	6	7	8	9	10	12	15	20	24	30	40	60	120	∞
$N_2 =$ 1	648	800	864	900	922	937	948	957	963	969	977	985	993	997	1001	1006	1010	1014	1018
2	38·51	39·00	39·16	39·25	39·30	39·33	39·36	39·37	39·39	39·40	39·42	39·43	39·45	39·46	39·46	39·47	39·48	39·49	39·50
3	17·44	16·04	15·44	15·10	14·88	14·74	14·62	14·54	14·47	14·42	14·34	14·25	14·17	14·12	14·08	14·04	13·99	13·95	13·90
4	12·22	10·65	9·98	9·60	9·36	9·20	9·07	8·98	8·90	8·84	8·75	8·66	8·56	8·51	8·46	8·41	8·36	8·31	8·26
5	10·01	8·43	7·76	7·39	7·15	6·98	6·85	6·76	6·68	6·62	6·52	6·43	6·33	6·28	6·23	6·18	6·12	6·07	6·02
6	8·81	7·26	6·60	6·23	5·99	5·82	5·70	5·60	5·52	5·46	5·37	5·27	5·17	5·12	5·07	5·01	4·96	4·90	4·85
7	8·07	6·54	5·89	5·52	5·29	5·12	4·99	4·90	4·82	4·76	4·67	4·57	4·47	4·42	4·36	4·31	4·25	4·20	4·14
8	7·57	6·06	5·42	5·05	4·82	4·65	4·53	4·43	4·36	4·30	4·20	4·10	4·00	3·95	3·89	3·84	3·78	3·73	3·67
9	7·21	5·71	5·08	4·72	4·48	4·32	4·20	4·10	4·03	3·96	3·87	3·77	3·67	3·61	3·56	3·51	3·45	3·39	3·33
10	6·94	5·46	4·83	4·47	4·24	4·07	3·95	3·85	3·78	3·72	3·62	3·52	3·42	3·37	3·31	3·26	3·20	3·14	3·08
12	6·55	5·10	4·47	4·12	3·89	3·73	3·61	3·51	3·44	3·37	3·28	3·18	3·07	3·02	2·96	2·91	2·85	2·79	2·72
15	6·20	4·76	4·15	3·80	3·58	3·41	3·29	3·20	3·12	3·06	2·96	2·86	2·76	2·70	2·64	2·58	2·52	2·46	2·40
20	5·87	4·46	3·86	3·51	3·29	3·13	3·01	2·91	2·84	2·77	2·68	2·57	2·46	2·41	2·35	2·29	2·22	2·16	2·09
24	5·72	4·32	3·72	3·38	3·15	2·99	2·87	2·78	2·70	2·64	2·54	2·44	2·33	2·27	2·21	2·15	2·08	2·01	1·94
30	5·57	4·18	3·59	3·25	3·03	2·87	2·75	2·65	2·57	2·51	2·41	2·31	2·20	2·14	2·07	2·01	1·94	1·87	1·79
40	5·42	4·05	3·46	3·13	2·90	2·74	2·62	2·53	2·45	2·39	2·29	2·18	2·07	2·01	1·94	1·88	1·80	1·72	1·64
60	5·29	3·93	3·34	3·01	2·79	2·63	2·51	2·41	2·33	2·27	2·17	2·06	1·94	1·88	1·82	1·74	1·67	1·58	1·48
120	5·15	3·80	3·23	2·89	2·67	2·52	2·39	2·30	2·22	2·16	2·05	1·94	1·82	1·76	1·69	1·61	1·53	1·43	1·31
∞	5·02	3·69	3·12	2·79	2·57	2·41	2·29	2·19	2·11	2·05	1·94	1·83	1·71	1·64	1·57	1·48	1·39	1·27	1·00

Abridged from Merrington, M. and C.M. Thompson. 1943. Tables of percentage points of the inverted beta (F) distribution. *Biometrika* 33:73–78.

Table A3. Mann–Whitney test.
5% significance level for two-tailed test.

$N_L =$	9	10	11	12	13	14	15	16	17	18	19	20
$N_S =$ **1**												
2	0	0	0	1	1	1	1	1	2	2	2	2
3	2	3	3	4	4	5	5	6	6	7	7	8
4	4	5	6	7	8	9	10	11	11	12	13	13
5	7	8	9	11	12	13	14	15	17	18	19	20
6	10	11	13	14	16	17	19	21	22	24	25	27
7	12	14	16	18	20	22	24	26	28	30	32	34
8	15	17	19	22	24	26	29	31	34	36	38	41
9	17	20	23	26	28	31	34	37	39	42	45	48
10	20	23	26	29	33	36	39	42	45	48	52	55
11	23	26	30	33	37	40	44	47	51	55	58	62
12	26	29	33	37	41	45	49	53	57	61	65	69
13	28	33	37	41	45	50	54	59	63	67	72	76
14	31	36	40	45	50	55	59	64	67	74	78	83
15	34	39	44	49	54	59	64	70	75	80	85	90
16	37	42	47	53	59	64	70	75	81	86	92	98
17	39	45	51	57	63	67	75	81	87	93	99	105
18	42	48	55	61	67	74	80	86	93	99	106	112
19	45	52	58	65	72	78	85	92	99	106	113	119
20	48	55	62	69	76	83	90	98	105	112	119	127

Adapted and abridged from Mann, H.B. and D.R. Whitney, 1947, On a test of whether one of two random variables is stochastically larger than the other. *Ann. of Math. Stat.* 18:52–54.

Table A4. Values of the test statistic required for 5% and 1% significance in the Kolmogorov–Smirnov two-sample test.

Sample sizes n	Significance level (per cent) 5	1
4	4	
5	4	5
6	5	6
7	5	6
8	5	6
9	6	7
10	6	7
11	6	8
12	6	8
13	7	8
14	7	8
15	7	9
16	7	9
17	8	9
18	8	10
19	8	10
20	8	10
21	8	10
22	9	11
23	9	11
24	9	11
25	9	11
26	9	11
27	9	12
28	10	12
29	10	12
30	10	12
35	11	13
40	11	14

From Campbell, R.C. 1974. Statistics for Biologists. 2nd Ed. Cambridge University Press, Cambridge, 385 pp.

Table A5$_1$. Critical values of r in the runs test.

Given in the bodies of Table A5$_1$ and Table A5$_2$ are various critical values of r for various values of n_1 and n_2. For the one-sample runs test, any value of r which is equal to or smaller than that shown in Table A5$_1$ or equal to or larger than that shown in Table A5$_2$ is significant at the .05 level. For the Wald–Wolfowitz two-sample runs test, any value of r which is equal to or smaller than that shown in Table A5$_1$ is significant at the .05 level.

n_1 \ n_2	2	3	4	5	6	7	8	9	10	11	12	13	14	15	16	17	18	19	20
2											2	2	2	2	2	2	2	2	2
3					2	2	2	2	2	2	2	2	2	3	3	3	3	3	3
4				2	2	2	3	3	3	3	3	3	3	3	4	4	4	4	4
5			2	2	3	3	3	3	3	4	4	4	4	4	4	4	5	5	5
6		2	2	3	3	3	3	4	4	4	4	5	5	5	5	5	5	6	6
7		2	2	3	3	3	4	4	5	5	5	5	5	6	6	6	6	6	6
8		2	3	3	3	4	4	5	5	5	6	6	6	6	6	7	7	7	7
9		2	3	3	4	4	5	5	5	6	6	6	7	7	7	7	8	8	8
10		2	3	3	4	5	5	5	6	6	7	7	7	7	8	8	8	8	9
11		2	3	4	4	5	5	6	6	7	7	7	8	8	8	9	9	9	9
12	2	2	3	4	4	5	6	6	7	7	7	8	8	8	9	9	9	10	10
13	2	2	3	4	5	5	6	6	7	7	8	8	9	9	9	10	10	10	10
14	2	2	3	4	5	5	6	7	7	8	8	9	9	9	10	10	10	11	11
15	2	3	3	4	5	6	6	7	7	8	8	9	9	10	10	11	11	11	12
16	2	3	4	4	5	6	6	7	8	8	9	9	10	10	11	11	11	12	12
17	2	3	4	4	5	6	7	7	8	9	9	10	10	11	11	11	12	12	13
18	2	3	4	5	5	6	7	8	8	9	9	10	10	11	11	12	12	13	13
19	2	3	4	5	6	6	7	8	8	9	10	10	11	11	12	12	13	13	13
20	2	3	4	5	6	6	7	8	9	9	10	10	11	12	12	13	13	13	14

Adapted from Swed, F.S. and C. Eisenhart. 1943. Tables for testing randomness of grouping in a sequence of alternatives. *Ann. Math. Stat.* 14:83–86.

Table A5$_2$. Critical values of r in the runs test *(Continued)*.[1]

n_1 \ n_2	2	3	4	5	6	7	8	9	10	11	12	13	14	15	16	17	18	19	20
2																			
3																			
4				9	9														
5			9	10	10	11	11												
6			9	10	11	12	12	13	13	13	13								
7				11	12	13	13	14	14	14	14	15	15	15					
8				11	12	13	14	14	15	15	16	16	16	16	17	17	17	17	17
9					13	14	14	15	16	16	16	17	17	18	18	18	18	18	18
10					13	14	15	16	16	17	17	18	18	18	19	19	19	20	20
11					13	14	15	16	17	17	18	19	19	19	20	20	20	21	21
12					13	14	16	16	17	18	19	19	20	20	21	21	21	22	22
13						15	16	17	18	19	19	20	20	21	21	22	22	23	23
14						15	16	17	18	19	20	20	21	22	22	23	23	23	24
15						15	16	18	18	19	20	21	22	22	23	23	24	24	25
16							17	18	19	20	21	21	22	23	23	24	25	25	25
17							17	18	19	20	21	22	23	23	24	25	25	26	26
18							17	18	19	20	21	22	23	24	25	25	26	26	27
19							17	18	20	21	22	23	23	24	25	26	26	27	27
20							17	18	20	21	22	23	24	25	25	26	27	27	28

1. From Campbell, R.C. 1974. Statistics for Biologists. 2nd Edition. Cambridge University Press, Cambridge, 385 pp.

Table A6. Critical values of chi square.

P	99	95	10	5	1	0·1
$\nu=1$	$0{\cdot}0^{3}157$	0·00393	2·71	3·84	6·63	10·83
2	0·0201	0·103	4·61	5·99	9·21	13·81
3	0·115	0·352	6·25	7·81	11·34	16·27
4	0·297	0·711	7·78	9·49	13·28	18·47
5	0·554	1·15	9·24	11·07	15·09	20·52
6	0·872	1·64	10·64	12·59	16·81	22·46
7	1·24	2·17	12·02	14·07	18·48	24·32
8	1·65	2·73	13·36	15·51	20·09	26·12
9	2·09	3·33	14·68	16·92	21·67	27·88
10	2·56	3·94	15·99	18·31	23·21	29·59
11	3·05	4·57	17·28	19·68	24·73	31·26
12	3·57	5·23	18·55	21·03	26·22	32·91
13	4·11	5·89	19·81	22·36	27·69	34·53
14	4·66	6·57	21·06	23·68	29·14	36·12
15	5·23	7·26	22·31	25·00	30·58	37·70
16	5·81	7·96	23·54	26·30	32·00	39·25
17	6·41	8·67	24·77	27·59	33·41	40·79
18	7·01	9·39	25·99	28·87	34·81	42·31
19	7·63	10·12	27·20	30·14	36·19	43·82
20	8·26	10·85	28·41	31·41	37·57	45·31
21	8·90	11·59	29·62	32·67	38·93	46·80
22	9·54	12·34	30·81	33·92	40·29	48·27
23	10·20	13·09	32·01	35·17	41·64	49·73
24	10·86	13·85	33·20	36·42	42·98	51·18
25	11·52	14·61	34·38	37·65	44·31	52·62
26	12·20	15·38	35·56	38·89	45·64	54·05
27	12·88	16·15	36·74	40·11	46·96	55·48
28	13·56	16·93	37·92	41·34	48·28	56·89
29	14·26	17·71	39·09	42·56	49·59	58·30
30	14·95	18·49	40·26	43·77	50·89	59·70
40	22·16	26·51	51·81	55·76	63·69	73·40
50	29·71	34·76	63·17	67·50	76·15	86·66
60	37·48	43·19	74·40	79·08	88·38	99·61
70	45·44	51·74	85·53	90·53	100·4	112·3
80	53·54	60·39	96·58	101·9	112·3	124·8
90	61·75	69·13	107·6	113·1	124·1	137·2
100	70·06	77·93	118·5	124·3	135·8	149·4

From R.C. Campbell. 1974. Statistics for Biologists. 2nd Ed. Cambridge University Press. London. 385 pp.

Table A7. Critical values of T in the Wilcoxon matched-pairs signed-ranks test.

N	Level of significance for one-tailed test		
	.025	.01	.005
	Level of significance for two-tailed test		
	.05	.02	.01
6	0	—	—
7	2	0	—
8	4	2	0
9	6	3	2
10	8	5	3
11	11	7	5
12	14	10	7
13	17	13	10
14	21	16	13
15	25	20	16
16	30	24	20
17	35	28	23
18	40	33	28
19	46	38	32
20	52	43	38
21	59	49	43
22	66	56	49
23	73	62	55
24	81	69	61
25	89	77	68

Abridged from Wilcoxon, F. 1949. Some rapid approximate statistical procedures. American Cyanamid Co. New York.

Table A8. Sign test.

L = frequency of the less frequent sign.
T = total frequency of *both* pluses and minuses.
Table value is the *probability* of obtaining L or fewer of the less frequent sign, out of a total of T pluses and minuses (probabilities are for a two-tailed test).

$L=$	0	1	2	3	4	5	6	7	8	9	10	11	12
$T=$ 5	062	376	1										
6	032	218	688	1									
7	016	124	454	1									
8	008	070	290	726	1								
9	004	040	180	508	1								
10	002	022	110	344	754	1							
11	000	012	066	226	548	1							
12	000	006	038	146	388	774	1						
13	000	004	022	092	266	582	1						
14	000	002	012	058	180	424	790	1					
15		000	008	036	118	302	608	1					
16		000	004	022	076	210	454	804	1				
17		000	002	016	050	144	332	630	1				
18		000	002	008	030	096	238	480	814	1			
19			000	004	020	064	168	360	648	1			
20			000	002	012	042	116	264	504	824	1		
21			000	002	008	026	078	180	384	664	1		
22				000	004	016	054	134	286	524	832	1	
23				000	002	010	017	094	210	404	678	1	
24				000	002	006	022	064	152	308	542	838	1
25					000	004	014	044	108	230	424	390	1

Abridged from Walker, H.M. and J. Lev. 1953. Statistical Inference. Holt, Rinehart and Winston, New York.

Table A9. Significance of Kendall's tau.

N = number of pairs of scores.
The τ values are the smallest values of τ significant at the 0.05 level for different values of N.

N	τ	N	τ
5	0·80	21	0·31
6	0·69	22	0·29
7	0·63	23	0·29
8	0·57	24	0·28
9	0·53	25	0·27
10	0·49	26	0·27
11	0·45	27	0·26
12	0·43	28	0·25
13	0·41	29	0·25
14	0·39	30	0·25
15	0·37		
16	0·35		
17	0·35		
18	0·33		
19	0·33		
20	0·31		

Adapted from Anderson, B.F. 1966. The Psychology Experiment. Brooks–Cole, Belmont, Calif.

Table A10. Table of critical values for Spearman's correlation coefficient, r_s.

N	Significance level (one-tailed test)	
	.05	.01
4	1.000	
5	.900	1.000
6	.829	.943
7	.714	.893
8	.643	.833
9	.600	.783
10	.564	.746
12	.506	.712
14	.456	.645
16	.425	.601
18	.399	.564
20	.377	.534
22	359	.508
24	.343	.485
26	.329	.465
28	.317	.448
30	.306	.432

Adapted from Olds, E.G. 1938. Distribution of sums of squares of rank differences for small numbers of individuals. *Ann. Math. Stat.* 9:133–148, and from Olds, E.G. 1949. The 5% significance levels for sums of squares of rank differences and a correction. *Ann. Math. Stat.* 20:117–118.

Appendix B
Selected Ethological Journals

The following are selected technical journals which publish articles in ethology. The years of publication are indicated following the journal title.

Journals containing large numbers of ethological articles:

Animal Behaviour (previously British Journal of Animal Behavior) 1953–
Animal Behaviour Monographs 1968–
Applied Animal Ethology 1974–
Behavioural Ecology and Sociobiology 1976–
Behaviour 1947–
Behavioural Processes 1976–
Biology of Behaviour 1976–
Insects of Socieux 1954–
Journal of Animal Behaviour 1911–1916
Primate Behavior 1970–
Sociobiology
Zeitschrift fur Tierpsychologie 1937–

Journals devoted to various aspects of animal behavior that frequently contain ethological articles:

Animal Learning and Behavior 1973–
Behavior Genetics 1970–
Behavior Research Methods and Instrumentation 1969–
Behavioral Biology (began as Communication in Behavioral Biology)
Behavioral Science 1956–
Brain, Behavior and Evolution 1968–
Comparative Psychology Monographs 1922–
Ecology 1920–
Evolution 1947–
Hormones and Behavior 1969–
Journal of Comparative and Physiological Psychology 1921–
Journal of Insect Physiology
Physiology and Behavior 1966–
The Behavioural and Brain Sciences 1978–

Journals devoted to animal groups which occasionally contain ethological articles:

Auk 1884–
Condor 1899–
Copeia 1913–
Folia Primatologica 1963–
Herpetologica 1936–
Journal of Herpetology 1968–
Journal of Mammalogy 1919–
Primates 1960–
Wilson Bulletin 1889–

There are numerous other technical journals in which ethological articles can occasionally be found. These are best found through the abstracting sources discussed on page 53.

Appendix C

Obtaining Funds

Most ethological research projects cost money, and that money is not always easy to come by. Large sums are not always necessary, and there is no direct correlation between the size of the budget and the quality and importance of the research. Many very worthwhile studies have been conducted on "shoestring" budgets. However, pursuit of funds generally becomes a reality for most professional ethologists. "Grantsmanship" is a skill which has too many dimensions to elucidate here. The best sources of information on writing successful proposals are experienced investigators and the following references:

Krathwohl, D.R. 1977. How to prepare a research proposal: Suggestions for those seeking funds for behavioral science research, 2nd ed. Syracuse Univ. Bookstore. 112 pp.

Lisk, D.J. 1971. Why research grant applications are turned down. BioScience 21(20):1025–1026.

MacIntyre, M. 1971. How to write a proposal. Education, Training and Research Services Corporation, Washington, D.C. 55 pp.

White, V.P. 1975. Grants: How to find out about them and what to do next. Plenum, New York. 354 pp.

Additional useful references are listed below:

The *Annual Register of Grant Support* (published by Marquis Who's Who, 4300 W. 62nd St., Indianapolis, Indiana 46206) is an annually revised directory of grant-support programs administered by government agencies, foundations, and by business, professional, and other organizations.

Most granters have application forms or preferred formats. Requests for this information can often be included with letters of inquiry, including an outline of your proposed research or a preproposal. The grantors can then respond as to whether or not they would consider a complete proposal on that topic.

1. Federal Agencies

Federal-agency grant programs are all listed in the *Catalog of Federal Domestic Assistance* which is available from the Superintendent of Documents, U.S. Government Printing Office, Washington, D.C. 20402.

Some of the federal agencies which will consider proposals for ethological research are listed below:

National Aeronautics and Space Administration. Grants from this agency to institutions have dropped off precipitously over the last several years. They have supported research on migration, navigation, and orientation. Information can be obtained from

NASA Headquarters Information Center
Washington, D.C. 20546

National Institutes of Health (NIH). This agency does fund ethological research within its Division of Research Grants. They fund research on animal behavior which may serve as a model for understanding human behavior. Proposals submitted to the Division are forwarded to the appropriate study section for review. Ethological proposals are likely to be reviewed in the Developmental Behavioral Sciences Study Section or the Experimental Psychology Study Section.

Division of Research Grants
National Institutes of Health
Bethesda, Maryland 20014

National Institute of Mental Health (NIMH). This agency funds research on principles underlying behavior. Like NIH, their objective is to promote studies of animal behavior which may serve as a model for understanding human behavior. Funding is through the Division of Extramural Research Programs, and animal ethology proposals are handled within the Behavioral Sciences Research Branch. Proposals are submitted to the address listed under NIH.

National Science Foundation. Ethological studies are funded through the Psychobiology Program of the Division of Behavioral and Neural Sciences and the Ecology and the Population Biology and Physiological Ecology Programs of the Division of Environmental Biology. Proposals are submitted to:

Central Processing Section
National Science Foundation
Washington, D.C. 20550

Further information can be obtained from the following paper written by the Director of the Psychobiology Program:

Stollnitz, F. 1976. NSF programs in the behavioral and neural sciences. Behav. Res. Methods Instrum. 8(2):65.

Smithsonian Institution. The Smithsonian Institution funds ethological studies through their Programs for Visiting Scholars, Scientists, and Students. The research is conducted at, or near, one of their installations:

National Zoological Park, Washington, D.C.
Smithsonian Tropical Research Institute, Balboa, Canal Zone
Convservation and Research Center, Front Royal, Virginia

Further information, including the publication *Smithsonian Opportunities for Research and Study in History, Art and Science*, can be obtained from:

Office of Academic Studies
Room S1356, Smithsonian Institution
Washington, D.C. 20560

United States Navy. The Navy funds research which is primarily relevant to their needs, some of which falls within the realm of psychological-behavioral studies. Further information can be found in

Young, J.L. 1976. Research support from the Office of Naval Research. Behav. Res. Methods Instrum. 8(2):66–67.

Inquiries and proposals should be addressed to

Office of Naval Research
Department of the Navy
Ballston Tower #1
800 N. Quincy St.
Arlington, Virginia 22217

A useful reference for U.S. government grants is

Urgo, L.A. 1972. A manual for obtaining government grants. Robert J. Corcoran Co., 40 Court St., Boston Massachusetts 02108. 20 pp.

2. Foundations and Other Organizations

There are a large number of foundations and organizations in the United States, some of which are receptive to proposals for funds for ethological research. The *Foundation Directory*, which lists about 2,500 foundations, lists their purpose and activities and is indexed by fields of interest. It is published by Columbia University Press, 136 S. Broadway, Irvington-on-Hudson, New York 10533.

The following is a list of selected foundations and organizations which provide limited funding for field studies in ethology. They should be contacted about their programs before a proposal is submitted.

American Museum of Natural History
Central Park West at 79th St.
New York, N.Y. 10024
Chapman Fund
Theodore Roosevelt Memorial Fund
American Philosophical Society

Conservation and Research Foundation
Box 1445
Connecticut College
New London, Connecticut 06320

Caesar Kleberg Foundation for Wildlife Conservation
535 San Antonio Bank and Trust Building
711 Navarro St.
San Antonio, Texas 78205

J.N. (Ding) Darling Foundation, Inc.
3663 Grand St.
Suite 608
Des Moines, Iowa 50312

Max McGraw Wildlife Foundation
P.O. Box 194
Dundee, Illinois 60118

National Academy of Sciences
Office of the Home Secretary
2101 Constitution Ave., N.W.
Washington, D.C. 20418
Marsh Fund

National Audubon Society
950 Third Ave.
New York, New York 10022

Committee for Research and Exploration
National Geographic Society
17th and M Streets, N.W.
Washington, D.C. 20036

The National Wildlife Federation Endowment, Inc.
1404 Oliver Building
Pittsburgh, Pennsylvania 15222

New York Zoological Society
The Zoological Park
New York, New York 10460

North American Wildlife Foundation
709 Wire Building
Washington, D.C. 20005

Renewable Natural Resources Foundation
5400 Grosvenor Lane
Bethesda, Maryland 20014

Rob and Bessie Welder Wildlife Foundation
P.O. Box 1400
Sinton, Texas 78387

Wildlife Management Institute
709 Wire Building
1000 Vermont Ave., NW
Washington, D.C. 20005

The World Wildlife Fund
1110 Morges, Switzerland

Useful references for writing foundation grant proposals are:

Dermer, J. 1972. How to write successful foundation presentations. Public Service Materials Center, 104 East 40th St., New York, N.Y. 10016. 80 pp.

Urgo, L.A. and R.J. Corcoran. 1971. A manual for obtaining foundation grants. Robert J. Corcoran Co., 40 Court St., Boston, Mass. 02108.

Assistance in Obtaining Funds

There are a few organizations which will assist a prospective researcher in finding funds for his research.

For a $250 annual fee the Funding Sources Clearinghouse, Inc., will attempt to find funding for your proposed project. Their address is: 260 Bancroff Way, Berkeley, CA 94704.

The Center for Field Research will assist in finding support for ethological studies. "*The Center arranges financial support for research investigators whose projects can constructively utilize nonspecialists in the field.* The Center is *not* the source of funds. Instead, it reviews and evaluates research proposals in a wide range of disciplines. The Center assigns those accepted to Earthwatch which, in turn, raises the funds from carefully selected nonspecialists who collectively finance the projects, in return for the opportunity to work as assistants to research scholars in the field" (page 5 in *A Guide to Funds and Volunteers for Field Research; The Center for Field Research,* 10 Juniper Road, Box 127, Belmont, Massachusetts 02178).

In summary, there is no shortcut to a well-conceived research project and a well-written proposal. Writing a proposal to study white elephants is one thing, trying to fund a proposal which is a white elephant is another.

Appendix D

The Study of Social Organizations

by GLEN MCBRIDE

Animal Behaviour Unit, University of Queensland, St. Lucia, Queensland, Australia

With contributions from:

G.W. Barlow, Zoology Department, University of California, Berkeley 94720, U.S.A.

M.C. Busnel, I.N.R.A. Laboratoire de Physiologie acoustique, C.N.R.Z. Domaine de Vilbert, Jouy-en-Josas 78, France.

C.C. Carpenter, Zoology Department, University of Oklahoma, Norman, Oklahoma 73069, U.S.A.

D.D. Dow, Zoology Department, University of Queensland, Australia.

P. Dwyer, Zoology Department, University of Queensland, Australia.

J.F. Eisenberg, National Zoological Park, Washington, D.C. 20009, U.S.A.

M.W. Fox, Psychology Department, Washington University, St. Louis, Missouri 63130, U.S.A.

K. Immelmann, Ethology Department, University of Bielefeld, P.O. Box 8640, 48 Bielefeld, West Germany.

J. Kikkawa, Zoology Department, University of Queensland, Australia.

H. Kummer, Zoologisches Institut der Universität Zürich, Aussenstation Oerliken, Birchstrasse 95, 8050 Zürich, Switzerland.

P. Leyhausen, Abteilung Lorenz, Max-Planck-Institut für Verhaltensphysiologie, 56 Wuppertal-Elberfeld, Boeltinger Weg 37, West Germany.

A.A. Myrberg, Jr., School of Marine and Atmospheric Science, 10 Rickenbacker Causeway, Miami, Florida 33149, U.S.A.

B. Nievergelt, Zoologisches Institut der Universität Zürich, Switzerland.

E.S. Reese, Zoology Department, University of Hawaii, 2538 The Mall, Honolulu, Hawaii 96822, U.S.A.

G. Richard, Lab. d'Ethologie, Université de Rennes, Rennes-Beaulieu 35, France.

W.M. Schleidt, Zoology Department, University of Maryland, College Park, Md 20742, U.S.A.

J.P. Scott, Center for Research on Social Behavior, Bowling Green, Ohio 43403, U.S.A.

History

During the XI and XII International Ethological Conferences, a group of people interested in the study of social organizations met to examine the problems associated with summarizing the knowledge available on the societies of animals. A committee was charged with the responsibility for preparing a questionnaire which would summarize all of the information available, or which it was thought should be available, to describe adequately the social organization of any species. We were also asked to publish the results of our study.

It was suggested that such a questionnaire would provide a valuable guide to those undertaking ethological field studies, and may provide a standard format for the presentation of data on animal societies.

Though many have contributed to the final draft presented, and some have filled in the questionnaire for individual species, the responsibility for any inadequacies is mine.

The questionnaire presented here is only a first attempt to make a formalized approach to the study of animal social systems. Readers and users are invited to send their comments to me—so that experience on its usefulness may continue to accumulate.

Introduction

Because there is no well-known and accepted model of the organization of animal societies, this questionnaire reflects the model of society held by the author, and built into the questions. This is inevitable, even though many views were sought in the development of the questionnaire, and incorporated in the questions. A brief account of the author's model is presented, firstly to reveal the bias built into the questions and to foster consistency in their interpretation, and secondly to draw comment and criticism, so that a general model of social systems may emerge, and thus modify the approach to their study.

Ecologically, a species is seen as a unit or subsystem within a community, harvesting energy and resources of specific types. The individuals are organized physiologically for the conversions of the chemical materials they harvest; they are equally well organized behaviourally to harvest, and also to distribute the particular resources available among the individuals of their species, socially. Animals live in that part of their species' niche determined by the social arrangements they have made with their neighbours. The matrix or net of such social arrangements is the society.

By natural selection, each species tends to maintain or "improve" the relationships with its community and environment; these relationships constitute the niche of a species. Improvement might occur when new productive relationships are developed, or when old-established relationships become more productive, perhaps by specialization. Alternatively losses of individuals or of resources to other species, predators, parasites, or competitors may be reduced. The evolutionary changes involved may modify individuals in the behaviour they use to deal with the environment, or by which they organize their societies, so that the latter evolve.

The patterns of societal organization seen in nature appear to be regular. The structural features of societies are seen most clearly, though often their functions are readily interpreted. There are some patterns of organization which are found repeatedly, and on analysis, all societies appear to combine variations in only four structural features. These are as follows:

1. Species organize differently for periods of time, into *social phases*, for example in the breeding and the non-breeding seasons;
2. There are usually several types of individuals, or *castes*, organized organically. Examples are infants, juveniles, males and females;
3. Animals normally aggregate into *groups* characteristic of the species, or they remain solitary;
4. The groups or individuals *disperse* in space in some regular pattern, such as territories or overlapping home ranges.

These four structural features provide the first four categories in the questionnaire. The remaining categories concern the organization of sexual, parental, and environmental behaviour, and the responses to changes in density.

Behaviour is the tool by which animals become moulded into the regular architectural pattern of societies. Without a systematic treatment of this social behaviour, a systematics of societies would be pointless.

Animals make behaviour, which is basically movements. Their sensory systems provide for the orientation of these movements and their adjustment by the surroundings. The movements are ordered into sequences which perform functional operations. Some sequences are ritualized into complex displays; others are ordered into *fixed action patterns*. While the fixity of the movements has recently been questioned, there is no doubt about the fixity, with redundancy, of their functional and communicative properties within any context. These movements or groups of movements, along with various attention-drawing or attention-hindering features of colour and form, are the simple behavioural tools; they are assembled by animals into behavioural sentences of various lengths and complexity. There appear to be rules of ordering and completion in the assembly process. The resulting sentences are best seen in actonic interactions, when the animal organizes its behaviour towards some non-responsive component of its environment, as in collecting food, digging a burrow, or building a nest. Here most of the behavioural sentences are completed.

When the behaviour is oriented to each other, then animals are dealing with a responsive component of their environment, in a social or interspecific interaction. In these situations, the ability to produce completed behavioural sentences is usually hindered by the responses of the interactant. In the same way, I may write in completed sentences, but in conversation, incompleted sentences are common. Incomplete sentences seem to be the essence of interaction. The outcome of interactions appears to be the resolution of the differing goals of the interactants. Each animal has the ability to anticipate the outcome of the other's behavioural sentences, and interrupts accordingly. Thus a blow may be parried, or a courtship sequence halted by a threat or withdrawal. The function of the interaction does not appear to include the completion of behavioural sentences; it is concerned with the building of adjustments or agreements between the interactants. Every such adjustment is a step in the formation of a social relationship, for they are remembered, and organize behaviour in future interactions between the pair. Each interaction thus has the effect of influencing behaviour in the next interaction or encounter. Thus predictability emerges as the animals build a relationship, which is a summary of the relevant parts of their experiences together.

All relationships divide the repertoires of behaviour each animal has into that which is included in and that which is excluded from future interactions between the pair. This summary is built into the perceptual system of each in such a way that a deviation would attract immediate attention. Since individuals do not use conscious choice in the process of including and excluding behaviour within their relationship, then one must assume that the summary system is preorganized, as well as the sorts of decisions being made. This preorganization may have occurred by evolutionary or ontogenetic processes, normally both.

A concrete example may help. All relationships may include spacing arrangements. Thus fights may be summarized as adjustments determining who "wins" where, whereupon we describe the relationship as territorial, with the boundaries learned: alternatively they may involve regular "wins" by a single individual over the range of the pair, so organizing the approach and orientation behaviour of the other; this gives the sort of relationship ethologists call dominant-subordinate. This in turn may involve the allocation of a variety of other behaviour between the two, leadership-follower, sentry duty, and attention structure, and this division of behaviour we call roles. Roles emphasise the individual aspect of the division of behaviour, while the relationship emphasises the process by which the division is made and maintained.

It is the regularity and predictability of the adjustments between individuals comprising social relationships which gives them their stability, and thus stable characteristics to societies in time, in space, and in the organization of the behaviour of individuals. Social relationships embody the glue, and the design features of animal societies.

It is within the matrix of social relationships that animals organize their behaviour toward the other components of their environments, so that every action is carried out within a social context. A solitary squirrel may bury nuts, but only in places about which it has developed relationships with neighbours, limiting their access. These arrangements will determine who shall dig up the nuts some months into the future. The digging up is organized before the burying. Any animal has some freedom to respond to its environment—but freedom means absence of constraints, and the source of constraints is most commonly social, built by arrangements with neighbours.

When any change occurs in the environment, such as drought, flood, or fire, then there are normally readjustments to be made between neighbours; readjustments also become necessary with changes in the density of conspecifics. When relationships are renegotiated, then the society changes, for societies are always as dynamic as are the relationships that shape them. A change in any one relationship usually means a change in others, for relationships are not independent. Consider the interdependence of dominant-subordinate relationships in the linear hierarchy of a small closed group.

Societies in this view are not fixed structures, but fixed systems. All levels of organization, and all of the control systems are the product of evolution by natural selection. Yet all of the societal information is carried only in the organization of the individuals, as their sensory system, their behavioural system or repertoire, and their summary or learning system.

A dynamic view of animal societies makes a typological summary of an animal society difficult. Yet either structure or change may be observed systematically, at every level of organization, with changes followed through to new stable structures. The changes begin in the way the individuals perceive and respond to each other, and proceed until new adjustments emerge. These become established among other changed relationships, so that a new societal structure emerges.

This is the structure of a process of change, whether the change is a regular event like the annual change in social phase, an occasional change like a dramatic change in density, or some quite novel change like the colonization of a new type of environment—perhaps as bizarre as a zoo! An account of the society of a species should aim to examine the processes of maintenance as well as change, at every level of organization, behavioural, interactional, relationship, group, and society. The questionnaire is organized to obtain such data.

Classification of Species: ....
Common Names: .... (language) ....

1. Social Organization in Time

In any species, there may be a number of separate and separable patterns of social organization, called social phases, each persisting for a period of time. These are the largest units of specialization, or social subsystems within the species.

Characteristically there is one social phase during the breeding season, and another for the remainder of the year. There may also be separate social phases for migration between these. Common social phases are: prebreeding, mating (with or separate from reproduction), migratory, moulting, non-breeding.

The smaller units of social specialization in time are the social subphases, found in gregarious animals. Here, the groups reorganize spatially for each of their functional activities, such as for moving, for feeding, resting, body care, sleep, and alarm. There is a separate social organization present in each of these units, or social subphases, and this is seen in different spatial and orienting responses to neighbours, and a different repertoire of interactions among the members.

While solitary animals may, and often do, synchronize their activities throughout the day, the subphases thus formed are less easily seen than in gregarious animals. The synchronization of subphase change in gregarious animals usually involves social facilitation, but solitary animals rely more on the time of day, though there may be communication between individuals at these times. Other solitary animals may still adopt periods of functional specialization in their behaviour, but alternate periods of activity with those of neighbours.

1.1 How many separate or separable social phases are observed? ....
1.2 List the names of the social phases observed. ....
1.3 How is the timing of phase change organized (*e.g.*, regular changes may depend upon day length, aided by social stimulation which may foster synchrony, synchronized but irregular changes may depend upon rain or temperature in adventitious or opportunistic breeding species)? ....

1.4 Is the phase change uniform over wide areas, or are there different phases in different areas (*e.g.*, non-synchronized breeding)? . . . .
1.5 How would you describe the main relationships in each phase? . . . .
1.6 Describe the patterns of relationship change between phases. . . . .
1.7 What categories of interaction are observed over the transition periods (*e.g.*, fighting in the winter flock, fights leading to territories, or to the breaking of bonds, courtship interaction)? . . . .
1.8 Describe the behaviour used in these interactions. . . . .
1.9 What are the regular cycles of activity (diurnal, tidal)? . . . .
1.10 Describe a typical activity cycle. . . . .
1.11 Is there synchrony between individuals in the changes of activity? . . . . All changes? . . . . Some changes? . . . . Which changes? . . . .
1.12 How is this synchrony organized? . . . .
1.13 What communicative calls or displays are observed in each subphase? . . . . or during each change of activity? . . . .
1.14 What caste changes occur between seasonal phases? (See below) . . . .

Each social phase is a separate social sub-system, perhaps with different castes, group structures and spacing patterns. Each social phase must now be treated separately as follows.

2. Specialized Types of Animals

Within most species of animals, there is considerable specialization of animals for functional activities. Much of the specialization is concerned with social behaviour, so that societal structures take particular forms because of the specialized types of animals they contain. The major form of animal specialization is into castes, or organically organized divisions of social labour. The common vertebrate castes include infants, juveniles, sub-adults (sometimes males and females are castes, fulfilling separate social as well as sexual functions), and there may be other specialized nonbreeding or subordinate castes, *e.g.* in the European ruff. The seasonal changes in social phase are generally associated with a change in seasonal castes, with the behaviour of the individuals organized quite differently for each seasonal caste, and of course, social phase. In invertebrates, there may be more than one larval caste, sex castes among the adults, and perhaps a range of specialized castes as in most ants and termites.

The functional incorporation of other species within the social organization of another species raises problems of organization here and may require separate treatment to handle all of the specialized relationships found, from nest parasites in social insects or birds, to commensals, or slave species. Yet it seems that the same types of relationships are also found between individuals of different castes.

For the purpose of description, the members of the other species may be treated as separate castes, to examine the nature of their relationships with the species being studied.

There are other specializations of animals beside castes. In time, the specialization of individuals responsible for the social subphase is the role, which involved an organization of the individuals' spacing relationships with its neighbours, and the repertoire of interactions available. Shorter still, an animal becomes highly specialized while it engages in a specific interaction.

Here, the emphasis will be on the specialization by caste, leaving the other specializations in time to be discussed around other features of their organization. Each social phase is treated separately.

2.1 What castes (and species) are present? . . . .
2.2 Are the sexes separable behaviourally as castes (*i.e.* the sexes may be organized as a social as well as a reproductive division of labour, or only as a sexual division of labour with otherwise equal social responses, in which case sex is not organized as a caste)?
2.3 Describe briefly the morphological differences between castes (*e.g.* size, weight, form). . . . .
2.4 What specializations for display are present in each caste (*e.g.* scent glands and marking behaviour, special vocal or vibratory apparatus, parts organized for visual displays)? . . . .
2.5 Are there differences in camouflage between castes? . . . .
2.6 Comment on the nature of the division of labour by caste. . . . .
2.7 What are the main relationships formed between individuals within castes? . . . . between castes? . . . .
2.8 What are the main interactions observed in the relationships described in 2.7? . . . .
2.9 How are the display specializations used in these interactions? . . . .
2.10 How and when are the display specializations used outside the interindividual interactions? . . . . To what part of the maintenance or change of social relationships do they contribute? . . . .

3. Group Structures

In each social phase, individuals may be solitary or there may be groups of two or more present, or regularly observed. The solitary individuals may be of any caste, and the groups may comprise individuals of one or more castes. The groups may be permanent or temporary, they may break up into subgroups or solitary individuals regularly or irregularly for varying periods; they may be fixed in space or mobile. The groups may be open to entry by outsiders, or to certain sorts of outsiders or they may be closed; the individuals may recognize each other as individuals or just as members by a group scent, or they may be anonymous. The aggregates may show internal structure with subgroups of various types. The different types of groups or subgroups or individuals may be similar, or have different and recognizable functions which complement each other within a social phase (or a subphase). The specialization within a group, within a subphase unit may be caste and role, or by subgroups.

In each social phase

3.1 What are the recognizable social units, individuals or aggregates? . . . .
3.2 Name each type of group (*e.g.* all male, all female, female-offspring, pair-offspring, multimale troop, harem, *etc.*). . . . .
3.3 Each type of group should now be classified separately. . . . .
3.4 What is the duration of the group (permanent, permanent throughout the social phase only, nightly, resting aggregations, aggregations around feed, *i.e.* for certain activities on subphases, regular or irregular gatherings, *etc.*)? . . . .
3.5 When does the group aggregate? . . . .
3.6 Group size and range? . . . .
3.7 Does the variability in group size correlate with any environmental conditions? . . . . What relationship? . . . .
3.8 Is the group composition constant throughout the phase? . . . .
3.9 Is the group size constant throughout the phase (except for deaths)? . . . .
3.10 When the group disperses, what are the dispersed units? . . . . How large? . . . . How many? . . . . What composition? . . . .
3.11 When does the group change compositions? . . . . What changes occur? . . . .
3.12 Are the young born or hatched in the group? . . . .
3.13 Do the young incorporate into the group? . . . . or leave it? . . . . At what age? . . . .
3.14 For anonymous groups, what are the aggregating stimuli (*e.g.* a food source, to individuals of the same size as in large fish shoals)? . . . . Is there a group identity (*e.g.* traditional attachment to a site, maintenance by a group scent, *etc.*)? . . . .
3.15 What outsiders may enter the group (all, all of the same size, all of one caste, none)? . . . .
3.16 For recognized groups, what is the pattern of recognition (of all members, of only a small subgroup within a larger aggregate, by rank only, by attachment to particular sites, *etc.*)? . . . .
3.17 What sensory modalities and stimuli seem important in recognition? . . . .
3.18 Is there a recognition interaction or part of an interaction? . . . . Describe the sequence. . . . .
3.19 How long is the process of integrating the outsider into the group? (immediate, some avoidance at first, a complex process-describe . . . .)
3.20 Describe the interactions when outsiders enter the group. . . . .
3.21 What appears to be the relationship between group members (*e.g.*

dominance-subordinate pair-bond, mating bond in harem, parent-offspring bond in families, *etc.*)? . . . .

3.22 What interactions were used in the formation of the relationships? . . . . Describe them. . . . .

3.23 What interactions are used to service the relationships (*e.g.* allofeeding, a trophallaxis, allopreening or allogrooming, threats and avoidances, *etc.*)? . . . . When? . . . . How frequently? . . . . Where? . . . .

3.24 Describe the interactions. . . . .

3.25 What cooperative or coordinated interactions are carried out by more than one individual? . . . .

3.26 What locating behaviours are used to maintain contact between group members (*e.g.* tail flicks or flashes, regular calls, body orientation so that nearest neighbours are visible, *etc.*)? . . . .

3.27 What range of social distances are maintained between group members? (Hediger defined social distance as the maximum distance an animal normally moves from the group. This varies with age in parent-offspring groups, and with activity subphase. The social distance expresses the aggregating forces and affects the pattern and extent of dispersion of the group.)

3.28 What "lost" behaviour is expressed by individuals beyond the social distance? . . . .

3.29 What rejoining behaviour is seen after "lost" behaviour? . . . .

3.30 Is there some regular interval between ontacts or contact signals—
When animals are close? . . . .
When widely dispersed? . . . .
What signals or contacts? . . . .
What intervals? . . . .
Do the signals vary with age? . . . .
Do the intervals? . . . .

3.31 What animals are dispersed? . . . . Do they use trails? . . . . Do they mark trails? . . . . How? . . . .

3.32 Is there occasional splitting of groups into two or more permanent groups? . . . . Describe process. . . . . Describe interactions seen. . . . . The behaviour used. . . . .

3.33 When groups disperse at end of phase, describe terminating process. . . . . The interactions seen. . . . . The behaviour used. . . . .

3.34 When group is permanent, describe the process of phase change. . . . . The interactions seen. . . . . The behaviour used. . . . .

3.35 What divisions of labour are apparent within the group? . . . .

3.36 Describe briefly any important short-term changes in behaviour within the group (*e.g.* with temperature change, rain, snow, alarm, *etc.*). . . . .

3.37 Describe briefly any long-term changes in behaviour within the group (*e.g.* as offspring grow). . . . .

3.38 Are there well organized subgroups (*e.g.* harems in a Gelada troop, pairs or pair families in geese, mother-daughter adult subgroups in waterbuffaloes, mother-offspring pairs, *etc.*)? . . . .

3.39 What is the size and composition of the subgroups? . . . .

3.40 Describe the relationship between members of the subgroup? . . . .
3.41 What interactions service the relationships within the subgroup? . . . .
3.42 Describe the relationships between individuals of different subgroups. . . . .
3.43 What interactions service the relationships between individuals of different subgroups? . . . .
3.44 How many subgroups are normally present? . . . .
3.45 General comments . . . .

4. The Organization of Spacing

Spacing patterns always arise as a result of the *distance-dependent behaviour* of individuals responding to each other as distance-dependent stimuli. The spacing patterns become stable as relationships form, expressing the behaviour which has shaped them. Though we may talk loosely about territories or personal spaces as though they were entities, the only reality is the regular form of the behaviour of the animals towards each other and in attachments to places. Different spacing between animals of the same and different castes reflects their different behaviour and relationships. When some animals form attachment relationships with each other, such as bonds or dominance-subordinate relationships, groups are formed. The shorter spacing between such individuals than between groups reflects these different relationships. Territorial neighbours form relationships which determine where each shall move, and also enable each to monitor the presence and whereabouts of the other. The relationships of animals commonly reflect priorities of access to fixed space, for all of or part of the time, or for different areas at different times, or just in the vicinity of neighbours. These space control features seem to be characteristic of all dominance relationships. When the restrictions on entry to a fixed area exclude all conspecifics, or members of one or more castes, or involve appeasement or submissive behaviour by intruders over the whole or part of the area, then there is a form of territorial behaviour. When there is a time sharing of the facilities of an area, with restrictions on approach to the resident individual or group, then there is some form of home range behaviour, often with dominance between the resident groups or "solitary" individuals. Naturally the avoidance properties of such spacing systems are arrayed along a scale from exclusive territorial systems through the overlapping territories to the fully overlapping home range systems.

Other groups show no site attachment and merge freely sometimes into aggregates of vast sizes, such as fish shoals (not small recognized groups with hierarchies, which are called schools). Within a group, the relationships may involve regular spacing, perhaps with polarized orientations, or may involve the flexible priority system of distance-dependent behaviours seen in dominance hierarchies.

Thus the three main spacing systems examined are territorial, home ranges (and combinations of the two), and the spacing between individuals within aggregates.

Where the spacing between individuals of more than one species is socially organized, the patterns of spacing and the behaviour observed should be described. The questions on castes should be used to include the other species.

Territories

4.1 Territorial behaviour is observed in individuals of the caste or castes or the groups described as . . . .

4.2 Are there interspecific territories? . . . . (Each type of territory should now be treated separately.) The caste or group holding the territory is . . . .

4.3 Does the animal or group remain on the territory at all times? . . . .

4.4 If not when is the territory occupied? . . . .

4.5 Are there rules for leaving the territory (*e.g.* by use of communal paths, or by flying high over neighboring territories)? . . . .

4.6 What is the function of the territory (nest site, mating and display, feeding, family care, roosting, *etc.*, or any combination of functions)? . . . .

4.7 Is the territory in a colony or aggregate of territories (*e.g.* colony of nesting birds, lek, *etc.*)? . . . . Are there boundaries (traditional, or limits to the size of the colony)? . . . .

4.8 Is the territory fixed in relation to a moving object (*e.g.* cattle egret and grazing cow, ant birds and army ants, *etc.*)? . . . .

4.9 Who "defends" territory? . . . .

4.10 Against whom? . . . .

4.11 Who may enter territory (subordinates, other castes, mates, offspring from previous broods, *etc.*)? . . . .

4.12 Are there territories within the territories (*e.g.* of subordinate males or of females within the territory of a male)? . . . .

4.13 Do the territories overlap? . . . . If so, what proportion of the range is exclusive territory? . . . .

4.14 What is the range of sizes of the territory? . . . . What is the distance between home sites? . . . . What may contribute to variation in territory size? . . . .

4.15 What is the range of density of individuals or groups? . . . .

4.16 Are there changes in territorial pattern when densities change (*e.g.* does the size change with density)? . . . .

4.17 Does the size of the area defended depend on the caste or species of the intruder? . . . .

4.18 What is the nature of the boundaries? . . . . Recognized equilibrium positions resulting from decreasing defence with increasing distance from home site? . . . . Or is defence equally intense over whole area? . . . .

Contiguous between neighbours? . . . . Often takes advantage of natural obstructions? . . . . Marked? . . . .

4.19 Where is territory "marked" (along boundaries, from fixed signalling sites, along paths, *etc.*)? . . . .

4.20 How is the territory marked? . . . . What sensory modalities are involved? . . . . By what behaviour? . . . . Are there special marking organs (*e.g.* special scent glands or apparently normal as by faeces, urine or saliva, or by some combination)? . . . .

4.21 When is the territory marked? . . . . Are there regular patterns in time? . . . . Regular display periods? . . . . Regular intervals between marking each "signpost"? . . . .

4.22 How is territorial intrusion detected? . . . . Is there regular monitoring behaviour, *e.g.* patrolling? . . . . Describe pattern. . . . .

4.23 Is there a recognizable territory centre? . . . . What form does it take (*e.g.* nest, burrow, lookout, *etc.*)? . . . .

4.24 Describe "off territory" or "intruding" behaviour by territorial animals. . . . .

4.25 Describe a typical interactive sequence (detect intruder, approach, engage, fight, submit, pursuit, *etc.*). . . . .

4.26 Describe any regular changes in behaviour or displays with the distance from the intruder. . . . .

4.27 Are different responses generated by other species besides conspecifics? . . . .

4.28 Is interspecific "territorial" behaviour mutual or by one species only (or is it anti-predator behaviour)? . . . .

4.29 What is the pattern of land use (*e.g.* by fixed paths, fixed activity sites, regular patterns of movement)? . . . .

4.30 How regularly is the space used (daily cycles, other cycles, irregular timing)? . . . .

4.31 Describe threat behaviour over a range of intensities (*e.g.* calls, displays, colour changes, or pheromones). . . .

4.32 Were the spacing patterns formed anew at the beginning of the phase? . . . .

4.33 Are the spacing patterns abandoned or changed at the end of the phase? . . . .

4.34 Describe a typical sequence of interactions between one pair of individuals during spacing relationship formation. . . . .

4.35 How many such relationships are formed in the establishment of the spacing pattern? . . . .

4.36 How many intrusions and interactions by non-neighbours occur during the formation of spacing relationships? . . . .

4.37 Describe typical interactions in the formation of relationships. . . . .

4.38 What behaviour is used during these interactions? . . . .

4.39 Can you show how the pattern of interactions involved in the formation of the spacing relationships develops into the patterns of servicing or maintaining interactions or into the pattern of monitoring each other's movements? . . . .

Home Ranges

This system involves attachment to and regular use of an area by established residents (individuals or groups) without exclusion of other residents. The relationship between such residents is expressed largely in avoidance of each other, organized in one of a number of possible ways. These include avoidance of dominants by subordinates within some sort of hierarchy with fixed rank relationships, or by a system of space and time sharing, either of which comprises a social organization of the solitary state (or isolated group). To enable fine controls over the avoidance patterns, the communicative system allows for regular monitoring of identity and location of neighbours, usually out of sight.

Questions 4.29 to 4.39 are appropriate to home ranges as well as to territorial behaviour, and should be answered first.

4.40 Ranges are used by resident individuals of the following castes . . . . or the groups of . . . . of size. . . . .
4.41 What non-resident individuals move through the area? . . . .
4.42 What are the sizes (and their variabilities) of the ranges used by each type of resident? . . . .
4.43 What is the normal variability in densities of individuals or groups? . . . .
4.44 What communicative patterns are used for monitoring the identity and location of resident neighbours? . . . . Describe the modality used and the behaviour and any special organs of emitter and receiver. . . . .
4.45 What behaviour is used to monitor the presence of non-resident intruders? . . . .
4.46 What interactions are used when non-residents are detected? . . . . Describe the sequence of interactions and behaviour. . . . .
4.47 What interactions occur when resident animals come in contact? . . . . Describe the sequence of behaviour for each type of resident (*e.g.* male-male, male-female, female-female, group-group, male-group, *etc.*). . . . .
4.48 What is the pattern of home range use (*e.g.* by paths and fixed sites)? . . . .
4.49 What are the fixed sites used for (*e.g.* defaecation, body care, sleep, *etc.*)? . . . .
4.50 What are the normal minimum distances of approach between individuals of each group or caste? . . . .
4.51 Do displays change with changes in the distance between individuals or groups? . . . .
4.52 Does there appear to be recognition between the approaching individuals or groups? . . . . Describe . . . .
4.53 Is there dominance behaviour seen between individuals or groups? Is it fixed or "peck right" dominance? . . . . Does dominance depend upon the place or time? . . . . or upon the distance from a home site? . . . . any other system of dominance-subordinate relationship? . . . .

4.54 Do the individuals or groups ever aggregate (*e.g.* at food or at roosting sites or for mobbing)? . . . .
4.55 Describe any interspecific relationships of a home range type. . . . .
4.56 Describe the behaviour and interactions associated with forming such interspecific relationships. . . . .
4.57 Describe any behavior or interactions associated with the maintenance of the interspecific relationships.
4.58 General comments. . . . .

5. Spacing Within Groups

Within groups, there is normally a wide range of distance-dependent behaviour. Usually Personal Distances are observed in most activities, and Individual Distances when the animals rest. Animals occupy Living Space within the group, limited by the Personal Distances of their neighbours and the Social Distances of the groups. Personal Distances and Social Distances normally change with the activity subphase, the time of the day, and with such conditions as rain or fog.

5.1 Is there regular spacing between individuals of the same caste? . . . . of different castes? . . . . Describe the spacing. . . . .
5.2 When spaced, is there a regular pattern of orientation (*e.g.* polarized)? . . . . only when moving? . . . . at what other times? . . . .
5.3 List the spacing distances between individuals of each caste or caste combination, the personal or minimum distances in each subphase including Individual Distances when resting. . . . . (The Social or maximum distances are described in 3.24.) Do these distances change with light intensity, rain or other environmental conditions? . . . . If so, how? . . . .
5.4 If contact animals, do they avoid close face to face orientation when resting? . . . . Describe. . . . .
5.5 What sensory modalities seem to be used to assess spacing distances? . . . .
5.6 Describe the behaviour which maintains spatial patterns when spacing becomes "too small", *i.e.* intrusions. . . . .
5.7 Are there dominance-subordinate relationships between the animals? . . . . of the same castes? . . . . of different castes? . . . .
5.8 Is there a single hierarchy or a separate one for each caste? . . . .
5.9 Is the hierarchy stable? . . . . If not, describe. . . . .
5.10 When a hierarchy is formed over a short period, as when a seasonal group assembles, describe the formation of relationships . . . . the interactions . . . . the behaviour. . . . .
5.11 Describe the interactions by which dominance-subordinate relationships are maintained. . . . .
5.12 Describe the range of threat behaviour. . . . .
5.13 Describe the range of submissive behaviours (*e.g.* flight—and how far,

avoidance, sexual presentations, infantile behaviours, *etc.*). . . . . (Submissive behaviour is defined as behaviour which terminates a fight or a threat interaction.)

5.14 Is there appeasement behaviour (seen when a subordinate approaches a dominant neighbour [temporary or permanent], without provoking a threat)? . . . . Describe. . . . . Are there other appeasing situations? . . . .

5.15 At what distance and/or orientation is submissive behaviour relaxed? . . . .

5.16 At what distance and/or orientation is appeasement behaviour shown? . . . .

5.17 Describe the formation of dominance of relationships among young animals. . . . .

5.18 Do the parents influence the status of their young? . . . . Describe the influences on the interaction by which the dominance-subordinate relationships are formed. . . . .

5.19 Are there special role behaviours associated with high and low ranks (*e.g.* leadership during movement, sentry duties while group feeds, boundary defence, *etc.*)? . . . . Describe. . . . .

5.20 Is there a pattern of dispersal related to rank (*e.g.* dominant at centre, subordinate at periphery)? . . . .

5.21 How are the castes distributed within the group? . . . .

5.22 General comments. . . . .

6. Sexual Behaviour

6.1 Is sexual behaviour seasonal? . . . . throughout the year? . . . . what pattern? . . . .

6.2 How do the sexes locate each other, at a distance? . . . . when close? . . . . identify each other? . . . .

6.3 What duration is the association between the mates (permanent, recurrent during each mating season, for a mating and breeding season, for a mating season only, shorter than a mating season, for the period of an oestrus or less, ephemeral, *etc.*)? . . . .

6.4 Are special mating bonds formed between the sexes? . . . . Describe. . . . .

6.5 Is there a succession of mates? . . . . Describe. . . . .

6.6 Describe prebond courtship. . . . .

6.7 Describe precopulatory courtship. . . . .

6.8 Describe the copulation interaction (*e.g.* fertilization external with no physical contact, cloacal contact, intromission, termination, *etc.*). . . . .

6.9 What special displays or sequences are used? . . . .

6.10 Describe any rejecting behaviour observed. . . . .

6.11 What is the division of labour in sexual interactions (*e.g.* who initiates each stage)? . . . .

6.12 Is there a rank order between mates? . . . . Does it change over time? . . . .

6.13 Is copulation more frequent than is necessary for fertilization? . . . . Does it recur throughout the period of the bond? . . . . throughout the year? . . . .

6.14 Describe bond-servicing interactions between the mates (*e.g.* allo-grooming, allofeeding *etc.*)? . . . . Comment on their evolutionary origins (*e.g.* derived from prebond courtship, copulations, from parent-offspring interactions). . . . .

6.15 Describe the interaction sequence leading to bond termination. . . . .

6.16 Describe the interactions seen in bond termination. . . . . the behaviour used. . . . .

6.17 Is mating monogamous, monogynous, polygamous, polygynous? . . . . Other? . . . . Describe. . . . .

6.18 Other comments. . . . .

7. Parent-Offspring Behaviour

Organisms may be divided reasonably well into two categories, those which have evolved family life, and those which have not. There are borderline cases where parents retain some care of their eggs until hatching, and others where young remain on a parental territory for a short time with no obvious attention to the parent.

Yet all invest much energy into the production of the next generation in a variety of ways. Non-family animals may migrate long distances to deposit their eggs in areas which have proven suitable for rearing, or may merely deposit eggs close to suitable food sources. Others like many wasps may house their eggs in artificial mud-shelled "eggs" with sufficient food to carry them through their larval stages. Still others replenish the food stocks, and so become family animals. Where eggs are produced and left, there is often camouflage. Throughout there is behaviour in carrying out these operations.

Family animals remain with their young for varying periods of time so that the young are in the presence of an experienced adult during their early development. Usually the behaviour of the parent then divides between attention to the world outside and attention to the young, while the attention of family young is usually directed from the world outside to the parent or parents. The rearing of the young is carried out in a societal subsystem, usually well incorporated within the society at large. Often the society at large organizes special aggregations or other societal structures for the period of family living.

7.1 Is reproduction sexual, asexual, parthenogenetic? . . . . Describe. . . . .

7.2 Is reproduction oviparous, ovoviviparous, viviparous? . . . . Describe. . . . .

7.3 Describe the sequence of behaviour leading up to oviposition or birth. . . . .

7.4 Describe oviposition. . . . . Where? . . . . When? . . . . How frequently? . . . . Are the eggs covered? . . . .
7.5 What is the significance of the particular places or times? . . . .
7.6 Is there any care or attention or defence of the eggs, or the area around the eggs? . . . . Describe. . . . .
7.7 Who cares for the eggs (mother, father, both parents, older siblings, members of another species, *etc.*)? . . . . Describe the care of eggs. . . . . including incubation where appropriate. . . . .
7.8 Is there a special nest, burrow, or other place in which the eggs are kept? . . . . Describe. . . . .
7.9 Where there is alternation of care for the eggs, what behavior and interactions are seen at the transfers (*e.g.* nest relief ceremonies)? . . . .
7.10 Are parents present at the hatching of the eggs? . . . . Which parent? . . . . Do they assist? . . . . Describe. . . . .
7.11 Describe the behaviour of parents and offspring after hatching in non-family animals (*e.g.* crocodiles or pythons, where the parents remain with the eggs through to hatching). . . . .
7.12 Describe the neonatal behaviour of the young non-family animal . . . . at hatching . . . . after metamorphosis. . . . .
7.13 Describe the social behaviour observed in the growing young of non-family animals. . . . .
7.14 What is the gestation or incubation length? . . . .
7.15 Is there delayed implantation? . . . . Discuss significance of delay in implantation. . . . .
7.16 What is the clutch, brood, or litter size? . . . . variation? . . . .
7.17 Where young are born, describe the behaviour leading up to parturition. . . . .
7.18 Where the young are born, describe the behaviour through parturition. . . . .
7.19 After birth or hatching, describe the first interactions between parent(s) and young. . . . . the first communicative behaviour between the two. . . . .
7.20 Does the parent(s) (or other adults) clean up the nest area after birth or parturition (*e.g.* remove egg shells and unhatched eggs, eat placenta, *etc.*)? . . . .
7.21 Rate the offspring along an altricial-precocial scale (using such characters as embryonic stage of development, locomotion, sight, fur or feathers, restriction within a larval cell, *etc.*). . . . .
7.22 How long is parental care maintained? . . . . What transitions (*e.g.* fledging or weaning)? . . . .
7.23 Is there "parental" care by siblings (*e.g.* in communal nesting mammals or birds, or the "social" insects)? . . . . By whom . . . . Describe. . . . .
7.24 Describe the interactions involved in parental care (*e.g.* body care, defence—including distraction displays—housekeeping at nest, supplying food to young—by bringing it, or leading the young to food, introducing them to food, feeding from bodily secretions or tropholaxis—supply

of warmth, removal of faeces or urine, retrieving young to self or to nest, carrying young). . . . .

7.25 Is an interindividual bond formed between parent(s) and young? . . . .
7.26 Describe the emergence of recognition within the bond (*i.e.* first an attachment to nest or a group scent, becoming discriminated into a recognition of the individuals). . . . . Describe the recognition behaviours. . .
7.27 Describe recognition behavior and organization between siblings. . . . .
7.28 Are all of the young hatched together or at intervals over a period (*i.e.* incubation begins with the first egg)? . . . .
7.29 Describe the interactions between the sibs (*e.g.* competitive, do the older young eventually eat their siblings? . . . . usually? . . . . or sometimes?) . . . .
7.30 Is there play behaviour? . . . . What types? . . . . How is it recognized? . . . . Are the parents always present? . . . . Is there noise? . . . . Describe behaviour, signals and metasignals, in typical sequences of solitary or social play. . . . .
7.31 At what age do the young leave the nest or site of birth? . . . . Do they accompany parents? . . . .
7.32 Does the offspring group remain together? . . . . with the parent(s)? . . . .
7.33 Is the attachment to a nest site retained? . . . . For how long? . . . . Describe. . . .
7.34 Describe the changes in allofeeding (*e.g.* at weaning or fledging, or after larval development). . . . .
7.35 What sorts of food do parents collect for young? . . . . How do they carry it? . . . . How far do they carry it? . . . .
7.36 How does the food-collecting behaviour affect parental roles (*e.g.* long hunting periods of raptors may contribute to division of labour whereby the female remains with the young and feeds them)? . . . .
7.37 Describe retrieval behaviour, including the signals used at each age. . . . . Is the retrieval to nest or to parent? . . . .
7.38 Who takes the active role in retrieval, the parent(s), young, or both? . . . . Are there changes? . . . .
7.39 Describe any contacting behaviour during movement. . . . .
7.40 Describe the interactions involved in the establishment and maintenance of the Social Distance between parents and young. . . . .
7.41 Describe the changes in the interactions associated with the breaking of the parent-offspring bond. . . . .
7.42 What risks are involved by parents in the care, especially the protection of their young? . . . .
7.43 What are the levels of mortality during the family care period? . . . . Discuss predation, parasitism and other losses. . . . . What are the predators or parasites of the young? . . . .
7.44 Is the mortality of the young affected by the age of the parent(s)? . . . . the social status of the parent(s)? . . . . Describe. . . . .
7.45 Describe any baby-sitting or creche behaviour. . . . .
7.46 General comments. . . . .

8. Ecological and Other Behaviour

So far, the questionnaire has focussed upon social behaviour and societal structure. Yet the animals are organized by their physiology and their behaviour to fit a niche within one or a restricted variety of ecosystems. Some aspects of the fitting into niches are largely passive, for example the depositing of urine and faeces. Yet those processes may also be organized, and even socially organized so as to profoundly affect the ecosystem. The use of communal dunging areas makes an important contribution to the distribution of nutrients and parasite eggs over an area, and also affects the communities of organisms which depend upon the faeces for a habitat.

Other organism-environment relations are not passive. Animals actively engage many components of their environment behaviourally in the harvest of energy, and in restraining the energy losses to other organisms, particularly to predators, sometimes the drain to parasites may be an intermediate stage leading to predation, and such behavioural devices as dunging areas may be important. By their behaviour, animals interact with components of their environment, with food sources, with predators, with shelter, within the context of their social system, yet in such a way as to build relationships with components of their surroundings. In so doing, they organize their perceptions of these components, build habituated maps of the parts, and arrange their attention structure to monitor such components, at least to detect changes. It is this incorporation of features of the surroundings into the animals' own organization which distinguishes environment from surroundings. Environments are parts of animals—of organism-environment relationships—and as such, are dynamic parts of the sensory-behavioural organization of animals.

Some questions are asked to uncover aspects of these relationships and many have already been asked. Emphasis is placed on the social aspects, though information on habitat may eventually contribute to the growing field of socioecology.

8.1 Describe the behaviour of defaecating and urinating. . . . .

8.2 What parasites or other organisms of importance to the species are associated with the excreta? . . . . How might the placement of the excreta contribute to the reinfestation of the host species by these organisms? . . . .

8.3 Describe the distribution of excreta within the habitat. . . . .

8.4 Describe social features of excreta distribution. . . . . What factors are involved (marking, depositing in camping areas, tradition in the use of dunging sites, *etc.*)? . . . .

8.5 Any general comments on excreta and associated behaviour? . . . .

8.6 What are the predators of the species? . . . . of each age? . . . . at each season? . . . . What time of day does most predation occur? . . . .
8.7 What defences does the organism have against these predators, with emphasis on behaviour? . . . .
8.8 Describe the intensity of predation with age. . . . . with hierarchial status . . . . with territorial status . . . . other social features of predation. . . . .
8.9 Describe alarm behaviour. . . . . Are there different types of alarms? . . . . What is the response of conspecifics to alarm displays? . . . . Describe sequence. . . . .
8.10 What are the main parasites or diseases? . . . . Is there any specific behaviour associated with the parasites? . . . . or the diseases? . . . .
8.11 Are there any social responses to diseased animals? . . . .
8.12 Is predation of sick animals high? . . . .
8.13 General comments on predation. . . . .
8.14 General comments on diseases and parasites. . . . .
8.15 Are any shelters built by the animals? . . . . Describe the form of the shelter and interactions concerned with construction and maintenance. . . . .
8.16 What other places of shelter used by the animals? . . . .
8.17 When does shelter seeking behaviour occur (season, time of day, storms, snow, *etc.*)? . . . .
8.18 What are the elements of the environment sheltered against (physical such as heat, dryness, light, *etc.*, or biological such as parasites or predators)? . . . .
8.19 What part does camouflage play in behaviour, and how does behaviour serve camouflage? . . . .
8.20 Does the organism have periods of dormancy (during drought, hibernation *etc.*)? . . . . How long? . . . . What preparation? . . . . Where does it occur? . . . . Any social features? . . . .
8.21 General comments on shelter behaviour? . . . .
8.22 What body care behaviour is observed? . . . . Describe sequence. . . . . Are there special organs involved? . . . . How are they used? . . . .
8.23 When is body care carried out? . . . . Lengths of periods? . . . . Frequency of periods? . . . . Time of day? . . . .
8.24 Are there any special types of body care (*e.g.* moulting)? . . . .
8.25 Where is body care carried out at the times mentioned? . . . . Are there special places? . . . . special "tools" (*e.g.* rubbing trees)? . . . .
8.26 What social behaviour accompanies body care? . . . .
8.27 General comments on body care behaviour. . . . .
8.28 What general category of feed is used (*e.g.* fructivores, insectivores, carnivores, scavenger, *etc.*)? . . . .
8.29 What food species are eaten? . . . . Is the range flexible? . . . . Is there a change throughout the year? . . . . with seasons? . . . . with droughts or floods? . . . . with age? . . . . with changes in density of food species? . . . .
8.30 Describe range of food-getting behaviour. . . . . What stimuli seem important? . . . . tactics used? . . . .

8.31 Is there any learning of food types? . . . . of food-getting behaviour? . . . . Describe acquisition? . . . .
8.32 What are the social features of food-getting behaviour? . . . .
8.33 Is there any food-hoarding behaviour? . . . . Describe the interactions. . . . . What social features? . . . . How is the control of the caches organized? . . . . What locating behaviour occurs at retrieval? . . . . Describe retrieval of food. . . . .
8.34 Describe any interspecific competition at food sources. . . . .
8.35 Describe any interspecific food sharing. . . . .
8.36 Describe any interspecific cooperation or arrangement in food-getting behaviour. . . . .
8.37 Does food-getting behaviour and food sources vary in different habitats or regions? . . . .
8.38 Are there traditional features of feeding behaviour? . . . .
8.39 Are new food sources ever incorporated into the repertoire of the species? . . . . Describe the conditions responsible for the new source. . . . . Describe the spread of the new source. . . . .
8.40 What parts of the animals' ranges are used for feeding? . . . .
8.41 Is there any hierarchy of food preferences (seen perhaps when food is plentiful and scarce)? . . . .
8.42 Are there differences in feeding behaviour between castes? . . . .
8.43 General comments on feeding behaviours. . . . .
8.44 Describe characteristic sequences of exploratory behaviour. . . . .
8.45 Describe the sequences of behaviour of an animal when a change occurs in its range. . . . .
8.46 What social features of exploratory behaviour are observed? . . . .
8.47 General comments on exploratory behaviour. . . . .
8.48 Are there periodic fluctuations in numbers? . . . . Regular or irregular? . . . .
8.49 Comment on cycles or periods. . . . .
8.50 Are there changes in density synchronized over large areas? . . . . Comment. . . . .
8.51 Describe changes in interindividual behaviour associated with changes in density. . . . . frequency of interaction . . . . types of interaction . . . . behaviour in same interactions? . . . .
8.52 Describe changes with density in group size or structure. . . . .
8.53 Describe changes with density in intergroup relationships. . . . .
8.54 Describe changes with density in spacing patterns between individuals . . . . between groups. . . . .
8.55 Describe changes in predation and disease. . . . .
8.56 Describe changes in inter- and intra-caste behaviour, interactions and relationships. . . . .
8.57 Describe changes in feeding behaviour. . . . .
8.58 Describe migratory patterns associated with density changes. . . . .
8.59 General comments on crowding. . . . .
8.60 General comments on density changes. . . . .
8.61 Describe briefly the habitat and ecological niche, referring particularly

to features which may be associated with the social organization of species. . . . .

8.62 Is behaviour different in different habitats? . . . . What features of the habitat seem responsible? . . . . for what differences in behaviour? . . . .

8.63 What is the range of habitats used? . . . . How is habitat assessed? . . . . selected? . . . . What are the preferences? . . . .

8.64 What resources may limit numbers? . . . . at which season? . . . . under what conditions? . . . .

8.65 General comments on habitats. . . . .

Summary

This paper presents a theoretical model of and a practical approach to the study of social systems. It was prepared after discussions at the XI and XII International Ethological Conferences, and each draft was presented to many ethologists for comment, and many responded.

The paper looks at the social organization within animal species, and at the way animals build, maintain, and change it by their behaviour. The questions asked move always from the individual behaviours, through the social interactions, to the social relationships and groups which are stable features of societies.

The main societal subsystems discussed are:

1. *Social phases*, or social structures which are maintained for periods of time;
2. Organic specialization by *castes*;
3. Social specialization in *groups*;
4. The *pattern of dispersal* of individuals or groups;
5. The social organization of sexual and parent-offspring behaviour;
6. The organization of behaviour in relation to the environment;
7. The dynamic aspects of animal societies, particularly with changes in density.

In each section, questions aim to draw out the pattern of behavioural organization, emphasizing the structural features of the theoretical model presented.

Index